SOIL MECHANICS FUNDAMENTALS

SOIL MECHANICS FUNDAMENTALS

Muni Budhu

Professor, Department of Civil Engineering and Engineering Mechanics
University of Arizona, USA

WILEY Blackwell

This edition first published 2015
© 2015 by John Wiley & Sons, Ltd

Registered office
John Wiley & Sons, Ltd, The Atrium, Southern Gate, Chichester, West Sussex, PO19 8SQ, United Kingdom.

For details of our global editorial offices, for customer services and for information about how to apply for permission to reuse the copyright material in this book please see our website at www.wiley.com/wiley-blackwell.

Library of Congress Cataloging-in-Publication Data

Budhu, M.
    Soil mechanics fundamentals / Muni Budhu. – Metric version.
        pages cm
Includes index.
    ISBN 978-1-119-01965-7 (paperback)
1. Soil mechanics.   I. Title.
TA710.B7654 2015b
624.1'5136–dc23

                                        2014046417

This book also appears in a Imperial measurement edition, ISBN 9780470577950.

A catalogue record for this book is available from the British Library.

Wiley also publishes its books in a variety of electronic formats. Some content that appears in print may not be available in electronic books.

Contents

About the Author

MUNIRAM (Muni) BUDHU is Professor of Civil Engineering and Engineering Mechanics at the University of Arizona, Tucson. He received his BSc (First Class Honors) in Civil Engineering from the University of the West Indies and his PhD in Soil Mechanics from Cambridge University, England. Prior to joining the University of Arizona, Dr. Budhu served on the faculty at the University of Guyana, Guyana; McMaster University, Canada; and the State University of New York at Buffalo. He spent sabbaticals as Visiting Professor at St. Catherine's College, Oxford University; Eidgenössische Technische Hochschule Zürich (Swiss Federal Institute of Technology, Zurich); and the University of Western Australia. He authored and co-authored many technical papers on various civil engineering and engineering mechanics topics including soil mechanics, foundation engineering, numerical modeling, hydraulic engineering, and engineering education. Dr. Budhu has developed interactive animations for learning various topics in soil mechanics and foundation engineering, fluid mechanics, statics, and interactive virtual labs. He is the co-founder of YourLabs, developer of a knowledge evaluation system (www.yourlabs.com). Dr. Budhu has authored two other textbooks, *Soil Mechanics and Foundations* and *Foundations and Earth Retaining Structures*. Both books are available from John Wiley & Sons (www.wiley.com).

Other Books by this Author

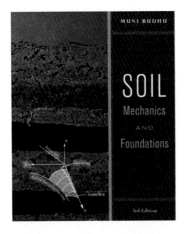

Soil Mechanics and Foundations,
3rd Edition
by Muni Budhu
ISBN: 978-0471-43117-6
An in-depth look at soil mechanics, including content for both an introductory soil mechanics and a foundations course. For students and other readers who wish to study the detailed mechanics connected with the fundamental concepts and principles. This textbbook includes critical state soils mechanics to provide a link between soil settlement and soil shear strength.

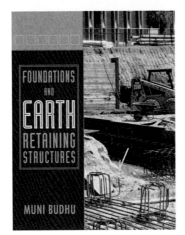

Foundations and Earth Retaining Structures
by Muni Budhu
ISBN: 978-0471-47012-0
Introduction to foundations and earth retaining structures, with fundamentals and practical applications of soil mechanics principles to the analysis and design of shallow and deep foundations, and earth retaining structures. In addition to a review of important soil mechanics concepts, this textbook discusses the uncertainties in geotechnical analysis and design, design philosophy and methodology, and design issues.

Website: www.wiley.com\go\budhu\soilmechanicsfundamentals

Preface

GOAL AND MOTIVATION

My intent in writing this textbook is to present accessible, clear, concise, and contemporary course content for a first course in soil mechanics to meet the needs of undergraduates not only in civil engineering but also in construction, mining, geological engineering, and related disciplines.

However, this textbook is not meant to be an engineering design manual nor a cookbook. It is structured to provide the user with a learning outcome that is a solid foundation on key soil mechanics principles for application in a later foundation engineering course and in engineering practice.

By studying with this textbook, students will acquire a contemporary understanding of the physical and mechanical properties of soils. They will be engaged in the presentation of these properties, in discussions and guidance on the fundamentals of soil mechanics. They will attain the problem-solving skills and background knowledge that will prepare them to think critically, make good decisions, and engage in lifelong learning.

PREREQUISITES

Students using this textbook are expected to have some background knowledge in Geology, Engineering Mechanics (Statics), and Mechanics of Materials.

UNITS

The primary unit of measure used in this textbook is the SI (International System) system of units. An imperial (US) units version version of this textbook is also available.

HALLMARK FEATURES

Contemporary methods: The text presents, discusses, and demonstrates contemporary ideas and methods of interpreting the physical and mechanical properties of soils that students will encounter as practicing engineers. In order to strike a balance between theory and practical applications for an introductory course in soil mechanics, the mechanics is kept to a minimum so that students can appreciate the background, assumptions, and limitations of the theories in use in the field.

The *implications of the key ideas* are discussed to provide students with an understanding of the context for the applications of these ideas.

A *modern explanation of soil behavior* is presented particularly in soil settlement and soil strength. These are foremost topics in the practice of geotechnical engineering. One-dimensional consolidation is presented in the context of soil settlement rather than as a separate topic (Chapter 7). The shear strength of soils is presented using contemporary thinking and approach. In particular, three popular failure criteria—Coulomb, Mohr-Coulomb, and Tresca—are discussed with regard to their applications and limitations. Students will be able to understand how to use these criteria to properly interpret soil test results and understand the differences between drained and undrained shear strength.

Pedagogy and design directed by modern learning theory: The content and presentation of the chapters are informed by modern theories of how students learn, especially with regard to metacognition.

Learning outcomes listed at the beginning of each chapter inform students what knowledge and skills they are expected to gain from the chapter. These form the bases for the problems at the end of each chapter. By measuring students' performance on the problems, an instructor can evaluate whether the learning outcomes have been satisfied.

Definitions of key terms at the beginning of each chapter define key terms and variables that will be used in the chapter.

Key points summaries throughout each chapter emphasize for students the most important points in the material they have just read.

Practical examples at the end of some chapters give students an opportunity to see how the prior and current principles are integrated to solve "real world type" problems. The students will learn how to find solutions for a "system" rather than a solution for a "component" of the system.

Consistent problem-solving strategy: Students generally have difficulty in translating a word problem into the steps and equations they need to use to solve it. They typically can't read a problem and understand what they need to do to solve it. This text provides and models consistent strategies to help students approach, analyze, and solve any problem. Example problems are solved by first developing a strategy and then stepping through the solution, identifying equations, and checking whether the results are reasonable as appropriate.

Three categories—*conceptual understanding, problem solving, and critical thinking and decision making*—of problems are delineated at the end of the chapter to assess students' knowledge mastery. These are not strict categories. In fact, the skills required in each category are intermixed. Problems within the *conceptual understanding* category are intended to assess understanding of key concepts and may contain problems to engage lateral thinking.

It is expected that the instructor may add additional problems as needed. Problems within the *problem-solving* category are intended to assess problem-solving skills and procedural fluency in the applications of the concepts and principles in the chapter. Problems within the *critical thinking and decision-making* category are intended to assess the student's analytical skills, lateral thinking, and ability to make good decisions. These problems have practical biases and require understanding of the fundamentals. Engineers are required to make decisions, often with limited data. Practical experience is a key contributor to good decisions. Because students will invariably not have the practical experience, they will have to use the fundamentals of soil mechanics, typical ranges of values for soils, and their cognitive skills to address problems within the *critical thinking and decision-making* category. The instructors can include additional materials to help the students develop critical thinking and decision-making skills.

Knowledge mastery assessment software. This textbook is integrated with YourLabs™ Knowledge Evaluation System (KES) (www.yourlabs.com). This system automatically grades students' solutions to the end of chapter problems. It allows students to answer the problems anywhere on any mobile device (smartphone, iPad, etc.) or any desktop computing device (PC, MAC, etc.). After answering each question in an assignment set by the instructor on KES, the student's answer (or answers to multi-parts problems) is compared to the correct answer (or answers in multi-parts problems) and scored. The student must step through the solution for each problem and answer preset queries to assess concept understanding, critical thinking, problem-solving skills, and procedural fluency. KES then analyzes the feedback from students immediately after submitting their responses and displays the analytics to the students and the instructor. The analytics inform the instructor what the students know and don't know, at what steps, and the types of mistakes made during problem solving. The instructor can re-teach what the students did not know in a timely manner and identify at-risks students. The analytics are also displayed to the student to self-reflect on his/her performance and take corrective action. Relevant instructional materials are linked to each problem, so the student can self-learn the materials either before or upon completion of the problem. Instructors can modify the questions and assets (links or embedded videos, images, customized instructional materials, etc.) and, at each step of the solution, add or delete solution steps or create a customized question. Each problem can be tagged with any standard required by academic or professional organizations. The analytics as well as students' scores are aggregated from the problem to assignment and to class or course levels.

GENESIS OF THIS BOOK

This textbook is an abridged version of the author's other textbook *Soil Mechanics and Foundations* (3rd ed., Wiley, 2011). The *Soil Mechanics and Foundations* textbook provides a more in-depth look at soil mechanics and includes content for both an introductory soil mechanics and a foundations course. For students and other readers who wish to study the detailed mechanics connected with the fundamental concepts and principles, they should consult the author's *Soil Mechanics and Foundations* textbook.

The present textbook, *Soil Mechanics Fundamentals*, arose from feedback from instructors' for a textbook similar to *Soil Mechanics and Foundations* that would cover just the essentials and appeal to a broad section of undergraduate students.

Acknowledgments

I am grateful to the many anonymous reviewers who offered valuable suggestions for improving this textbook. Ibrahim Adiyaman, my former graduate student at University of Arizona, Tucson, worked tirelessly on the Solutions Manual.

Madeleine Metcalfe and Harriet Konishi of John Wiley & Sons were especially helpful in getting this book completed.

Notes for Students and Instructors

WHAT IS SOIL MECHANICS AND WHY IS IT IMPORTANT?

Soil mechanics is the study of the response of soils to loads. These loads may come from human-made structures (e.g., buildings), gravity (earth pressures), and natural phenomena (e.g., earthquake). Soils are natural, complex materials consisting of solids, liquids, and gases. To study soil behavior, we have to couple concepts in solid mechanics (e.g., statics) and fluid mechanics. However, these mechanics are insufficient to obtain a complete understanding of soil behavior because of the uncertainties of the applied loads, the vagaries of natural forces, and the intricate, natural distribution of different soil types. We have to utilize these mechanics with simplifying assumptions and call on experience to make decisions (judgment) on soil behavior.

A good understanding of soil behavior is necessary for us to analyze and design support systems (foundations) for infrastructures (e.g., roads and highways, pipelines, bridges, tunnels, embankments), energy systems (e.g., hydroelectric power stations, wind turbines, solar supports, geothermal and nuclear plants) and environmental systems (e.g., solid waste disposal, reservoirs, water treatment and water distribution systems, flood protection systems). The stability and life of any of these systems depend on the stability, strength, and deformation of soils. If the soil fails, these systems founded on or within it will fail or be impaired, regardless of how well these systems are designed. Thus, successful civil engineering projects are heavily dependent on our understanding of soil behavior. The iconic structures shown in Figure 1 would not exist if soil mechanics was not applied successfully.

PURPOSES OF THIS BOOK

This book is intended to provide the reader with a prefatory understanding of the properties and behavior of soils for later applications to foundation analysis and design.

LEARNING OUTCOMES

When you complete studying this textbook you should be able to:

Figure 1 (a) Willis tower (formerly the Sears Tower) in Chicago, (b) Empire State Building in New York City, and (c) Hoover Dam at the border of Arizona and Nevada.

- Describe soils and determine their physical characteristics such as grain size, water content, void ratio, and unit weight.
- Classify soils.
- Determine the compaction of soils and be able to specify and monitor field compaction.
- Understand the importance of soil investigations and be able to plan and conduct a soil investigation.
- Understand one- and two-dimensional flow of water through soils and be able to determine hydraulic conductivity, porewater pressures, and seepage stresses.
- Understand how stresses are distributed within soils from surface loads and the limitations in calculating these stresses.
- Understand the concept of effective stress and be able to calculate total and effective stresses, and porewater pressures.
- Be able to determine consolidation parameters and calculate one-dimensional consolidation settlement.
- Be able to discriminate between "drained" and "undrained" conditions.
- Understand the stress–strain response of soils.
- Determine soil strength parameters from soil tests, for example, the friction angle and undrained shear strength.

ASSESSMENT

Students will be assessed on how well they absorb and use the fundamentals of soil mechanics through problems at the end of the chapter. These problems assess concept understanding, critical thinking, and problem-solving skills. The problems in this textbook are coordinated with the YourLabs™ Knowledge Evaluation System (see the Preface for more detail).

WEBSITE

Additional materials are available at www.wiley.com\go\budhu\soilmechanicsfundamentals.

Additional support materials are available on the book's companion website at www.wiley. com\go\budhu\soilmechanicsfundamentals.

DESCRIPTION OF CHAPTERS

The sequencing of the chapters is such that the pre-knowledge required in a chapter is covered in previous chapters. This is difficult for soil mechanics because many of the concepts covered in the chapters are linked. Wherever necessary, identification is given of the later chapter in which a concept is discussed more fully.

Chapter 1 covers soil composition and particle sizes. It describes soil types and explains the differences between fine-grained and coarse-grained soils.

Chapter 2 introduces the physical soil parameters, and explains how these parameters are determined from standard tests and their usage in soil classification.

Chapter 3 discusses the purpose, planning, and execution of a soils investigation. It describes the types of common in situ testing devices and laboratory tests to determine physical and mechanical soil parameters.

Chapter 4 discusses both the one-dimensional and two-dimensional flows of water through soils. It shows how water flows through soil can be analyzed using Darcy's law and Laplace's equation. Procedures for drawing flownets and interpreting flowrate, porewater pressures, and seepage condition are covered.

Chapter 5 describes soil compaction and explains why it is important to specify and monitor soil compaction in the field.

Chapter 6 is about the amount and distribution of stresses in soils from surface loads. Boussinesq's solutions for common surface loads on a semi-infinite soil mass are presented and limitations of their use are described. The concept of effective stress is explained with and without the influence of seepage stresses.

Chapter 7 discusses soil settlement. It explains how to estimate the settlement of coarse-grained soils based on the assumption of elastic behavior. It covers the limitations of using elasticity and the difficulties of making reliable predictions of settlement. Also, the discussion covers the basic concept of soil consolidation, the determination of consolidation parameters, and methods to calculate primary consolidation settlement and secondary compression.

Chapter 8 brings the discussion to the shear strength of soils. Soils are treated using the contemporary idealization of them as dilatant-frictional materials rather than their conventional idealization as cohesive-frictional materials. Typical stress–strain responses of coarse-grained and fine-grained soils are presented and discussed. The chapter discusses the implications of drained and undrained conditions, cohesion, soil suction, and cementation on the shear resistance of soils. Interpretations and limitations of using the Coulomb, Mohr–Coulomb, and Tresca failure criteria are considered as well.

Appendix A presents the derivation of a solution for the one-dimensional consolidation theory as proposed by Karl Terzaghi (1925).

Appendix B describes the procedure to determine the stress state using Mohr's circle. It is intended as a brief review in order to assist the student in drawing Mohr's circles to interpret soil failure using the Mohr–Coulomb failure criterion.

Appendix C provides a collection of frequently used tables taken from the various chapters to allow for easy access to tables listing values of typical soil parameters and with information summaries.

Appendix D provides a collection of equations used in this textbook. It can be copied and used for assignments and examinations.

For instructors who wish to introduce additional materials in their lectures or examinations, a special chapter (Chapter 9 [Imperial Units only]) is available at www.wiley .com\go\budhu\soilmechanicsfundamentals. Chapter 9 presents some common applications of soil mechanics. It is intended for students who will not move forward to a course in Foundation Engineering. These applications include simple shallow and deep foundations, lateral earth pressures on simple retaining walls, and the stability of infinite slopes. Simple soil profiles are used in these applications to satisfy a key assumption (homogeneous soil) in the interpretation of shear strength.

Notation, Abbreviations, Unit Notation, and Conversion Factors

NOTATION

Note: A prime (') after notation for stress denotes effective stress.

A	Area
B	Width
c_{cm}	Cementation strength
c_o	Cohesion or shear strength from intermolecular forces
c_t	Soil tension
C	Apparent undrained shear strength or apparent cohesion
C_c	Compression index
C_r	Recompression index
C_v	Vertical coefficient of consolidation
C_α	Secondary compression index
CC	Coefficient of curvature
CI	Consistency index
CPT	Cone penetrometer test
CSL	Critical state line
Cu	Uniformity coefficient
D	Diameter
D_r	Relative density
D_{10}	Effective particle size
D_{50}	Average particle diameter
e	Void ratio
E	Modulus of elasticity
E_{sec}	Secant modulus
G_s	Specific gravity
h_p	Pressure head
h_z	Elevation head
H	Height
H_{dr}	Drainage path
H_o	Height
i	Hydraulic gradient

I_d	Density index
k	Hydraulic conductivity for saturated soils
k_z	Hydraulic conductivity in vertical direction for saturated soils
K_a	Active lateral earth pressure coefficient
K_o	Lateral earth pressure coefficient at rest
K_p	Passive lateral earth pressure coefficient
L	Length
LI	Liquidity index
LL	Liquid limit
LS	Linear shrinkage
m_v	Modulus of volume compressibility
n	Porosity
N	Standard penetration number
NCL	Normal consolidation line
OCR	Overconsolidation ratio with respect to vertical effective stress
q	Flow rate
q_s	Surface stress
q_z	Flow rate in vertical direction
Q	Flow, quantity of flow, and also vertical load
R_d	Unit weight ratio or density ratio
R_T	Temperature correction factor
s_u	Undrained shear strength
S	Degree of saturation
SF	Swell factor
SI	Shrinkage index
SL	Shrinkage limit
SPT	Standard penetration test
SR	Shrinkage ratio
S_t	Sensitivity
u	Porewater pressure
u_a	Pore air pressure
U	Average degree of consolidation
URL	Unloading/reloading line
v	Velocity
v_s	Seepage velocity
V	Volume
V'	Specific volume
V_a	Volume of air
V_s	Volume of solid
V_w	Volume of water
w	Water content
w_{opt}	Optimum water content
W	Weight
W_a	Weight of air
W_s	Weight of solid
W_w	Weight of water
z	Depth
α	Dilation angle
α_p	Peak dilation angle
ε_p	Volumetric strain
ε_z	Normal strain

ϕ'	Generic friction angle
ϕ'_{cs}	Critical state friction angle
ϕ'_p	Peak friction angle
ϕ'_r	Residual friction angle
γ	Bulk unit weight
γ'	Effective unit weight
γ_d	Dry unit weight
$\gamma_{d(max)}$	Maximum dry unit weight
γ_{sat}	Saturated unit weight
γ_w	Unit weight of water
γ_{zx}	Shear strain
μ	Viscosity
ν	Poisson's ratio
ρ_e	Elastic settlement
ρ_{pc}	Primary consolidation
ρ_{sc}	Secondary consolidation settlement
ρ_t	Total settlement
σ	Normal stress
τ	Shear stress
τ_{cs}	Critical state shear strength
τ_f	Shear strength at failure
τ_p	Peak shear strength
τ_r	Residual shear strength
ξ_o	Apparent friction angle

ABBREVIATIONS

AASHTO	American Association of State Highway and Transportation Officials
ASTM	American Society for Testing and Materials
USCS	Unified Soil Classification System
USGS	United States Geological Service

UNIT NOTATION AND CONVERSION FACTORS

Pa	Pascal
kPa	kiloPascal (1000 Pa)
MPa	megaPascal (1000 kPa)
mm	millimeter
cm	centimeter (10 mm)
m	meter (1000 mm or 100 cm)
km	kilometers (1000 m)
hectare	10,000 m^2
in.	inch
ksf	kips per square foot
lb	pounds
pcf	pounds per cubic foot
psf	pounds per square foot

1.00 kip = 1000 pounds (lb)
1.00 ksf = 1000 pounds per square foot (psf)

US Customary Units		SI Units

Length

1.00 in.	=	2.54 cm
1.00 ft	=	30.5 cm

Mass and Weight

1.00 lb	=	454 g
1.00 lb	=	4.46 N
1 kip	=	1000 lb

Area

1.00 in.^{2}	=	6.45 cm^2
1.00 ft^2	=	0.0929 m^2

Volume

1.00 mL	=	1.00 cm^3
1.00 L	=	1000 cm^3
1.00 ft^3	=	0.0283 m^3
1.00 in.^{3}	=	16.4 cm^3

Temperature

°F	=	1.8(°C) + 32
°C	=	(°F − 32)/1.8

Pressure

1.00 psi	=	6.895 kPa
1.00 psi	=	144 psf
1.00 ksi	=	1000 psi

Unit Weight and Mass Density

1.00 pcf	=	16.0 kg/m^3
1.00 pcf	=	0.157 kN/m^3

Unit weight of fresh water = 9.81 kN/m^3 or 62.4 pounds per cubic foot (pcf)
Unit weight of salted water = 10.1 kN/m^3 or 64 pounds per cubic foot (pcf)

Universal Constants

g	=	9.81 m/s^2
g	=	32.2 ft/s^2

Chapter 1
Composition and Particle Sizes of Soils

1.1 INTRODUCTION

The purpose of this chapter is to introduce you to the composition and particle sizes of soils. Soils are complex, natural materials, and soils vary widely. The composition and particle sizes of soils influence the load-bearing and settlement characteristics of soils.

Learning outcomes

When you complete this chapter, you should be able to do the following:

- Understand and describe the formation of soils.
- Understand and describe the composition of soils.
- Determine particle size distribution of a soil mass.
- Interpret grading curves.

1.2 DEFINITIONS OF KEY TERMS

Minerals are chemical elements that constitute rocks.

Rocks are the aggregation of minerals into a hard mass.

Soils are materials that are derived from the weathering of rocks.

Effective particle size (D_{10}) is the average particle diameter of the soil at the 10th percentile; that is, 10% of the particles are smaller than this size (diameter).

Average particle diameter (D_{50}) is the average particle diameter of the soil.

Uniformity coefficient (Cu) is a numerical measure of uniformity (majority of grains are approximately the same size).

Coefficient of curvature (CC) is a measure of the shape of the particle distribution curve (other terms used are the coefficient of gradation and the coefficient of concavity).

Soil Mechanics Fundamentals, First Edition. Muni Budhu.
© 2015 John Wiley & Sons, Ltd. Published 2015 by John Wiley & Sons, Ltd.
Companion website: www.wiley.com\go\budhu\soilmechanicsfundamentals

1.3 COMPOSITION OF SOILS

1.3.1 Soil formation

Engineering soils are formed from the physical and chemical weathering of rocks. Soils may also contain organic matter from the decomposition of plants and animals. In this textbook, we will focus on soils that have insignificant amounts of organic content. Physical weathering involves reduction of size without any change in the original composition of the parent rock. The main agents responsible for this process are exfoliation, unloading, erosion, freezing, and thawing. Chemical weathering causes both reductions in size and chemical alteration of the original parent rock. The main agents responsible for chemical weathering are hydration, carbonation, and oxidation. Often chemical and physical weathering takes place in concert.

Soils that remain at the site of weathering are called residual soils. These soils retain many of the elements that comprise the parent rock. Alluvial soils, also called fluvial soils, are soils that were transported by rivers and streams. The composition of these soils depends on the environment under which they were transported and is often different from the parent rock. The profile of alluvial soils usually consists of layers of different soils. Much of our construction activity has been and is occurring in and on alluvial soils.

1.3.2 Soil types

Gravels, sands, silts, and clays are used to identify specific textures in soils. We will refer to these soil textures as soil types; that is, sand is one soil type, clay is another. Texture refers to the appearance or feel of a soil. Sands and gravels are grouped together as coarse-grained soils. Clays and silts are fine-grained soils. Coarse-grained soils feel gritty and hard. Fine-grained soils feel smooth. The coarseness of soils is determined from knowing the distribution of particle sizes, which is the primary means of classifying coarse-grained soils. To characterize fine-grained soils, we need further information on the types of minerals present and their contents. The response of fine-grained soils to loads, known as the mechanical behavior, depends on the type of predominant minerals present.

Currently, many soil descriptions and soil types are in usage. A few of these are listed below.

- *Alluvial soils* are fine sediments that have been eroded from rock and transported by water, and have settled on river- and streambeds.
- *Calcareous soil* contains calcium carbonate and effervesces when treated with hydrochloric acid.
- *Caliche* consists of gravel, sand, and clay cemented together by calcium carbonate.
- *Collovial soils* (collovium) are soils found at the base of mountains that have been eroded by the combination of water and gravity.
- *Eolian* soils are sand-sized particles deposited by wind.
- *Expansive soils* are clays that undergo large volume changes from cycles of wetting and drying.
- *Glacial soils* are mixed soils consisting of rock debris, sand, silt, clays, and boulders.
- *Glacial till* is a soil that consists mainly of coarse particles.

- *Glacial clays* are soils that were deposited in ancient lakes and subsequently frozen. The thawing of these lakes has revealed soil profiles of neatly stratified silt and clay, sometimes called varved clay. The silt layer is light in color and was deposited during summer periods, while the thinner, dark clay layer was deposited during winter periods.
- *Gypsum* is calcium sulfate formed under heat and pressure from sediments in ocean brine.
- *Lacustrine* soils are mostly silts and clays deposited in glacial lake waters.
- *Lateritic* soils are residual soils that are cemented with iron oxides and are found in tropical regions.
- *Loam* is a mixture of sand, silt, and clay that may contain organic material.
- *Loess* is a wind-blown, uniform, fine-grained soil.
- *Marine soils* are sand, silts, and clays deposited in salt or brackish water.
- *Marl* (marlstone) is a mud (see definition of mud below) cemented by calcium carbonate or lime.
- *Mud* is clay and silt mixed with water into a viscous fluid.

1.3.3 Soil minerals

Minerals are crystalline materials and make up the solids constituent of a soil. Minerals are classified according to chemical composition and structure. Most minerals of interest to geotechnical engineers are composed of oxygen and silicon, two of the most abundant elements on earth.

Quartz (a common mineral in rocks) is the principal mineral of coarse-grained soils. Quartz is hard and composed of silicon dioxide (SiO_2) in colored, colorless, and transparent hexagonal crystals. The particles of coarse-grained soil are thus naturally angular. Weathering, especially by water, can alter the angular shape to a rounded one.

Clay minerals are made up of phyllosilicates, which are parallel sheets of silicates. Silicates are a group of minerals with a structural unit called the silica tetrahedron. A central silica cation (positively charged ion) is surrounded by four oxygen anions (negatively charged ions), one at each corner of the tetrahedron (Figure 1.1a). The charge on a single tetrahedron is -4, and to achieve a neutral charge, cations must be added or single tetrahedrons must be linked to each other sharing oxygen ions. Silicate minerals are formed by the addition of cations and interactions of tetrahedrons. Silica tetrahedrons combine to form sheets, called silicate sheets or laminae, which are thin layers of silica tetrahedrons in which three oxygen ions are shared between adjacent tetrahedrons (Figure 1.1b). Silicate sheets may contain other structural units such as alumina sheets. Alumina sheets are formed by combination of alumina minerals, which consists of an aluminum ion surrounded by six oxygen or hydroxyl atoms in an octahedron (Figure 1.1c, d).

The mineral particles of fine-grained soils are platy. The main groups of crystalline materials that make up fine-grained soils, principally clays, are the minerals kaolinite, illite, and montmorillonite. These minerals are the products from weathering of feldspar and muscovite mica, families of rock-forming silicate minerals that are abundant on the Earth's surface. Kaolinite has a structure that consists of one silica sheet and one alumina sheet bonded together into a layer about 0.72 nm thick and stacked repeatedly (Figure 1.2a). The layers are held together by hydrogen bonds. Tightly stacked layers result from numerous hydrogen bonds. Kaolinite is common in clays in humid tropical regions. Illite consists of repeated layers of one alumina sheet sandwiched by two silicate sheets (Figure 1.2b). The layers, each of thickness 0.96 nm, are held together by potassium ions.

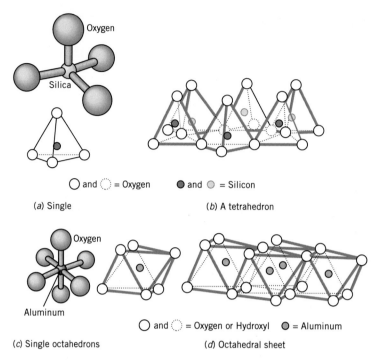

○ and ◌ = Oxygen   ● and ◔ = Silicon

(a) Single   (b) A tetrahedron

○ and ◌ = Oxygen or Hydroxyl   ● = Aluminum

(c) Single octahedrons   (d) Octahedral sheet

Figure 1.1   (a) Silica tetrahedrons, (b) silica sheets, (c) single aluminum octahedrons, and (d) aluminum sheets.

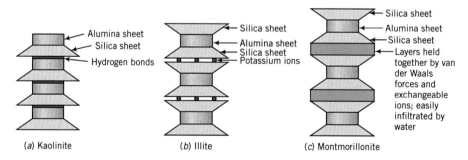

(a) Kaolinite   (b) Illite   (c) Montmorillonite

Figure 1.2   Structure of (a) kaolinite, (b) illite, and (c) montmorillonite.

Montmorillonite has a structure similar to illite, but the layers are held together by weak van der Waals forces. Montmorillonite belongs to the smectite clay family. It is an aluminum smectite with a small amount of Al^{+3} replaced by Mg^{2+}. This causes a charge inequity that is balanced by exchangeable cations Na^+ or Ca^{2+} and oriented water (Figure 1.2c). Additional water can easily enter the bond and separate the layers in montmorillonite, causing swelling. If the predominant exchangeable cation is Ca^{2+} (calcium smectite), there are two water layers, whereas if it is Na^+ (sodium smectite), there is usually only one water layer. Sodium smectite can absorb enough water to cause the particles to fully separate. Calcium smectites do not usually absorb enough water to cause particle separation because of their divalent cations. Montmorillonite is often called a swelling or expansive clay. Worldwide, it is responsible for billions of dollars in damages to structures (on ground and below ground).

1.3.4 *Surface forces and adsorbed water*

If we subdivide a body, the ratio of its surface area to its volume increases. For example, a cube with sides of 1 cm has a surface area of 6 cm^2. If we subdivide this cube into smaller cubes with sides of 1 mm, the original volume is unchanged, but the surface area increases to 60 cm^2. The surface area per unit mass (specific surface) of sands is typically 0.01 m^2 per gram, whereas for clays it is as high as 1000 m^2 per gram (montmorillonite). The specific surface of kaolinite ranges from 10 to 20 m^2 per gram, while that of illite ranges from 65 to 100 m^2 per gram. The surface area of 45 grams of illite is equivalent to the area of a football field. Because of the large surface areas of fine-grained soils, surface forces significantly influence their behavior compared to coarse-grained soils. The clay–water interaction coupled with the large surface areas results in clays having larger water-holding capacity in a large number of smaller pore spaces compared with coarse-grained soils.

The surface charges on the particles of fine-grained soils are negative (anions). These negative surface charges attract cations and the positively charged side of water molecules from surrounding water. Consequently, a thin film or layer of water, called adsorbed water, is bonded to the mineral surfaces. The thin film or layer of water is known as the diffuse double layer (Figure 1.3). The largest concentration of cations occurs at the mineral surface and decreases exponentially with distance away from the surface (Figure 1.3).

Surface forces on clay particles are of two types. One type, called attracting forces, is due to London–van der Waals forces. These forces are far-reaching and decrease in inverse proportion to l^2 (l is the distance between two particles). The other type, called repelling forces, is due to the diffuse double layer. Around each particle is an ionic cloud. When two particles are far apart, the electric charge on each is neutralized by equal and opposite charge of the ionic cloud around it. When the particles move closer together such that the clouds mutually penetrate each other, the negative charges on the particles cause repulsion.

Drying of most soils, with few exceptions (e.g., gypsum), using an oven for which the standard temperature is 105 $\pm$ 5°C cannot remove the adsorbed water. The adsorbed water influences the way a soil behaves. For example, plasticity in soils, which we will deal with in Chapter 4, is attributed to the adsorbed water. Toxic chemicals that seep into the ground contaminate soil and groundwater. Knowledge of the surface chemistry of fine-grained soils is important in understanding the migration, sequestration, rerelease, and ultimate removal of toxic compounds from soils.

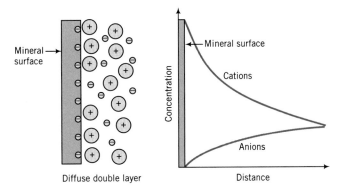

Figure 1.3 Diffuse double layer.

1.3.5   *Soil fabric*

Soil (minerals) particles are assumed to be rigid. During deposition, the mineral particles are arranged into structural frameworks that we call soil fabric (Figure 1.4). Each particle is in random contact with neighboring particles. The environment under which deposition occurs influences the structural framework that is formed. In particular, the electrochemical environment has the greatest influence on the kind of soil fabric that is formed during deposition of fine-grained soils.

Two common types of soil fabric—flocculated and dispersed—are formed during soil deposition of fine-grained soils, as shown schematically in Figure 1.4. A flocculated structure, formed in a saltwater environment, results when many particles tend to orient parallel to one another. A flocculated structure, formed in a freshwater environment, results when many particles tend to orient perpendicular to one another. A dispersed structure occurs when a majority of the particles orient parallel to one another.

Any loading (tectonic or otherwise) during or after deposition permanently alters the soil fabric or structural arrangement in a way that is unique to that particular loading condition. Consequently, the history of loading and changes in the environment is imprinted in the soil fabric. The soil fabric is the brain; it retains the memory of the birth of the soil and subsequent changes that occur.

The spaces between the mineral particles are called voids, which may be filled with liquids (essentially water), gases (essentially air), and cementitious materials (e.g., calcium carbonate). Voids occupy a large proportion of the soil volume. Interconnected voids form the passageway through which water flows in and out of soils. If we change the volume of voids, we will cause the soil to either compress (settle) or expand (dilate). Loads applied by a building, for example, will cause the mineral particles to be forced closer together, reducing the volume of voids and changing the orientation of the structural framework.

Consequently, the building settles. The amount of settlement depends on how much we compress the volume of voids. The rate at which the settlement occurs depends on the interconnectivity of the voids. Free water, not the adsorbed water, and/or air trapped in the voids must be forced out for settlement to occur. The decrease in volume, which results in settlement of buildings and other structures, is usually very slow (almost ceaseless) in fine-grained

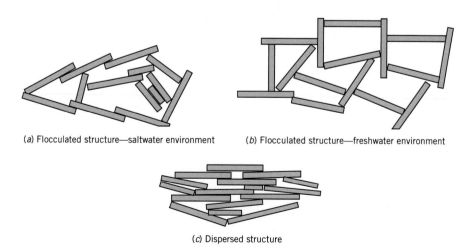

(*a*) Flocculated structure—saltwater environment   (*b*) Flocculated structure—freshwater environment

(*c*) Dispersed structure

Figure 1.4   Soil fabric.

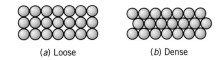

(a) Loose   (b) Dense

Figure 1.5   Loose and dense packing of spheres.

soils because these soils have large surface areas compared with coarse-grained soils. The larger surface areas provide greater resistance to the flow of water through the voids.

If the rigid (mostly quartz) particles of coarse-grained soils can be approximated by spheres, then the loosest packing (maximum void spaces) would occur when the spheres are stacked one on top of another (Figure 1.5a). The densest packing would occur when the spheres are packed in a staggered pattern, as shown in Figure 1.5b. Real coarse-grained soils consist of an assortment of particle sizes and shapes, and consequently, the packing is random. From your physics course, mass is volume multiplied by density. The density of soil particles is approximately 2.7 grams/cm^3. For spherical soil particles of diameter D (cm), the mass is $2.7 \times (\pi D^3/6)$. So the number of particles per gram of soil is $0.7/D^3$. Thus, a single gram of a fine sand of diameter 0.015 cm would consist of about 207,400 particles.

Key points

1. Soils are derived from the weathering of rocks and are broadly described by terms such as gravels, sands, silts, and clays.
2. Physical weathering causes reduction in size of the parent rock without change in its composition.
3. Chemical weathering causes reduction in size and chemical composition that differs from the parent rock.
4. Gravels and sands are coarse-grained soils; silts and clays are fine-grained soils.
5. Coarse-grained soils are composed mainly of quartz.
6. Clays are composed of three main types of minerals: kaolinite, illite, and montmorillonite.
7. The clay minerals consist of silica and alumina sheets that are combined to form layers. The bonds between layers play a very important role in the mechanical behavior of clays. The bond between the layers in montmorillonite is very weak compared with kaolinite and illite. Water can easily enter between the layers in montmorillonite, causing swelling.
8. A thin layer of water, called adsorbed water, is bonded to the mineral surfaces of soils. This layer significantly influences the physical and mechanical characteristics of fine-grained soils.

1.4 DETERMINATION OF PARTICLE SIZE

1.4.1 *Particle size of coarse-grained soils*

The distribution of particle sizes or average grain diameter of coarse-grained soils—gravels and sands—is obtained by screening a known weight of the soil through a stack of sieves of progressively finer mesh size. A typical stack of sieves is shown in Figure 1.6.

Sieve no.	Opening (mm)
3/8″	9.53
4	4.75
10	2
20	0.85
40	0.425
100	0.15
200	0.075

Figure 1.6 Stack of US sieves.

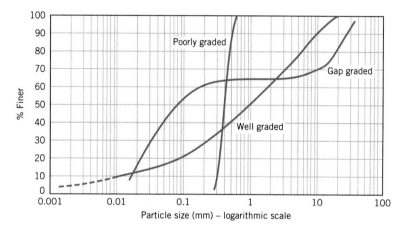

Figure 1.7 Particle size distribution curves.

In the United States, each sieve is identified by either a number that corresponds to the number of square holes per linear inch of mesh or the size of the opening. Large sieve (mesh) openings (75 mm to 9.5 mm) are designated by the sieve opening size, while smaller sieve sizes are designated by numbers. Other countries define sieves based on their actual opening in either millimeters or microns. The particle diameter in the screening process, often called sieve analysis, is the maximum dimension of a particle that will pass through the square hole of a particular mesh. A known weight of dry soil is placed on the largest sieve (the top sieve), and the nest of sieves is then placed on a vibrator, called a sieve shaker, and shaken. The nest of sieves is dismantled, one sieve at a time. The soil retained on each sieve is weighed, and the percentage of soil retained on each sieve is calculated. The results are plotted on a graph of percentage of particles finer than a given sieve size (not the percentage retained) as the ordinate versus the logarithm of the particle sizes, shown in Figure 1.7. The resulting plot is called a particle size distribution curve, or simply, the gradation curve. Engineers have found it convenient to use a logarithmic scale for particle size because the ratio of particle sizes from the largest to the smallest in a soil can be greater than 10^4.

Let W_i be the weight of soil retained on the ith sieve from the top of the nest of sieves and W be the total soil weight. The percentage weight retained is

$$\% \text{ retained on } i\text{th sieve} = \frac{W_i}{W} \times 100 \tag{1.1}$$

The percentage finer is

$$\% \text{ finer than } i\text{th sieve} = 100 - \sum_{i=1}^{i} (\% \text{ retained on } i\text{th sieve}) \tag{1.2}$$

The particles less than 0.075 mm (passing the No. 200 sieve) are collectively called fines. The fines content (usually greater than 35%) can significantly influence the engineering properties and behavior of a soil.

1.4.2 *Particle size of fine-grained soils*

The screening process cannot be used for fine-grained soils—silts and clays—because of their extremely small size. The common laboratory method used to determine the size distribution of fine-grained soils is a hydrometer test (Figure 1.8). The hydrometer test involves mixing a small amount of soil into a suspension and observing how the suspension settles in time. Larger particles will settle quickly, followed by smaller particles. When the hydrometer is lowered into the suspension, it will sink until the buoyancy force is sufficient to balance the weight of the hydrometer.

The length of the hydrometer projecting above the suspension is a function of the density, so it is possible to calibrate the hydrometer to read the density of the suspension at different times. The calibration of the hydrometer is affected by temperature and the specific gravity of the suspended solids. A correction factor must be applied to the hydrometer reading based on the test temperatures used.

Typically, a hydrometer test is conducted by taking a small quantity of a dry, fine-grained soil (approximately 50 grams) and thoroughly mixing it with distilled water to form a paste. The paste is placed in a 1000 mL (=1 liter = 1000 cm^3) glass cylinder, and distilled water is added to bring the level to the 1000 mL mark. The glass cylinder is then repeatedly shaken and inverted before being placed in a constant-temperature bath. A hydrometer is placed in the glass cylinder and a clock is simultaneously started. At different times, the hydrometer

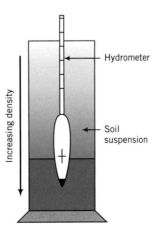

Figure 1.8 Hydrometer in soil-water suspension.

is read. The diameter D (mm) of the particle at time t_D (minute) is calculated from Stokes's law as

$$D = \sqrt{\frac{30\mu z}{980(G_s - 1)t_D}} = K\sqrt{\frac{z}{t_D}} = K\sqrt{v_{set}} \tag{1.3}$$

where μ is the viscosity of water [0.01 Poise at 20°C; 10 Poise = 1 Pascal second (Pa.s) = 1000 centiPoise], z is the effective depth (cm) of the hydrometer, G_s is the specific gravity of the soil particles, and $K = \sqrt{30\mu/980(G_s - 1)}$ is a parameter that depends on temperature and the specific gravity of the soil particles, and v_{set} is the settling velocity. For most soils, $G_s \approx 2.7$. At a temperature of 20°C and for $G_s = 2.7$, $K = 0.01341$.

In the application of Stokes's law, the particles are assumed to be free-falling spheres with no collision. But the mineral particles of fine-grained soils are platelike, and collision of particles during sedimentation is unavoidable. Also, Stokes's law is valid only for laminar flow with Reynolds number (Re $= vD\gamma_w/\mu g$, where v is velocity, D is the diameter of the particle, γ_w is the unit weight of water, μ is the viscosity of water at 20°C, and g is the acceleration due to gravity) smaller than 1. Laminar flow prevails for particle sizes in the range 0.001 mm $< D <$ 0.1 mm. By using the material of average particle size $<$ 0.075 mm, laminar flow is automatically satisfied for particles greater than 0.001 mm. Particles smaller than 0.001 mm are colloids. Electrostatic forces influence the motion of colloids, and Stokes's law is not valid. Brownian motion describes the random movement of colloids.

It is important to distinguish silts from clays because, apart from particle size differences, they have different strength and deformation properties. Silts have lower strength than clays and absorb smaller amounts of water to become "liquid like." Silts tend to dry and become powdery, whereas clays become brittle on drying.

The results of the hydrometer test suffice for most geotechnical engineering needs. For more accurate size distribution measurements in fine-grained soils, other, more sophisticated methods are available (e.g., light-scattering methods). The dashed line in Figure 1.7 shows a typical particle size distribution for fine-grained soils.

1.5   CHARACTERIZATION OF SOILS BASED ON PARTICLE SIZE

The grading curve is used for textural classification of soils. Various classification systems have evolved over the years to describe soils based on their particle size distribution. Each system was developed for a specific engineering purpose. In the United States, the popular systems are the Unified Soil Classification System (USCS), the American Society for Testing and Materials (ASTM) system (a modification of the USCS system), and the American Association of State Highway and Transportation Officials (AASHTO) system (Figure 1.9). Other countries such as those in Europe use the Euro-Standards. We will discuss soil classification in more detail in Chapter 2.

In this textbook, we will use the USCS system because this is sufficiently general for educational purposes. We will modify it slightly by delimiting clays as having particles less than 0.002 mm. Soils are separated into two categories. One category is coarse-grained soils, which are thus delineated if more than 50% of the soil is greater than 0.075 mm. The other category is fine-grained soils, which are thus delineated if more than 50% of the soil is finer than 0.075 mm. Coarse-grained soils are subdivided into gravels and sands, while fine-grained soils are divided into silts and clays. Each soil type—gravel, sand, silt, and clay—is identified by grain size, as shown in Table 1.1. Clays have particle sizes less than 0.002 mm.

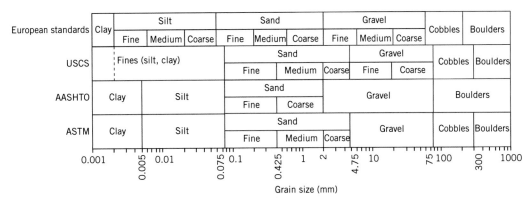

Figure 1.9 Comparison of four systems describing soil types based on particle size.

Table 1.1 Soil types, descriptions, and average grain sizes.

Category	Soil type	Symbol	Description	Grain size, D
Coarse-grained	Gravel	G	Rounded and/or angular bulky hard rock, coarsely divided	Coarse: >75 mm Fine: 4.75 mm–19 mm
	Sand	S	Rounded and/or angular hard rock, finely divided	Coarse: 2.0 mm–4.75 mm Medium: 0.425 mm–2.0 mm Fine: 0.075 mm–0.425 mm
Fine-grained (also called fines)	Silt	M	Particle size between clay and sand; nonplastic or very slightly plastic; exhibits little or no strength when dried; easily brushed off when dried	0.002 mm–0.075 mm
	Clay	C	Particles are smooth and mostly clay minerals; greasy and sticky when wet; exhibits plasticity and significant strength when dried; water reduces strength	<0.002 mm

These sizes are established for convenience. Sand, which consists mainly of quartz minerals, can be grounded to a powder with particle sizes less than 0.002 mm, but the powder will not behave like a clay. Clays have high specific surface area, which is the surface area of the particles divided by the mass. Real soils consist of a mixture of particle sizes.

The amount of fines (materials with particle sizes < 0.075 mm) can considerably influence the response of a soil to loads. For example, a soil containing more than 35% of fines is likely to behave like a fine-grained soil. Fines content less than 5% has little or no influence on the soil behavior. Thus knowledge of the fines content in a soil is critical to understanding how that soil can be used as a construction material or as a foundation for a structure. The selection of a soil for a particular use may depend on the assortment of particles it contains. Two coefficients have been defined to provide guidance on distinguishing soils based on the distribution of the particles.

One of these is a numerical measure of uniformity, called the *uniformity coefficient, Cu*, defined as

$$Cu = \frac{D_{60}}{D_{10}} \tag{1.4}$$

where D_{60} is the diameter of the soil particles for which 60% of the particles are finer, and D_{10} is the diameter of the soil particles for which 10% of the particles are finer. Both of these diameters are obtained from the grading curve.

The other coefficient is the *coefficient of curvature*, CC (other terms used are the coefficient of gradation and the coefficient of concavity), defined as

$$CC = \frac{(D_{30})^2}{D_{10}D_{60}} \tag{1.5}$$

where D_{30} is the diameter of the soil particles for which 30% of the particles are finer. The average particle diameter is D_{50}.

The minimum value of *Cu* is 1 and corresponds to an assemblage of particles of the same size. The shape of the gradation curve indicates the range of particles in a soil. The gradation curve for a poorly graded soil is almost vertical (Figure 1.7). Humps in the gradation curve indicate two or more poorly graded soils. A well-graded soil is indicated by a flat curve (Figure 1.7). The absence of certain grain sizes, termed gap graded, is diagnosed by a sudden change of slope in the particle size distribution curve, as shown in Figure 1.7.

Poorly graded soils are sorted by water (e.g., beach sands) or by wind. Gap-graded soils are also sorted by water, but certain sizes were not transported. Well graded soils are produced by bulk transport processes (e.g., glacial till). The uniformity coefficient and the coefficient of concavity are strictly applicable to coarse-grained soils. The limits of uniformity coefficient and the coefficient of concavity to characterize well graded and poorly graded are as follows:

Well graded	gravel content > sand content	$Cu \geq 4;\ 1 \leq CC \leq 3$
	sand content > gravel content	$Cu \geq 6;\ 1 \leq CC \leq 3$
Poorly graded	gravel content > sand content	$Cu < 4;\ CC < 1$ or $CC > 3$
	sand content > gravel content	$Cu < 6;\ CC < 1$ or $CC > 3$

Gap graded soils are outside the limits of *Cu* and CC for well-graded and poorly graded soils.

The diameter D_{10} is called the effective size of the soil and was described by Allen Hazen (1892) in connection with his work on soil filters. The effective size is the diameter of an artificial sphere that will produce approximately the same effect as an irregularly shaped particle. The effective size is particularly important in regulating the flow of water through soils, and can dictate the mechanical behavior of soils since the coarser fractions may not be in effective contact with each other; that is, they float in a matrix of finer particles. The higher the D_{10} value, the coarser is the soil and the better are the drainage characteristics.

Particle size analyses have many uses in engineering. They are used to select aggregates for concrete, soils for the construction of dams and highways, soils as filters for drainage, and soils as material for grouting and chemical injection. In Chapter 2, you will learn about how the particle size distribution is used with other physical properties of soils in a classification system designed to help you select soils for particular applications.

Key points

1. A sieve analysis is used to determine the grain size distribution of coarse-grained soils.
2. For fine-grained soils, a hydrometer analysis is used to find the particle size distribution.
3. Particle size distribution is represented on a semi-logarithmic plot of percentage finer (ordinate, arithmetic scale) versus particle size (abscissa, logarithmic scale).
4. The particle size distribution plot is used to delineate the different soil textures (percentages of gravel, sand, silt, and clay) in a soil.
5. The effective size, D_{10}, is the diameter of the particles of which 10% of the soil is finer. D_{10} is an important value in regulating flow through soils and can significantly influence the mechanical behavior of soils.
6. D_{50} is the average grain size diameter of the soil.
7. Two coefficients—the uniformity coefficient and the coefficient of curvature—are used to characterize the particle size distribution. Poorly graded soils have steep gradation curves. Well graded soils are indicated by relatively flat particle distribution curves and have uniformity coefficients >4, coefficients of curvature between 1 and 3. Gap-graded soils are indicated by one or more humps on the gradation curves.

EXAMPLE 1.1 *Calculation of Percentage Finer than a Given Sieve in a Sieve Analysis Test*

A particle analysis test was conducted on a dry soil. The total mass used in test was 500 grams. All 500 grams are greater than 9.5 mm. The total mass of particles greater than 0.075 mm was 220 grams. Determine the percentage of coarse-grained and fine-grained soil particles.

Strategy Calculate the cumulative percentage greater than 0.075 mm, and then subtract it from 100 to get percentage finer than. Use Table 1.1 to guide you to get the amount of each soil category.

Solution 1.1

Step 1: Determine percentage greater than 0.075 mm.

Mass of soil particles greater than 0.075 mm, $M_r = 220$ grams.

Total mass, $M_t = 500$ grams.

% of soil particles greater than 0.075 mm =

$$\frac{M_r}{M_t} \times 100 = \frac{220}{500} \times 100 = 44\%$$

Step 2: Determine percentage finer than 0.075 mm.

% finer than 0.075 mm = 100 − 44 = 56%.

Step 3: Determine % coarse-grained and fine-grained particles.

% coarse-grained soil particles = % of particles greater than 0.075 mm = 44%.

% fine-grained soil particles = % of particles finer than 0.075 mm = 56%.

EXAMPLE 1.2 *Calculating Particle Size Distribution and Interpretation of Soil Type from a Sieve Analysis Test*

A sieve analysis test was conducted using 650 grams of soil. The results are as follows.

Sieve opening (mm)	9.53	4.75	2	0.85	0.425	0.15	0.075	Pan
Mass retained (grams)	0	53	76	73	142	85.4	120.5	99.8

Determine (a) the amount of coarse-grained and fine-grained soils, and (b) the amount of each soil type based on the USCS system.

Strategy Calculate the percentage finer and plot the gradation curve. Extract the amount of coarse-grained soil (particle sizes ≥ 0.075 mm) and the amount of fine-grained soil (particle sizes < 0.075 mm). Use Table 1.1 to guide you to get the amount of each soil type.

Solution 1.2

Step 1: Set up a table or a spreadsheet to do the calculations.

A	B	C	D	E
Sieve opening (mm)	Mass retained (grams), M_r	% Retained $(100 \times M_r/M_t)$	Σ (% Retained) (Σ column D)	% Finer (100 − column E)
9.53	0	0.0	0.0	100.0
4.75	53.0	8.2	add ⟶ 8.2	91.8
2.00	76.0	11.7 ⟵ ⟶ 19.9		80.1
0.85	73.0	11.2	31.1	68.9
0.425	142.0	21.9	52.9	47.1
0.15	85.4	13.1	66.1	33.9
0.075	120.5	18.5	check 84.6 ⟶	15.4
	99.8	15.4		15.4
Total mass = M_t =	649.7	100.0		

Note: In the sieve analysis test, some mass is lost because particles are stuck in the sieves. Use the sum of the mass after the test. You should always check that the sum of the soil retained on all sieves plus the pan is equal to 100% (column C in the table).

Step 2: Plot grading curve.

See Figure E1.2.

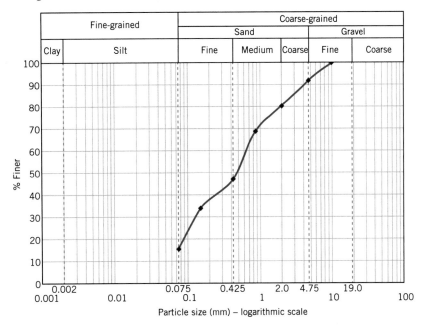

Figure E1.2

Step 3: Extract soil type.

(a) The amount of fine-grained soil is the percentage finer than 0.075 mm. The amount of coarse-grained soil is the percentage coarser than 0.075 mm.

$$\% \text{ fine-grained soil} = 15.4\%.$$

$$\% \text{ coarse-grained soil} = 100 - 15.4 = 84.6\%.$$

Check answer: % fine-grained soil + % coarse-grained soil must be 100%.

That is: 15.4 + 84.6 = 100%.

(b)
$$\text{Fine gravel (\%)} = 8.2$$

$$\text{Total gravel (\%)} = 8.2$$

$$\text{Coarse sand (\%)} = 11.7$$

$$\text{Medium sand (\%)} = 33.0$$

$$\text{Fine sand (\%)} = 31.7$$

$$\text{Total sand (\%)} = 76.4$$

$$\text{Silt} + \text{clay (\%)} = 15.4$$

Check answer: total gravel (%) + total sand (%) + silt (%) + clay (%) must equal 100%

That is: 8.2 + 76.4 + 15.4 = 100%

EXAMPLE 1.3 *Interpreting Sieve Analysis Data*

A sample of a dry, coarse-grained material of mass 500 grams was shaken through a nest of sieves, and the following results were as given in the table below.

(a) Plot the particle size distribution (gradation) curve.

(b) Determine (1) the effective size, (2) the average particle size, (3) the uniformity coefficient, and (4) the coefficient of curvature.

(c) Determine the textural composition of the soil (the amount of gravel, sand, etc.).

Sieve opening (mm)	Mass retained (grams)
4.75	0
2.00	14.8
0.85	98.0
0.425	90.1
0.15	181.9
0.075	108.8
	6.1

Strategy The best way to solve this type of problem is to make a table to carry out the calculations and then plot a gradation curve. Total mass (M) of dry sample used is 500 grams, but on summing the masses of the retained soil in column 2 we obtain 499.7 grams. The reduction in mass is due to losses mainly from a small quantity of soil that is stuck in the meshes of the sieves. You should use the "after sieving" total mass of 499.7 grams in the calculations.

Solution 1.3

Step 1: Tabulate data to obtain percentage finer.

See the table below.

Mass retained (grams), M_r	% Retained $(M_r/M) \times 100$	Σ% retained	% Finer
0	0	0	100 − 0 = 100
14.8	3.0	add ⟶ 3.0	100 − 3.0 = 97.0
98.0	19.6 ◄	⟶ 22.6	100 − 22.6 = 77.4
90.1	18.0	40.6	100 − 40.6 = 59.4
181.9	36.4	77.0	100 − 77 = 23.0
108.8	21.8	98.8	100 − 98.8 = 1.2
6.1	1.2	check	
Total mass M_t = 499.7	100.0		

Step 2: Plot the gradation curve.

See Figure E1.3 for a plot of the gradation curve.

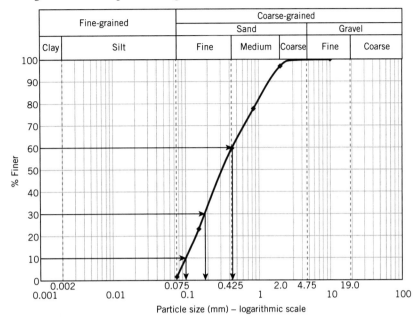

Figure E1.3

Step 3: Extract the effective size.

$$\text{Effective size} = D_{10} = 0.1 \text{ mm}$$

Step 4: Extract percentages of gravel, sand, silt, and clay.

$$\text{Gravel} = 0\%$$

$$\text{Sand} = 98.8\%$$

$$\text{Silt and clay} = 1.2\%$$

Check answer: gravel (%) + sand (%) + silt (%) + clay (%) must equal 100%.

That is: $0 + 98.8 + 1.2 = 100\%$

Step 5: Calculate Cu and CC.

$$Cu = \frac{D_{60}}{D_{10}} = \frac{0.45}{0.1} = 4.5$$

$$CC = \frac{(D_{30})^2}{D_{10}D_{60}} = \frac{0.18^2}{0.1 \times 0.45} = 7.2$$

EXAMPLE 1.4 *Calculation of Particle Diameter from Hydrometer Test Data*

After a time of 1 minute in a hydrometer test, the effective depth was 0.8 cm. The average temperature measured was 20°C and the specific gravity of the soil particles was 2.7, calculate the diameter of the particles using Stokes's law. Are these silt or clay particles?

Strategy This is a straightforward application of Equation (1.3).

Solution 1.4

Step 1: Calculate the particle diameter using Stokes's law.

$z = 0.8$ cm and $t_D = 1$ minute. For the temperature and specific gravity of the soil particles, $K = 0.01341$

$$D = K\sqrt{\frac{z}{t_D}} = 0.01341\sqrt{\frac{0.8}{1}} = 0.012 \text{ mm}$$

Step 2: Identify the soil type.

Silt particles have sizes between 0.075 mm and 0.002 mm.

Therefore, the soil particles belong to the silt fraction of the soil.

EXAMPLE 1.5 *Interpreting Hydrometer Analysis*

Sixty-five grams of the soil finer than 0.075 mm (passing the No. 200 sieve) in Example 1.2 was used to conduct a hydrometer test. The results are shown in the table below. What are the amounts of clays and silts in the soil?

Time (min)	Hydrometer reading (gram/liter)	Temperature (°C)	Corrected distance of fall (cm)	Grain size (mm)	% Finer by weight
1	40.0	22.5	8.90	0.0396	82.2
2	34.0	22.5	9.21	0.0285	68.8
3	32.0	22.0	9.96	0.0243	64.2
4	30.0	22.0	10.29	0.0214	59.7
8	27.0	22.0	10.96	0.0156	53.1
15	25.0	21.5	11.17	0.0116	48.4
30	23.0	21.5	11.45	0.0083	43.9
60	21.0	21.5	11.96	0.0060	39.5
240	17.0	20.0	12.45	0.0031	30.0
900	14.0	19.0	13.10	0.0017	22.9

Solution 1.5

Step 1: Plot percentage finer versus particle size (log scale).

See Figure E1.5.

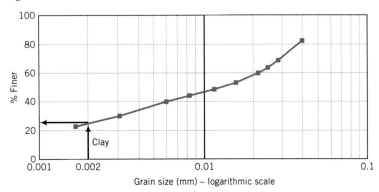

Figure E1.5

Step 2: Extract percentage finer than 0.002 mm.

$$\% \text{ finer than } 0.002 \text{ mm} = 25\%.$$

$$\% \text{ of clay and silt from Example } 1.2 = 15.4\%.$$

$$\% \text{ clay in the soil in Example } 1.2 \text{ is } (25/100) \times 15.4 = 3.9\%.$$

$$\% \text{ silt} = 15.4 - 3.9 = 11.5\%.$$

Check answer: silt (%) + clay (%) must equal 15.4%:

$$11.5 + 3.9 = 15.4\%$$

1.6 COMPARISON OF COARSE-GRAINED AND FINE-GRAINED SOILS FOR ENGINEERING USE

Coarse-grained soils have good load-bearing capacities and good drainage qualities, and their strength. They are practically incompressible when dense, but significant volume changes can occur when they are loose. Vibrations accentuate volume changes in loose, coarse-grained soils by rearranging the soil fabric into a dense configuration. Coarse-grained soils with angular particles have higher strengths, higher compressibilities, and lower densities than coarse-grained soils with rounded particles. The engineering properties of coarse-grained soils are controlled mainly by the grain size of the particles and their structural arrangement. Changes in moisture conditions do not significantly affect the volume change under static loading.

Coarse-grained soils are generally described as free draining. However, the term free draining means that the soil allows free passage of water in a relatively short time (a few minutes). Fines content (silts and clays) can significantly alter the flow conditions in these soils. Gravel, boulders, and coarse sands with fines content less than 5% are free draining. Fine sand, especially if it exists as a thick layer, is not free draining.

Fine-grained soils have poor load-bearing capacities compared with coarse-grained soils. Fine-grained soils are practically impermeable (not free draining), change volume and strength with variations in moisture conditions, and are susceptible to frost. Mineralogical factors rather than grain size control the engineering properties of fine-grained soils. Thin layers of fine-grained soils, even within thick deposits of coarse-grained soils, have been responsible for many geotechnical failures, and therefore, you need to pay special attention to fine-grained soils.

Key points

1. Fine-grained soils have much larger surface areas than coarse-grained soils and are responsible for the major physical and mechanical differences between coarse-grained and fine-grained soils.
2. The engineering properties of fine-grained soils depend mainly on mineralogical factors.
3. Coarse-grained soils have good load-bearing capacities and good drainage qualities. Changes in moisture conditions do not significantly affect the volume-change characteristics under static loading.
4. Fine-grained soils have low load-bearing capacities and poor drainage qualities. Changes in moisture conditions strongly influence the volume-change characteristics and strength of fine-grained soils.

1.7 SUMMARY

Soils are derived from the weathering of rocks by physical and chemical processes. The main groups of soils for engineering purposes from these processes are coarse-grained soils —sand and gravels—and fine-grained soils—silts and clays. Particle size is sufficient to identify coarse-grained soils. Fine-grained soils require mineralogical characterization in addition to particle size for identification. Coarse-grained and fine-grained soils have different engineering properties. Moisture content changes strongly influence the behavior of fine-grained soils. Moisture content changes do not significantly influence the behavior of coarse-grained soils under static loading.

EXERCISES

Concept understanding

1.1 Describe the processes responsible for the formation of soils from rock.

1.2 (a) What is a mineral?

　　(b) Describe the differences among the three main soil minerals.

　　(c) Which mineral group is most important for soils and why?

1.3 Which of the three main clay minerals undergo large volume change in contact with water and why?

1.4 (a) What is soil fabric?

(b) What is the name for the spaces between mineral particles?

(c) Why are the spaces between mineral particles important to geoengineers?

(d) Explain the differences between a flocculated and a dispersed structure?

1.5 Describe the differences among alluvial, collovial, glacial, and lateritic soils.

1.6 (a) What are the two types of surface forces in clayey soils?

(b) What is adsorbed water?

(c) Can you remove the adsorbed water by oven drying at 105°C? Explain.

1.7 What is the shape of the hole in a standard sieve?

1.8 What tests would you specify to determine the grain size of a sand that contains fine-grained soils?

Problem solving

1.9 In a sieve analysis test, the amount of soil retained on all sieves with 0.425 mm opening and above is 100 grams. The total mass used in the test is 500 grams.

(a) Determine the percentage of the soil greater than 0.425 mm (No. 40 sieve).

(b) Determine the percentage finer than 0.425 mm.

1.10 The data from a particle size analysis on a sample of a dry soil at a depth of 0.5 m near a mountain range (colluvium) are given in the table below.

Sieve opening (mm)	9.53	4.75	2.0	0.84	0.425	0.15	0.075	Pan
Mass retained (grams)	0	31	38	58	126	120	68	58

(a) What is the total mass of the soil retained on all sieves including the pan?

(b) If the total mass used at the start of the test is 500 grams, what is the percentage loss? Explain why this loss occurred in the test.

(c) Plot the particle size distribution curve.

(d) What are the percentages of coarse-grained and fine-grained soils in the sample.

1.11 The effective depth measured in a hydrometer test after 8 minutes is 1 cm. (a) Determine the average particle size if K is 0.01341, and (b) identify the soil type (e.g., silt or clay) corresponding to the average particle size.

1.12 The following results were obtained from sieve analyses of two soils.

Sieve opening (mm)	Mass (grams)	
	Soil A	Soil B
4.75	0	0
2.00	20.2	48.1
0.85	25.7	219.5
0.425	60.4	67.3
0.15	98.1	137.2
0.075	127.2	22.1
	168.2	5.6

Hydrometer tests on these soils gave the following results.

Particle size (mm)	% finer	
	Soil A	Soil B
0.05	22.6	1.0
0.01	13.8	0.9
0.005	12.2	0.8
0.002	5	0.5

(a) Plot the gradation curve for each soil on the same graph.

(b) How much coarse-grained and fine-grained soils are in each soil?

(c) What are the percentages of clay and silt in each soil according to USCS?

(d) Determine D_{10} for each soil.

(e) Determine the uniformity coefficient and the coefficient of concavity for each soil.

(f) Describe the gradation curve (e.g., well graded) for each soil?

Critical thinking and decision making

1.13 Why do geoengineers plot particle distribution curves on a semi-log scale with particle size on the abscissa (logarithmic scale) versus percentage finer on the ordinate (arithmetic scale)? Is there any theoretical justification for this? Would the shape of the grain size graph be different if arithmetic rather than semi-log scale is used?

1.14 If a soil consists of sand and fines, would drying the soil and then sieving it through a standard stack of sieves give accurate results on the fines content? Justify your answer.

1.15 If you have to select a soil for a roadway that requires good drainage qualities, what soil type would you select and why?

1.16 A house foundation consists of a concrete slab casted on a clay soil. The homeowner planted vegetation near one side of the foundation and watered it regularly, sometimes excessively. She noticed that this side of the foundation curled upward, the concrete slab cracked and several cracks appeared on the wall. What do you think is likely the predominant mineral in the clay soil? Justify your answer.

Chapter 2
Phase Relationships, Physical Soil States, and Soil Classification

2.1 INTRODUCTION

Soils are naturally complex, multiphase materials. They are generally a matrix of an assortment of particles (solids), fluids, and gases. Each influences the behavior of the soil mass as a whole. Unless we understand the composition of a soil mass, we will be unable to estimate how it will behave under loads and how we can use it as a construction material. Geoengineers have devised classification systems based on the results of simple, quick soil tests. These classifications help us make decisions about the suitability of particular types of soils for typical geoengineering systems.

In this chapter, we will dismantle soil into three constituents and examine how the proportions of each constituent characterize soils. We will also briefly describe standard tests to determine the physical states of soils. The results of these tests and determination of particle size distribution (Chapter 1) allow us to classify soils.

Learning outcomes

When you complete this chapter, you should be able to do the following:

- Determine the proportions of the main constituents in a soil.
- Understand how water changes the states of soils, particularly fine-grained soils.
- Determine index parameters of soils.
- Classify soils.

2.2 DEFINITIONS OF KEY TERMS

Water content (*w*) is the ratio of the weight of water to the weight of solids.
Void ratio (*e*) is the ratio of the volume of void spaces to the volume of solids.

Soil Mechanics Fundamentals, First Edition. Muni Budhu.
© 2015 John Wiley & Sons, Ltd. Published 2015 by John Wiley & Sons, Ltd.
Companion website: www.wiley.com\go\budhu\soilmechanicsfundamentals

Porosity (n) is the ratio of the volume of voids to the total volume of soil.

Degree of saturation (S) is the ratio of the volume of water to the volume of voids.

Bulk unit weight (γ) is the weight density, that is, the weight of a soil per unit volume.

Saturated unit weight (γ$_{sat}$) is the weight of a saturated soil per unit volume.

Dry unit weight (γ$_d$) is the weight of a dry soil per unit volume.

Effective unit weight (γ') is the weight of a saturated soil submerged in water per unit volume.

Relative density (D$_r$) is an index that quantifies the degree of packing between the loosest and densest state of coarse-grained soils.

Density index (I$_d$) is a similar measure (not identical) to relative density.

Unit weight ratio or density ratio (R$_d$) is the ratio of the unit weight of the soil to that of water.

Swell factor (SF) is the ratio of the volume of excavated material to the volume of in situ material (sometimes called borrow pit material or bank material).

Liquid limit (LL) is the water content at which a soil changes from a plastic state to a liquid state.

Plastic limit (PL) is the water content at which a soil changes from a semisolid to a plastic state.

Shrinkage limit (SL) is the water content at which a soil changes from a solid to a semisolid state without further change in volume.

Plasticity index (PI) is the range of water content for which a soil will behave as a plastic material (deformation without cracking).

Liquidity index (LI) is a measure of soil strength using the Atterberg limits (liquid and plastic limits based on test data).

Shrinkage index (SI) is the range of water content for which a soil will behave as a semisolid (deformation with cracking).

2.3 PHASE RELATIONSHIPS

Soil is composed of solids, liquids, and gases (Figure 2.1a). The solid phase may be minerals, organic matter, or both. The spaces between the solids (soil particles) are called voids. Water is often the predominant liquid and air is the predominant gas. We will use the terms water and air instead of liquid and gas. The soil water is called porewater and plays a very important role in the behavior of soils under load. If all the voids are filled by water, the soil is

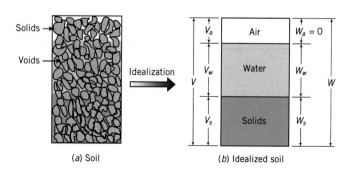

(a) Soil (b) Idealized soil

Figure 2.1 Soil phases.

saturated. Otherwise, the soil is unsaturated. If all the voids are filled with air, the soil is said to be dry.

We can idealize the three phases of soil, as shown in Figure 2.1b. The physical parameters of soils are influenced by the relative proportions of each of these phases. The total volume of the soil is the sum of the volume of solids (V_s), volume of water (V_w), and volume of air (V_a); that is,

$$V = V_s + V_w + V_a = V_s + V_v \qquad (2.1)$$

where

$$V_v = V_w + V_a$$

is the volume of voids. The weight of the soil is the sum of the weight of solids (W_s) and the weight of water (W_w). The weight of air is negligible. Thus,

$$W = W_s + W_w \qquad (2.2)$$

The following definitions have been established to describe the proportion of each constituent in a soil. Each equation can be presented with different variables. The most popular and convenient forms are given. You should try to memorize these relationships. When you work on these relationships, think about a bread dough in which you have to reconstruct the amount of the constituent ingredients, for example, the amount of flour or water. If you add too much water to a bread dough, it becomes softer and more malleable. The same phenomenon occurs in fine-grained soils.

1. **Water content** (*w*) is the ratio, often expressed as a percentage, of the weight of water to the weight of solids:

$$w = \frac{W_w}{W_s} \times 100\% \qquad (2.3)$$

   The water content of a soil is found by weighing a sample of the soil and then placing it in an oven at $110 \pm 5°C$ until the weight of the sample remains constant; that is, all the absorbed water is driven out. For most soils, a constant weight is achieved in about 24 hours. The soil is removed from the oven, cooled, and then weighed. It is a common mistake to use the total weight in the denominator. Remember, it is the weight (or mass) of the solids. Some fine-grained soils may contain appreciable amounts of adsorbed water that cannot be removed by drying at $110 \pm 5°C$. If organic matter is present in a soil, it may oxidize and decompose at $110 \pm 5°C$. Thus, the weight loss may not be entirely due to the loss of water.

2. **Void ratio** (*e*) is the ratio of the volume of void space to the volume of solids. The void ratio is usually expressed as a decimal quantity:

$$e = \frac{V_v}{V_s} \qquad (2.4)$$

3. **Specific volume** (*V'*) is the volume of soil per unit volume of solids:

$$V' = \frac{V}{V_s} = 1 + e \qquad (2.5)$$

   This equation is useful in relating volumes.

4. **Porosity (*n*)** is the ratio of the volume of voids to the total volume. Porosity is usually expressed as a percentage:

$$n = \frac{V_v}{V} \qquad (2.6)$$

Porosity and void ratio are related by the expression

$$n = \frac{e}{1+e} \qquad (2.7)$$

Let us prove Equation (2.7). We will start with the basic definition given in Equation (2.6), and then we algebraically manipulate it to get Equation (2.7). The total volume is decomposed into the volume of solids and the volume of voids, and then both the numerator and denominator are divided by the volume of solids; that is,

$$n = \frac{V_v}{V} = \frac{V_v}{V_s + V_v} = \frac{V_v / V_s}{V_s / V_s + V_v / V_s} = \frac{e}{1+e}$$

The porosity of soils can vary widely. If the particles of coarse-grained soils were spheres, the maximum and minimum porosities would be 48% and 26%, respectively. This is equivalent to maximum and minimum void ratios of 0.91 and 0.35, respectively. The void ratios of real coarse-grained soils vary between 1 and 0.3. Clay soils often have void ratios greater than 1.

5. **Specific gravity (*G_s*)** is the ratio of the weight of the soil solids to the weight of water of equal volume:

$$G_s = \frac{W_s}{V_s \gamma_w} \qquad (2.8)$$

where $\gamma_w = 9.81 \, kN/m^3$ is the unit weight of water. The specific gravity of soils ranges from approximately 2.3 to 2.8; the lower range (2.3 to 2.5) are for silt particles with traces of organic material. For most problems, G_s can be assumed, with little error, to be equal to 2.7.

   Two types of container are used to determine the specific gravity for soil particles less than 4.75 mm (No. 4 sieve). One is a volumetric flask (at least 100 mL) that is used for coarse-grained soils. The other is a 50-mL density bottle (stoppered bottle) that is used for fine-grained soils. The container is weighed and a small quantity of dry soil is placed in it. The mass of the container and the dry soil is determined. De-aired water is added to the soil in the container. The container is then agitated to remove air bubbles. When all air bubbles have been removed, the container is filled with de-aired water. The mass of container, soil, and water is determined. The contents of the container are discarded and the container is thoroughly cleaned. De-aired water is added to fill the container and the mass of the container and water is determined.

   Let M_1 be the mass of the oven-dried soil, M_2 be the mass of the container and water, and M_3 be the mass of the container, oven-dried soil, and water. The mass of water displaced by the soil particles is $M_4 = M_1 + M_2 - M_3$, and $G_s = M_1/M_4$.

6. **Degree of saturation (*S*)** is the ratio, often expressed as a percentage, of the volume of water to the volume of voids:

$$S = \frac{V_w}{V_v} = \frac{wG_s}{e} \quad \text{or} \quad Se = wG_s \qquad (2.9)$$

If $S = 1$ or 100%, the soil is saturated. If $S = 0$, the soil is bone dry. It is practically impossible to obtain a soil with $S = 0$. The water content in Equation (2.9) is a decimal quantity (water content of 10% is $w = 0.1$ in the equation).

7. **Unit weight** is the weight of a soil per unit volume. We will use the term **bulk unit weight**, γ, to denote unit weight:

$$\gamma = \frac{W}{V} = \left(\frac{G_s + Se}{1+e}\right)\gamma_w \tag{2.10}$$

Special cases

(a) Saturated unit weight $(S = 1)$:

$$\gamma_{sat} = \left(\frac{G_s + e}{1+e}\right)\gamma_w \tag{2.11}$$

(b) Dry unit weight $(S = 0)$:

$$\gamma_d = \frac{W_s}{V} = \left(\frac{G_s}{1+e}\right)\gamma_w = \frac{\gamma}{1+w} = G_s\gamma_w(1-n) \tag{2.12}$$

(c) Effective or buoyant unit weight is the weight of a saturated soil, surrounded by water, per unit volume of soil:

$$\gamma' = \gamma_{sat} - \gamma_w = \left(\frac{G_s - 1}{1+e}\right)\gamma_w \tag{2.13}$$

The equations for unit weights and other relationships in this section can be written in different forms for convenience. For example, the bulk unit weight can be written as $\gamma = \gamma_d + nS\gamma_w$. This is convenient if γ_d, n, and S are given.

Typical values of unit weight of soils are given in Table 2.1. This and other tables of ranges of typical soil values in this textbook are based on observed and reported values for various soil types. They are intended for guidance.

Let us consider the limits of unit weights for soils. We will use the ratio of the soil's unit weight to that of water, which for a saturated soil is (γ_{sat}/γ_w). This ratio is a dimensionless quantity that we will label as R_d. Thus, R_d indicates how much soil is heavier than water per unit volume (Table 2.1). We will label R_d as the unit weight ratio or density ratio. If $e = 0$, then the soil has no voids. It is now an incompressible solid. In this case, $R_d = \gamma_{sat}/\gamma_w = \gamma/\gamma_w = \gamma_d/\gamma_w = G_s$, which is the specific gravity of the soil solids. This is the upper limit R_d. If G_s is 1, then the soil is theoretically water, which is unlikely because the soil solids will confer a unit weight greater than 1. In the extreme (but unlikely) case of the soil becoming water, $\gamma_{sat}/\gamma_w = \gamma/\gamma_w = 1$. The actual lower limit of R_d for soils will correspond

Table 2.1 Typical values of unit weight for soils.

Soil type	γ_{sat} (kN/m³)	R_d	γ_d (kN/m³)	R_d
Gravel	20–22	2.04–2.24	15–17	1.52–1.73
Sand	18–20	1.84–2.04	13–16	1.33–1.63
Silt	18–20	1.84–2.04	14–18	1.43–1.84
Clay	16–22	1.63–2.24	14–21	1.43–2.15

to the maximum void ratio. In the case of clays, the lower limit of R_d will correspond to the void ratio at the water content for which the clay becomes a viscous fluid (see Section 2.4). Therefore, the theoretical limiting conditions for a saturated soil are $G_s \geq \gamma_{sat}/\gamma_w > 1$ or $G_s \geq R_d > 1$. You can use the limits of R_d to judge the reasonableness of soil unit weights.

Geoengineers are particularly concerned with strength and settlement (deformation) of soils. These limits provide a range of R_d for strength and settlement consideration. At or near the upper limit, the soil will have the highest strength and will undergo small settlement. Near the lower limit, the soil will have the lowest strength and will undergo the highest settlement.

8. **Relative density** (D_r) is an index that indicates the degree of packing between the loosest and densest possible state of coarse-grained soils as determined by experiments:

$$D_r = \frac{e_{max} - e}{e_{max} - e_{min}} \tag{2.14}$$

where e_{max} is the maximum void ratio (loosest condition), e_{min} is the minimum void ratio (densest condition), and e is the current void ratio.

The relative density can also be written as

$$D_r = \frac{\gamma_d - (\gamma_d)_{min}}{(\gamma_d)_{max} - (\gamma_d)_{min}} \left\{ \frac{(\gamma_d)_{max}}{\gamma_d} \right\} \tag{2.15}$$

The maximum void ratio is obtained by pouring dry sand, for example, into a mold of volume (V) 2830 cm^3 using a funnel. The sand that fills the mold is weighed. If the weight of the sand is W, then, by combining Equations (2.10) and (2.12), we get $e_{max} = G_s \gamma_w (V/W) - 1$. One method of determining the minimum void ratio is by vibrating the sand with a weight imposing a vertical stress of 14 kPa on top of the sand. Vibration occurs for 8 minutes at a frequency of 3600 Hz and amplitude of 0.33 mm. From the weight of the sand (W_1) and the volume (V_1) occupied by it after vibration, we can calculate the minimum void ratio using $e_{min} = G_s \gamma_w (V_1/W_1) - 1$.

The maximum void ratio is a basic soil property, but the minimum void ratio is not—it depends on the method used to obtain it. Theoretically, the denominator, $(e_{max} - e_{min})$, of Equation (2.14) for a given soil should be a constant. But, in practice, it is not because the minimum void ratio obtained by the laboratory method described above is not necessarily the minimum void ratio that the soil can achieve naturally.

In practice, geoengineers have correlated relative density with various parameters for coarse-grained soils (essentially sand). These correlations are often weak (low coefficient of regression). However, they serve as guidance for preliminary assessment of earthworks, foundations and vibrations of sand. A description of sand based on relative density and porosity is given in Table 2.2. A dense sand (relative density between 70% and 85%) is likely to be stronger and will settle less than a loose sand.

9. **Density index** (I_d) is a similar measure (not identical) to relative density:

$$I_d = \frac{\gamma_d - (\gamma_d)_{min}}{(\gamma_d)_{max} - (\gamma_d)_{min}} \tag{2.16}$$

From Equation (2.15), the relation between D_r and I_d is

$$D_r = I_d \left\{ \frac{(\gamma_d)_{max}}{\gamma_d} \right\} \qquad (2.17)$$

Table 2.2 Description of coarse-grained soils based on relative density and porosity.

D_r (%)	Porosity, n (%)	Description
0–20	100–80	Very loose
20–40	80–60	Loose
40–70	60–30	Medium dense
70–85	30–15	Dense
85–100	<15	Very dense

It is easier to calculate and measure dry density than void ratio, so Equations (2.16) and (2.17) are preferable to Equation (2.14).

10. *Swell factor* (*SF*) or free swell factor is the ratio of the volume of excavated material to the volume of in situ material (sometimes called borrow pit material or bank material):

$$SF = \frac{\text{Volume of excavated material}}{\text{Volume of in situ material}} \times 100 (\%) \qquad (2.18)$$

Free swell ranges for some clay minerals are shown in Table 2.3.

Table 2.3 Ranges of free swell for some clay minerals.

Clay minerals	Free swell (%)
Calcium montmorillonite (Ca-smectite)	45–145
Sodium montmorillonite (Na-smectite)	1400–1600
Illite	15–120
Kaolinite	5–60

EXAMPLE 2.1   *Calculating Water Content*

A wet clay soil and its container weigh 1.10 N. After the wet clay soil and its container was placed in an oven at 110°C for 24 hours, the weight reduced to 0.89 N. If the container weighs 0.22 N, calculate the water content of the clay soil.

Strategy Calculate the water content and then the dry weight of the soil. Then use Equation (2.3). You could also make a sketch of a phase diagram to give you a visual.

Solution 2.1

Step 1: Determine the water content.

See phase diagram, Figure E2.1.

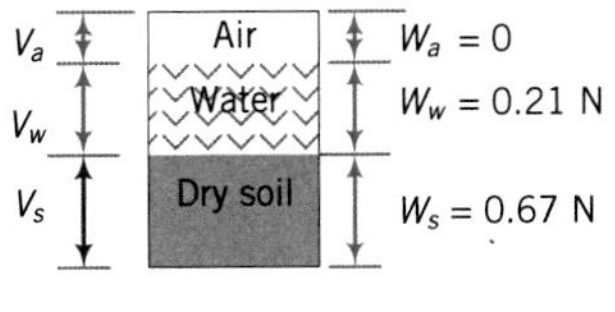

Figure E2.1

Setting up a table as shown below helps you to work logical and keeps the calculation tidy.

W_c = weight of container	0.22 N
W_{wc} = weight of wet soil and container	1.10 N
W_{dc} = weight of dry soil and container	0.89 N
W_d = weight of dry soil = $W_{dc} - W_c$	0.67 N
W_w = weight of water = $W_{wc} - W_{dc}$	0.21 N
w = water content = $100 \times W_w/W_d = 100 \times 0.21/0.67$	31.3 %

Step 2: Check answer.

You can check your answer by using other methods of calculation and/or back calculate to ensure that you used the correct parameters (or variables or constants) and values. In this case, the key parameters are the weight of water and the weight of dry soil.

Check weight of dry soil, $W_d = W_w/w = 0.21/0.313 = 0.67 \text{N}$.

You should always recheck your calculation for water content because water content (directly or indirectly) affects the values of other phase parameters such as void ratio and unit weights.

EXAMPLE 2.2   *Specific Gravity of a Coarse-Grained Soil*

A specific gravity test was conducted on a sand. The data are as shown below. Calculate the specific gravity.

Mass of oven-dried sand	= 92.6 grams
Mass of flask and water	= 663.2 grams
Mass of flask, oven-dried sand, and water	= 722.4 grams

Strategy Prepare a table of the data and carry out the calculations as given in Section 2.3 (item 5: specific gravity).

Solution 2.2

Step 1: Determine the specific gravity.

M_1 = mass of oven-dried soil	= 92.6 grams
M_2 = mass of flask and water	= 663.2 grams
M_3 = mass of flask, oven-dried soil, and water	= 722.4 grams
M_4 = mass of water displaced by soil particles = $M_1 + M_2 - M_3$	= 33.4 grams
Specific gravity, $G_s = M_1/M_4$	= 2.77 grams

Step 2: Check if the answer is reasonable.

The range of specific gravity for most soil types is 2.3 to 2.8. The value 2.77 is within that range.

EXAMPLE 2.3   *Calculation of Void Ratio and Porosity*

A container of volume 0.003m^3 weighs 8.9 N. Dry sand was poured to fill the container. The container and the sand weigh 53.4 N. Calculate (a) the void ratio of the sand, (b) the porosity of the sand, and (c) describe the packing (loose or dense) of the soil. Assume that $G_s = 2.7$.

Strategy Since you know the volume and the dry unit weight, you can calculate the dry unit weight and then find e using Equation (2.12). The porosity can be found using the void ratio–porosity relationship.

Solution 2.3

See phase diagram, Figure E2.3.

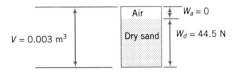

Figure E2.3

Step 1: Calculate the weight of dry sand.

$$\text{Weight of sand and container} = 53.4 \, \text{N}$$

$$\text{Weight of container} = 8.9 \, \text{N}$$

$$\text{Weight of dry sand, } W_d = 53.4 - 8.9 = 44.5 \, \text{N}$$

Step 2: Calculate dry unit weight.

$$\gamma_d = \frac{W_d}{V} = \frac{44.5}{0.003} = 14833 \, \text{N/m}^3 = 14.83 \, \text{kN/m}^3$$

Check reasonableness of the answer. From Table 2.1, this value of dry unit weight is reasonable for sand.

Step 3: Calculate the void ratio.

Equation (2.12): $\gamma_d = \dfrac{W_s}{V} = \dfrac{G_s}{1+e} \gamma_w$

Solving for e, we get

$$e = G_s \frac{\gamma_w}{\gamma_d} - 1 = 2.7 \frac{9.81}{14.83} - 1 = 0.786$$

Check if the answer is correct. Use a different method to calculate the void ratio.

Volume of solids, $V_s = \dfrac{W_d}{G_s \gamma_w} = \dfrac{44.5}{2.7 \times 9810} = 0.00168 \, \text{m}^3$

Volume of voids, $V_v = 0.003 - 0.00168 = 0.00132 \, \text{m}^3$

$e = \dfrac{V_v}{V_s} = \dfrac{0.00132}{0.00168} = 0.786$, which is equal to the value of e calculated above.

Step 4: Calculate the porosity.

Equation (2.7): $n = \dfrac{e}{1+e} = \dfrac{0.786}{1+0.786} = 0.44 = 44\%$

Check if the answer is correct and reasonable. Check using a different method.

$$\text{Total volume of soil, } V_t = 0.003 \text{ m}^3$$

$$\text{Volume of voids, } V_v = 0.00132 \text{ m}^3$$

$n = V_v/V_t = 0.00132/0.003 = 0.44 = 44\%$, which is equal to the value of n calculated above. This value of porosity is reasonable since in practice the range of n is $0 < n < 1$ (see Table 2.2)

Step 5: Describe the soil.

Table 2.2: For $n = 44\%$, the sand is medium dense.

EXAMPLE 2.4 *Calculating Soil Constituents*

A sample of saturated clay and its container weight 5.78 N. The clay in its container was placed in an oven for 24 hours at 105°C. The weight reduced to a constant value of 4.49 N. The weight of the container is 0.89 N. If $G_s = 2.7$, determine the (a) water content, (b) void ratio, (c) bulk unit weight, (d) dry unit weight, and (e) effective unit weight.

Strategy Write down what is given and then use the appropriate equations to find the unknowns. You are given the weight of the natural soil, sometimes called the wet weight and the dry weight of the soil. The difference between these will give the weight of water. You can find the water content by using Equation (2.3). You are also given a saturated soil, which means that $S = 1$.

Solution 2.4

Step 1: Write down what is given.

$$\text{Weight of wet (saturated) sample} + \text{container} = 5.78 \text{ N}$$
$$\text{Weight of dry sample} + \text{container} = 4.89 \text{ N}$$

Step 2: Determine the weight of water and the weight of dry soil.

$$\text{Weight of water: } W_w = 5.78 - 4.89 = 0.89 \text{ N}$$
$$\text{Weight of dry soil: } W_d = 4.89 - 0.89 = 4.00 \text{ N}$$

Step 3: Determine the water content.

$$w = \frac{W_w}{W_d} \times 100 = \frac{0.89}{4.00} \times 100 = 22.3\%$$

Note: The denominator is the weight of solids, not the total weight. You should always recheck that you have calculated w correctly because it will cause the values of the other soil parameters that dependent on it to be incorrect.

Step 4: Determine the void ratio.

$$e = \frac{wG_s}{S} = \frac{0.223 \times 2.7}{1} = 0.6$$

Step 5: Determine the bulk unit weight.

$$\gamma = \frac{W}{V} = \frac{G_s \gamma_w (1+w)}{1+e} \quad \text{(see Example 2.2)}$$

$$\gamma = \frac{2.7 \times 9.81(1+0.223)}{1+0.6} = 20.25 \text{ kN/m}^3$$

In this case the soil is saturated, so the bulk unit weight is equal to the saturated unit weight.

Step 6: Determine the dry unit weight.

$$\gamma_d = \left(\frac{\gamma}{1+w}\right) = \frac{20.25}{1+0.223} = 16.56 \, \text{kN/m}^3$$

Step 7: Determine the effective unit weight.

$$\gamma' = \gamma_{sat} - \gamma_w = 20.25 - 9.81 = 10.44 \, \text{kN/m}^3$$

Step 8: Check if the answers are reasonable.

From Table 2.1, the unit weights calculated are within the range for soils. So the answers are reasonable.

EXAMPLE 2.5 *Calculation of Water Content of an Unsaturated Soil*

The void space in a sand taken near a river consists of 80% air and 20% water. The dry unit weight is $\gamma_d = 15 \, \text{kN/m}^3$ and $G_s = 2.7$. Determine the water content.

Strategy From the amount of air and water in the voids, you can calculate the void ratio from Equation (2.12). Then use Equation (2.9) to find the water content.

Solution 2.5

Step 1: Calculate the void ratio from the dry unit weight.

$$\gamma_d = \frac{G_s \gamma_w}{1+e}$$

Solving for e, we get $e = \dfrac{G_s \gamma_w}{\gamma_d} - 1 = \dfrac{2.7 \times 9.81}{15} - 1 = 0.766$

Step 2: Calculate the water content.

$$Se = wG_s$$
$$w = Se/G_s$$

We need to find the degree of saturation.

The degree of saturation is the ratio of the volume of water to the volume of voids. Since the volume of water is 20% of the void volume, the degree of saturation is 20%, that is, $S = 0.2$.

$$w = \frac{Se}{G_s} = \frac{0.2 \times 0.766}{2.7} = 0.057 = 5.7\%$$

Check the answer.

You could substitute $e = wG_s/S$ in the equation for dry unit weight in step 1 and find w directly instead of finding e first. We can, however, try another method.

$$\text{Volume of solids in 1 m}^3 \text{ of soil, } V_s = \frac{W_d}{G_s \gamma_w} = \frac{15}{2.7 \times 9.81} = 0.566 \, \text{m}^3$$

Note: You are given that one cubic meter of dry soil weights 15 kN, that is, dry unit weight.

$$\text{Volume of voids in 1 m}^3 \text{ of soil, } V_v = \text{total volume} - \text{volume of solids}$$
$$= 1 - 0.566 = 0.434 \, \text{m}^3$$

$e = \dfrac{V_v}{V_s} = \dfrac{0.434}{0.566} = 0.766$, which is equal to the value of e calculated above.

Recheck the calculations for w with $e = 0.766$ and $S = 0.2$ gives $w = 5.7\%$ (see step 2)

EXAMPLE 2.6   *Determination of Aggregate Requirement for a Roadway*

Aggregates from a gravel pit are required for constructing a road embankment. The porosity of the aggregates at the gravel pit is 60%. The desired porosity of the compacted aggregates in the embankment is 25%. For a section of the embankment 7.5 m wide × 0.6 m compacted thickness × 1 km long, (a) calculate the volume of aggregates required. (b) Just before the construction of the embankment, a re-check of the porosity of the aggregates at the gravel pit was conducted. The porosity was 65% rather than 60% (an error of +8.3%), what extra percentage of aggregates would be required to construct the embankment. Assume a swell factor of 1.

Strategy The simplest way is to find a relationship between the n and the volume of the aggregates. The swell factor is 1, so no volume correction is required.

Solution 2.6

Step 1: Calculate the volume of the embankment.

$$V = 7.5 \times 0.6 \times 1 \times 1000 = 4500 \text{ m}^3 \quad \text{(Note: 1 km = 1000 m)}$$

Step 2: Calculate the volume of aggregate required.

Let V_{gp} = volume of aggregate required from the gravel pit, and V_{emb} = volume of the embankment. The subscript gp denotes gravel pit and the subscript emb denotes embankment.

$$\frac{V_{gp}}{V_{emb}} = \frac{1 + e_{gp}}{1 + e_{emb}}$$

We can now substitute $e = n/(1-n)$ in the equation above.

$$\therefore \frac{V_{gp}}{V_{emb}} = \frac{1 + \dfrac{n_{gp}}{1 - n_{gp}}}{1 + \dfrac{n_{emb}}{1 - n_{emb}}} = \frac{\dfrac{1 - n_{gp} + n_{gp}}{1 - n_{gp}}}{\dfrac{1 - n_{emb} + n_{emb}}{1 - n_{emb}}} = \frac{1 - n_{emb}}{1 - n_{gp}}$$

$$= \frac{1 - 0.25}{1 - 0.6} = 1.875$$

$$V_{gp} = 1.875 \times V = 1.875 \times 4500 = 8437.5 \text{ m}^3$$

Note: $\dfrac{1 + e_{gp}}{1 + e_{emb}} = \dfrac{(\gamma_d)_{emb}}{(\gamma_d)_{gp}}$, which is the ratio of dry unit weights

Step 3: Calculate the extra volume of aggregate required.

$$\frac{V_{gp}}{V_{emb}} = \frac{1 - 0.25}{1 - 0.65} = 2.14$$

$$V_{gp} = 2.14 \times V = 2.14 \times 4500 = 9630 \text{ m}^3$$

extra volume required $= \Delta V_{gp} = 9630 - 8437.5 = 1192.5 \text{ m}^3$

$$\text{\% extra volume} = \frac{1192.5}{8437.5} \times 100 = 14.1\% \text{ or } \frac{(2.14 - 1.875)}{1.875} \times 100 = 14.1\%$$

Therefore, an 8.3% error in the initial porosity can cause an extra aggregate cost of at least 14%.

EXAMPLE 2.7 *Application of Soil Constituent Relationships to a Practical Problem*

An embankment for a highway is to be constructed from a sandy clay compacted to a dry unit weight of 18.5 kN/m³. The sandy clay has to be trucked to the site from a borrow pit (a site at which soils are available for construction use). The bulk unit weight of the sandy clay in the borrow pit is 16 kN/m³, and its natural water content is 8%. Calculate the volume of sandy clay from the borrow pit required for 1 cubic meter of embankment. The swell factor is 1.1 (10% free swell). Assume that $G_s = 2.7$.

Strategy This problem can be solved in many ways. Perhaps the easiest is to use the ratio of the dry unit weight of the compacted soil to dry unit weight of the borrow pit soil (see Example 2.6).

Solution 2.7

Step 1: Find the dry unit weight of the borrow pit soil.

$$\gamma_d = \frac{\gamma}{1+w} = \frac{16}{1+0.08} = 14.8 \text{ kN/m}^3$$

Step 2: Find the volume of borrow pit soil required per cubic foot of embankment.

Let V_{bp} = volume of aggregate required from the borrow pit, and V_{emb} = volume of the embankment. The subscript *bp* denotes borrow pit and the subscript *emb* denotes embankment. Without consideration of swell factor,

$$\frac{V_{bp}}{V_{emb}} = \frac{(\gamma_d)_{emb}}{(\gamma_d)_{bp}} = \frac{18.5}{14.8} = 1.25$$
$$V_{bp} = V_{emb} \times 1.25 = 1 \times 1.25 = 1.25 \text{ m}^3$$

With consideration of swell factor,

$$\text{Volume required} = SF \times V_{bp} = 1.1 \times 1.25 = 1.375 \approx 1.4 \text{ m}^3$$

Step 2: Check if the answer is reasonable.

Since the dry unit weight in the borrow pit is lower than that in the embankment, the volume of soil from the borrow pit to be transported to construct the embankment will be greater than 1 m³. The answer is reasonable.

EXAMPLE 2.8 *Application of Soil Constituent Relationships to a Practical Problem*

The borrow soil in Example 2.7 is to be compacted at a water content of 12% to attain a dry unit weight of 18.5 kN/m³. Determine the volume of water required per cubic meter of embankment, assuming 10% loss of water during transportation. Neglect swell.

Strategy Since water content is related to the weight of solids and not the total weight, we need to use the data given to find the weight of solids.

Solution 2.8

Step 1: Determine the weight of solids per unit volume of borrow pit soil.

From step 1 of Example 2.7, the weight of solids per unit volume, $W_d = 14.8$ kN.

Step 2: Determine the amount of water required.

With 10% loss during transportation, the water content of the soil on delivery
$= 0.9 \times 8 = 7.2\%$

Additional water $= 12 - 7.2 = 4.8\%$

Weight of water for $1 \, \text{m}^3$ of soil $= W_w = w \, W_d = 0.048 \times 14.8 = 0.71 \, \text{kN}$

$$V_w = \frac{W_w}{\gamma_w} = \frac{0.71}{9.81} = 0.072 \, \text{m}^3$$

What's next ... Water significantly influences the strength and deformation of fine-grained soils. In the next section, we discuss how water changes the state of fine-grained soils.

2.4 PHYSICAL STATES AND INDEX PARAMETERS OF FINE-GRAINED SOILS

The physical and mechanical behavior of fine-grained soils is linked to four distinct states: solid, semisolid, plastic, and liquid, in order of increasing water content. Let us consider a soil initially in a liquid state that is allowed to dry uniformly. If we plot a diagram of volume versus water content as shown in Figure 2.2, we can locate the original liquid state as point A. As the soil dries, its water content reduces and, consequently, so does its volume (see Figure 2.2b).

At point B, the soil becomes so stiff that it can no longer flow as a liquid. The boundary water content at point B is called the liquid limit; it is denoted by *LL*. As the soil continues to dry, there is a range of water content at which the soil can be molded into any desired shape without rupture. The soil at this state is said to exhibit plastic behavior: the ability to deform continuously without rupture. But if drying is continued beyond the range of water content for plastic behavior, the soil becomes a semisolid. The soil cannot be molded

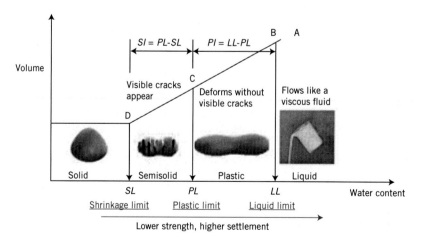

Figure 2.2 Changes in soil states as a function of soil volume and water content.

now without visible cracks appearing. The water content at which the soil changes from a plastic to a semisolid is known as the plastic limit, denoted by *PL*, point C. The range of water contents over which the soil deforms plastically is known as the plasticity index, *PI*:

$$PI = LL - PL \tag{2.19}$$

As the soil continues to dry, it comes to a final state called the solid state. At this state, no further volume change occurs because nearly all the water in the soil has been removed. The water content at which the soil changes from a semisolid to a solid is called the shrinkage limit, denoted by *SL*, point D. The shrinkage limit is useful for the determination of the swelling and shrinking capacity of soils. The range of water content from the plastic limit to the shrinkage limit for which the soil behaves as a semisolid is called the shrinkage index (*SI*),

$$SI = PL - SL \tag{2.20}$$

The shrinkage limit and shrinkage index are important parameters to evaluate the seasonable effects on fine-grained soils. For example, wet clay slopes will shrink during drying in, say, the summer period and cracks may develop at the top of these slopes.In subsequent rainfall, the cracks will acts as conduits for the water to add water pressure (hydrostatic pressure; see Chapter 4) over the depth of the cracks and the soil at the top of the slope will soften (reduce strength). This could lead to slope instability or failure.

We have changed the state of fine-grained soils by changing the water content. Since design geoengineers are primarily interested in the strength and deformation of soils, we can associate specific strength characteristics with each of the soil states. At one extreme, the liquid state, the soil has the lowest strength and the largest deformation. At the other extreme, the solid state, the soil has the largest strength and the lowest deformation. A measure of soil strength using the Atterberg limits is known as the liquidity index (*LI*) and is expressed as

$$LI = \frac{w - PL}{PI} \tag{2.21}$$

The liquidity index is the ratio of the difference in water content between the natural or in situ water content of a soil and its plastic limit to its plasticity index. Table 2.4 shows a description of soil strength based on values of *LI*.

The plasticity index, the liquidity index and shrinkage index are called index parameters. Swedish soil scientist Albert Atterberg (1911) developed tests to determine the index parameters. However, these tests, especially the tests for the liquid limit (see Section 2.5) do not necessarily correspond to the water contents at which the transition from one soil state to another occurs (Figure 2.2). The tests devised by Atterberg and refined later (Arthur Casagrande, 1932) are convenient for engineering (practical) purposes rather than for scientific proof. The index parameters from these tests are called Atterberg limits and are often simply stated as liquid limit, plastic limit, and shrinkage limit.

Table 2.4 Description of the strength of fine-grained soils based on liquidity index.

Values of *LI*	Description of soil strength
$LI < 0$	Semisolid state: high strength, brittle (sudden) fracture is expected
$0 < LI < 1$	Plastic state: intermediate strength, soil deforms like a plastic material
$LI > 1$	Liquid state: low strength, soil deforms like a viscous fluid

Typical range of values of Atterberg limits for soils are shown in Table 2.5. As a reminder, these values in the table are for guidance. These limits depend on the type of predominant mineral in the soil. If montmorillonite is the predominant mineral, the liquid limit can exceed 100%. Why? Recall that the bond between the layers in montmorillonite is weak and large amounts of water can easily infiltrate the spaces between the layers. In the case of kaolinite, the layers are held relatively tightly and water cannot easily infiltrate between the layers in comparison with montmorillonite. Therefore, you can expect the liquid and plastic limits for kaolinite to be, in general, much lower than those for either montmorillonite or illite.

Soil consistency or simply consistency is analogous to viscosity in liquids and indicates internal resistance to forces that tend to deform the soil. The internal resistance may come from inter-particle forces (cohesion or adhesion), cementation, inter-particle friction, and soil suction. Terms such as stiff, hard, firm, plastic, soft, and very soft are often used to describe consistency. Consistency changes with water content. A measure of consistency is provided by the consistency index defined as

$$CI = \frac{LL - w}{LL - PL} = \frac{LL - w}{PI} \tag{2.22}$$

The description in Table 2.6 does not apply to expansive and collapsible soils.

Alec Skempton (1953) showed that for soils with a particular mineralogy, the plasticity index is linearly related to the amount of the clay fraction. He coined a term called activity (A) to describe the importance of the clay fractions on the plasticity index. The equation for A is

$$A = \frac{PI}{\text{Clay fraction (\%)}} \tag{2.23}$$

Table 2.5 Typical Atterberg limits for soils.

Soil type	LL (%)	PL (%)	PI (%)
Sand		Nonplastic	
Silt	30–40	20–25	10–15
Clay	40–150	25–50	15–100
Minerals			
Kaolinite	50–60	30–40	10–25
Illite	95–120	50–60	50–70
Montmorillonite	290–710	50–100	200–660

Table 2.6 Description of fine-grained soils based on consistency index.

Description	CI
Very soft (ooze out of finger when squeezed)	<0.25
Soft (easily molded by finger)	0.25–0.50
Firm or medium (can be molded using strong finger pressure)	0.50–0.75
Stiff (finger pressure dents soil)	0.75–1.00
Very stiff (finger pressure barely dents soil, but soil cracks under significant pressure)	>1

Table 2.7 Activity of clay-rich soils.

Description	Activity, A
Inactive	<0.75
Normal	0.75–1.25
Active	1.25–2
Very (highly) active (e.g., montmorillonite or bentonite)	>6
Minerals	
Kaolinite	0.3–0.5
Illite	0.5–1.3
Na-montmorillonite	4–7
Ca-montmorillonite	0.5–2.0

The clay fraction in Equation (2.23) is the amount of particles less than $2\,\mu m$. Activity is one of the factors used in identifying expansive or swelling soils. Typical values of activity are given in Table 2.7.

EXAMPLE 2.9 *Calculations of Plasticity Index, Liquidity Index, and Activity*

A fine-grained soil has a liquid limit of 300% and a plastic limit of 55%. The natural water content of the soil in the field is 80% and the clay content is 60%.

(a) Determine the plasticity index, the liquidity index, and the activity.

(b) Describe the soil state in the field.

(c) What is the predominant mineral in this soil?

(d) If this soil were under a concrete slab used as a foundation for a building and water were to seep into it from watering of a lawn, what would you expect to happen to the foundation?

Strategy Use Equations (2.19), (2.21), and (2.23) and Table 2.4, Table 2.5, and Table 2.7 to solve this problem.

Solution 2.9

Step 1: Calculate the plasticity index, liquidity index, and activity.

(a) $PI = LL - PL = 300 - 55 = 245\%$

$$LI = \frac{w - PL}{PI} = \frac{80 - 55}{245} = 0.1$$

$$A = \frac{PI}{\text{Clay fraction }(\%)} = \frac{245}{60} = 4.1$$

Step 2: Determine the state of the soil in the field.

(b) Based on Table 2.4, the soil with $LI = 0.1$ is at the low end of the plastic state.

Step 3: Determine the predominant mineral.

(c) From Table 2.5 and Table 2.6, the predominant mineral is montmorillonite (most likely, Na-montmorillonite).

Step 4: Determine the consequences of water seeping into the soil.

(d) Seepage from lawn watering will cause the soil to expand (montmorillonite is an expansive soil). Because the water content in the montmorillonite will not increase uniformly under the foundation, the expansion will not be uniform. More expansion will occur at the edge of the slab because the water content will be greater there. Consequently, the concrete foundation will curl upward at the edge and most likely crack. Construction on expansive soils requires special attention to water management issues such as drainage and landscape. Generally, plants and lawns should be at least 3 m away from the edge of the foundation and the land should be sculpted to drain water away from the foundation.

2.5 DETERMINATION OF THE LIQUID, PLASTIC, AND SHRINKAGE LIMITS

2.5.1 Casagrande's cup method

The liquid limit is determined from an apparatus (Figure 2.3) that consists of a semispherical brass cup that is repeatedly dropped onto a hard rubber base from a height of 10 mm by a cam-operated mechanism. Casagrande (1932) developed this apparatus, and the procedure for the test is called the Casagrande cup method.

A dry powder of the soil is mixed with distilled water into a paste and placed in the cup to a thickness of about 12.5 mm. The soil surface is smoothed and a groove is cut into the soil using a standard grooving tool. The crank operating the cam is turned at a rate of two revolutions per second, and the number of blows required to close the groove over a length of 12.5 mm is counted and recorded. A specimen of soil within the closed portion is extracted for determination of the water content. The liquid limit is defined as the water content at which the groove cut into the soil will close over a distance of 12.5 mm following 25 blows. This is difficult to achieve in a single test. Four or more tests at different water

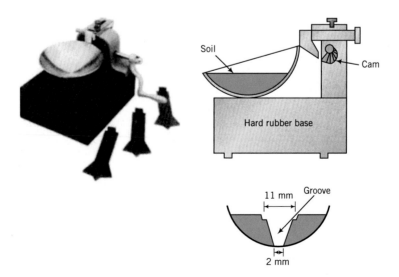

Figure 2.3 Cup apparatus for the determination of liquid limit. (Photo courtesy of Geotest.)

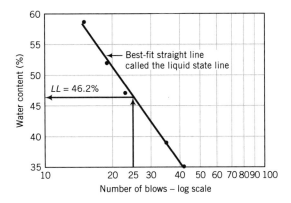

Figure 2.4   Typical liquid limit results from Casagrande's cup method.

Figure 2.5   Soil at plastic limit.

contents are usually required for terminal blows (number of blows to close the groove over a distance of 12.5 mm) usually ranging from 10 to 40. The results are presented in a plot of water content (ordinate, arithmetic scale) versus terminal blows (abscissa, logarithmic scale) as shown in Figure 2.4.

The best-fit straight line to the data points, usually called the flow line, is drawn. We will call this line the liquid state line to distinguish it from flow lines used in describing the flow of water through soils. The liquid limit is read from the graph as the water content on the liquid state line corresponding to 25 blows.

The cup method of determining the liquid limit has many shortcomings. Two of these are:

1. The tendency of soils of low plasticity to slide and to liquefy with shock in the cup rather than to flow plastically.
2. Sensitivity to operator technique and to small differences in apparatus.

2.5.2 Plastic limit test

The plastic limit is determined by rolling a small clay sample into threads and finding the water content at which threads of approximately 3 mm diameter will just start to crumble (Figure 2.5). Two or more determinations are made, and the average water content is reported as the plastic limit.

2.5.3 *Fall cone method to determine liquid and plastic limits*

A fall cone test, popular in Europe and Asia, appears to offer a more accurate (less prone to operator's errors) method of determining both the liquid and plastic limits. In the fall cone test (Figure 2.6), a cone with an apex angle of 30° and total mass of 80 grams is suspended above, but just in contact with, the soil sample. The cone is permitted to fall freely for a period of 5 seconds. The water content corresponding to a cone penetration of 20 mm defines the liquid limit.

The sample preparation is similar to the cup method except that the sample container in the fall cone test has a different shape and size (Figure 2.6). Four or more tests at different water contents are also required because of the difficulty of achieving the liquid limit from a single test. The results are plotted as water content (ordinate, logarithmic scale) versus penetration (abscissa, logarithmic scale), and the best-fit straight line (liquid state line) linking the data points is drawn (Figure 2.7). The liquid limit is read from the plot as the water content on the liquid state line corresponding to a penetration of 20 mm.

The plastic limit is found by projecting the best fit-straight line backward to intersect the water content axis at a depth of penetration of 1 mm. The water content at this depth of penetration (1 mm) is C. The plastic limit is given as (Feng, 2000):

$$PL = C(2)^m \qquad (2.24)$$

where *m* is the slope (taken as positive) of the best-fit straight line. If you use a spreadsheet program, you can obtain C and *m* from a power trend line function that gives the best-fit equation.

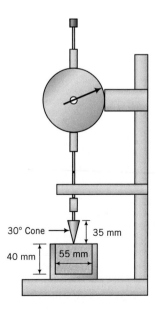

Figure 2.6 Fall cone apparatus.

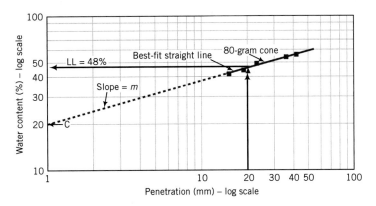

Figure 2.7 Typical fall cone test results.

2.5.4 Shrinkage limit

The shrinkage limit is determined as follows. A mass of wet soil, M_1, is placed in a porcelain dish 44 mm diameter and 12 mm high and then oven-dried. The volume of oven-dried soil is determined by using mercury to occupy the vacant spaces caused by shrinkage. The mass of the mercury is determined, and the volume decrease caused by shrinkage can be calculated from the known density of mercury. The shrinkage limit is calculated from

$$SL = \left(\frac{M_1 - M_2}{M_2} - \frac{V_1 - V_2}{M_2}\frac{\gamma_w}{g}\right)\times 100 = \left(w - \frac{V_1 - V_2}{M_2}\frac{\gamma_w}{g}\right)\times 100 \qquad (2.25)$$

where M_1 is the mass of the wet soil, M_2 is the mass of the oven-dried soil, w is water content (not in mass of mercury percentage), V_1 is the volume of wet soil, V_2 (= mass of mercury/density of mercury) is the volume of the oven-dried density of mercury soil, and g is the acceleration due to gravity ($9.81\,\text{m/s}^2$).

The linear shrinkage ratio, LS, is

$$LS = 1 - \sqrt[3]{\frac{V_2}{V_1}} \qquad (2.26)$$

The shrinkage ratio is

$$SR = \frac{M_2 g}{V_2 \gamma_w} \qquad (2.27)$$

The shrinkage limit can be estimated from the liquid limit and plasticity index by the following empirical expression:

$$SL = 46.4\left(\frac{LL + 45.5}{PI + 46.4}\right) - 43.5 \qquad (2.28)$$

where LL and PI are percentages.

Key points

1. Fine-grained soils can exist in one of four states: solid, semisolid, plastic, or liquid.
2. Water is the agent that is responsible for changing the states of soils.
3. A soil gets weaker if its water content increases.
4. Three limits are defined based on the water content that causes a change of state. These are the liquid limit—the water content that caused the soil to change from a liquid to a plastic state; the plastic limit—the water content that caused the soil to change from a plastic to a semisolid; and the shrinkage limit—the water content that caused the soil to change from a semisolid to a solid state. Water contents at approximately these limits are found from laboratory tests.
5. The plasticity index defines the range of water content for which the soil behaves like a plastic material.
6. The liquidity index gives a qualitative measure of strength.
7. The soil strength is lowest at the liquid state and highest at the solid state.

EXAMPLE 2.10 *Interpreting Liquid Limit Data from Casagrande's Cup Device*

A liquid limit test, conducted on a soil sample in the cup device, gave the following results:

Number of blows	10	19	23	27	40
Water content (%)	60.0	45.2	39.8	36.5	25.2

Two determinations for the plastic limit gave water contents of 20.3% and 20.8%. Determine (**a**) the liquid limit and plastic limit, (**b**) the plasticity index, (**c**) the liquidity index if the natural water content is 27.4%, (**d**) the void ratio at the liquid limit if $G_s = 2.7$ and (**e**) estimate the shrinkage limit. If the soil were to be loaded to failure, would you expect a brittle failure?

Strategy To get the liquid limit, you must make a semi-logarithmic plot of water content versus number of blows. Use the data to make your plot; then extract the liquid limit (water content on the liquid state line corresponding to 25 blows). Two determinations of the plastic limit were made, and the differences in the results were small. So, use the average value of water content as the plastic limit.

Solution 2.10

Step 1: Plot the data.
See Figure E2.10.

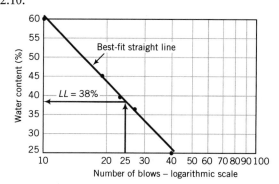

Figure E2.10

Step 2: Extract the liquid limit.

The water content on the liquid state line corresponding to a terminal blow of 25 gives the liquid limit.

$$LL = 38\%$$

Step 3: Calculate the plastic limit.

The plastic limit is

$$PL = \frac{20.3 + 20.8}{2} = 20.6\%$$

Step 4: Calculate PI.

$$PI = LL - PL = 38 - 20.6 = 17.4\%$$

The LL, PL, and PI are reasonable for typical soils (Table 2.5).

Step 5: Calculate LI.

$$LI = \frac{(w - PL)}{PI} = \frac{27.4 - 20.6}{17.4} = 0.39$$

Step 6: Calculate the void ratio.

Assume the soil is saturated at the liquid limit. For a saturated soil, $e = wG_s$. Thus,

$$e_{LL} = LLG_s = 0.38 \times 2.7 = 1.03$$

Step 7: Estimate the shrinkage limit.

$$SL = 46.4 \left(\frac{LL + 45.5}{PI + 46.4} \right) - 43.5 = 46.4 \left(\frac{38 + 45.5}{17.4 + 46.4} \right) - 43.5 = 17.2\%$$

Step 8: Estimate type of failure.

Brittle failure is not expected, as the soil is in a plastic state ($0 < LI < 1$).

EXAMPLE 2.11   *Interpreting Fall Cone Data*

The results of a fall cone test are shown in the table below.

Cone mass	80-gram cone				
Penetration (mm)	5.5	7.8	14.8	22	32
Water content (%)	39.0	44.8	52.5	60.3	67

Determine (a) the liquid limit, (b) the plastic limit, (c) the plasticity index, and (d) the liquidity index if the natural water content is 46%.

Strategy Adopt the same strategy as in Example 2.10. Make a plot of water content versus penetration, both at logarithmic scale. Use the data to make your plot, then extract the liquid limit (water content on the liquid state line corresponding to 20 mm). Find the water content difference between the two liquid state lines at any fixed penetration. Use this value to determine the plastic limit from Equation (2.24).

Solution 2.11

Step 1: Plot the data.

See Figure E2.11.

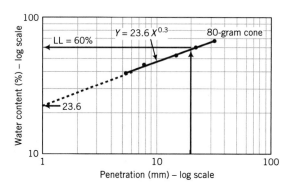

Figure E2.11 Plot of fall cone results.

Step 2: Extract the liquid limit.

$$LL = 60\%$$

Step 3: Determine the plastic limit.

The best-fit straight line for the 80-gram cone is $Y = 23.6X^{0.3}$ where Y is water content and X is penetration. Therefore, C = 23.6 and $m = 0.3$. From Equation (2.24):

$$PL = C(2)^m = 23.6(2)^{0.3}$$
$$= 29\%$$

Step 4: Determine PI.

$$PI = LL - PL = 60 - 29 = 31\%$$

Step 5: Determine LI.

$$LI = \frac{w - PL}{PI} = \frac{46 - 29}{31} = 0.55$$

EXAMPLE 2.12 *Determination of the Shrinkage Limit*

The following results were recorded in a shrinkage limit test using mercury. Determine the shrinkage limit.

Mass of container	17.5 grams
Mass of wet soil and container	78.1 grams
Mass of dish	130.0 grams
Mass of dish and displaced mercury	462.0 grams
Mass of dry soil and container	62.4 grams

Strategy Use a table to conduct the calculation based on Equation (2.25).

Solution 2.12

Step 1: Set up a table or, better yet, use a spreadsheet to carry out the calculations.

M_c = mass of container	17.5 grams
M_{wc} = mass of wet soil and container	78.1 grams
M_d = mass of dish	130.0 grams
M_{dm} = mass of dish and displaced mercury	462.0 grams
ρ_m = density of mercury	13.6 grams/cm^3
M_{dc} = mass of dry soil and container	62.4 grams
V_1 = volume of wet soil	32.4 cm^3
V_2 = volume of oven-dried soil	22.4 cm^3
M_1 = mass of wet soil	60.6 grams
M_2 = mass of dry soil	46.9 grams
Shrinkage limit = $[((M_1 - M_2)/M_2) - ((V_1 - V_2)/M_2)\gamma_w/g]100$	12.1%

What's next ... We now know how to obtain some basic soil data—particle size and indexes—from quick, simple tests. The question that arises is: What do we do with these data? Engineers have been using them to get a first impression on the use and possible performance of a soil for a particular purpose such as a construction material for an embankment. This is currently achieved by classification systems. Next, we will study a few of these systems.

2.6 SOIL CLASSIFICATION SCHEMES

A classification scheme provides a method of identifying soils in a particular group that would likely exhibit similar characteristics. Soil classification is used to specify a certain soil type that is best suited for a given application. Also, the classification scheme can be used to establish a soil profile along a desired cross section of a soil mass. There are several classification schemes available. Each was devised for a specific use in different countries and localities. In this textbook, we will use the Unified Soil Classification System (USCS) that was originally developed for use in airfield construction but was later modified for general use.

2.6.1 *The Unified Soil Classification System (USCS)*

The USCS uses symbols for the particle size groups. These symbols and their representations are G = gravel, S = sand, M = silt, and C = clay. These are combined with other symbols expressing gradation characteristics—W for well graded and P for poorly graded—and plasticity characteristics—H for high and L for low, and a symbol O, indicating the presence of organic material. A typical classification of CL means a clay soil with low plasticity, while SP means a poorly graded sand. The degree of plasticity (high or low) is based on ranges in plasticity index. Different countries and localities have different ranges. For this textbook, we will use the plasticity chart (Figure 2.8) as a guide on the degree of plasticity. From this chart, low plasticity is when $LL < 50\%$ and $PI < 37\%$; high plasticity is when $LL > 50\%$ and $PI > 37\%$.

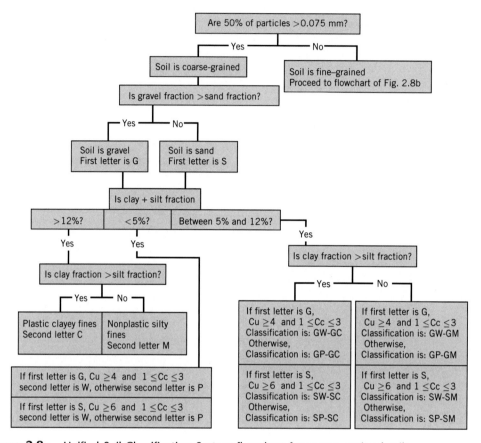

Figure 2.8a Unified Soil Classification System flowchart for coarse-grained soils.

Soils are classified by group symbols and group names. For example, we can have a soil with a group symbol, SW-SM, and group name, which describes the soil, as "well-graded sand with silt" if the gravel content is less than 15%. A flowchart to classify soils based on the USCS is shown in Figure 2.8a–b.

2.6.2 Plasticity chart

Experimental results from soils tested from different parts of the world were plotted on a graph of plasticity index (ordinate) versus liquid limit (abscissa). It was found that clays, silts, and organic soils lie in distinct regions of the graph. A line defined by the equation

$$PI = 0.73(LL - 20)\%, \quad PI \geq 4 \tag{2.29}$$

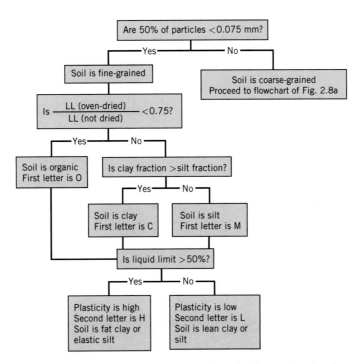

Figure 2.8b Unified Soil Classification System flowchart for fine-grained soils.

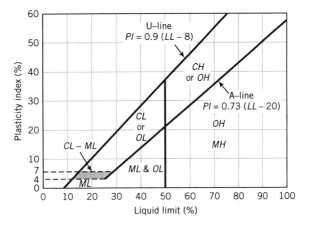

Figure 2.9 Plasticity chart.

called the "A-line," delineates the boundaries between clays (above the line) and silts and organic soils (below the line), as shown in Figure 2.9. A second line, the U-line, expressed as $PI = 0.9(LL - 8)\%$, defines the upper limit of the correlation between plasticity index and liquid limit. If the results of your soil tests fall above the U-line, you should be suspicious of your results and repeat your tests.

Table 2.8 Engineering Use Chart (after Wagner, 1957).

Typical names of soil groups	Group symbols	Important properties			
		Permeability when compacted	Shearing strength when compacted and saturated	Compressibility when compacted and saturated	Workability as a construction material
Well-graded gravels, gravel-sand mixtures, little or no fines	GW	Pervious	Excellent	Negligible	Excellent
Poorly graded gravels, gravel–sand mixtures, little or no fines	GP	Very Pervious	Good	Negligible	Good
Silty gravels, poorly graded gravel-sand-silt mixtures	GM	Semipervious to impervious	Good	Negligible	Good
Clayey gravels, poorly graded gravel-sand-clay mixtures	GC	Impervious	Good to fair	Very low	Good
Well-graded sands, gravelly sands, little or no fines	SW	Pervious	Excellent	Negligible	Excellent
Poorly graded sands, gravelly sands, little or no fines	SP	Pervious	Good	Very low	Fair
Silty sands, poorly graded sand-silt mixtures	SM	Semipervious to impervious	Good	Low	Fair
Clayey sands, poorly graded sand-clay mixtures	SC	Impervious	Good to fair	Low	Good
Inorganic silts and very fine sands, rock flour, silty, or clayey fine sands with slight plasticity	ML	Semipervious to impervious	Fair	Medium	Fair
Inorganic clays of low to medium plasticity, gravelly clays, sandy clays, silky clays, lean clays	CL	Impervious	Fair	Medium	Good to fair
Organic silts and organic silt-clays of low plasticity	OL	Semipervious to impervious	Poor	Medium	Fair
Inorganic silts, micaceous or diatomaceous fine sandy, or silty soils, elastic silts	MH	Semipervious to impervious	Fair to poor	High	Poor
Inorganic clays of high plasticity, fat clays	CH	Impervious	Poor	High	Poor
Organic clays of medium to high plasticity	OH	Impervious	Poor	High	Poor
Peat and other highly organic soils	Pt	–	–	–	–

2.7 ENGINEERING USE CHART

You may ask: How do I use a soil classification to select a soil for a particular type of construction, for example, a dam? Geotechnical engineers have prepared charts based on experience to assist you in selecting a soil for a particular construction purpose. One such chart is shown in Table 2.8.

The numerical values 1 to 14 are ratings, with 1 being the best. The chart should only be used to provide guidance and to make a preliminary assessment of the suitability of a soil for a particular use. You should not rely on such descriptions as "excellent" shear strength or "negligible" compressibility (settlement) to make final design and construction decisions. In Chapters 7 and 8 we will deal with more reliable methods to determine settlement and strength.

								Roadways		
Rolled earth dams			Canal sections		Foundations		Fills			
Homogeneous embankment	Core	Shell	Erosion resistance	Compacted earth lining	Seepage important	Seepage not important	Frost heave not possible	Frost heave possible	Surfacing
–	–	1	1	–	–	1	1	1	3
–	–	2	2	–	–	3	3	3	–
2	4	–	4	4	1	4	4	9	5
1	1	–	3	1	2	6	5	5	1
–	–	3 if gravelly	6	–	–	2	2	2	4
–	–	4 if gravelly	7 if gravelly	–	–	5	6	4	–
4	5	–	8 if gravelly	5 erosion critical	3	7	8	10	6
3	2	–	5	2	4	8	7	6	2
6	6	–	–	6 erosion critical	6	9	10	11	–
5	3	–	9	3	5	10	9	7	7
8	8	–	–	7 erosion critical	7	11	11	12	–
9	9	–	–	–	8	12	12	13	–
7	7	–	10	8 volume change critical	9	13	13	8	–
10	10	–	–	–	10	14	14	14	–
–	–	–	–	–	–	–	–	–	–

Relative desirability for various uses

EXAMPLE 2.13 *Soil Classification According to USCS*

Particle size analyses were carried out on two soils—soil A and soil B—and the particle size distribution curves are shown in Figure E2.13. The Atterberg limits for the two soils are as shown below:

Soil	LL	PL
A	26% (oven-dried; assume same for not dried)	18 %
B	Nonplastic	

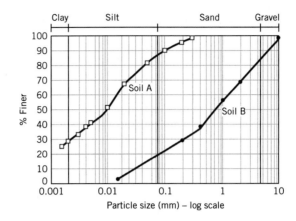

Figure E2.13

(a) Classify these soils according to USCS.

(b) Is either of the soils organic?

(c) In a preliminary assessment, which of the two soils is a better material for the core of a rolled earth dam?

Strategy If you examine the flowcharts of Figure 2.8a, b, you will notice that you need to identify the various soil types based on texture: for example, the percentage of gravel or sand. Use the particle size distribution curve to extract the different percentages of each soil type, and then follow the flowchart. To determine whether your soil is organic or inorganic, plot your Atterberg limits on the plasticity chart and check whether the limits fall within an inorganic or organic soil region.

Solution 2.13

Step 1: Determine the percentages of each soil type from the particle size distribution curve.

Constituent	Soil A	Soil B
Percentage of particles greater than 0.075 mm	12	80
Gravel fraction (%)	0	16
Sand fraction (%)	12	64
Silt fraction (%) (between 0.075 and 0.005 mm)	47	20
Clay fraction (%) (<0.002 mm)	29	0

Step 2: Use the flowchart (Figure 2.8a,b) to classify the soils.

Soil A: % finer than sieve 0.075 mm (% fines) = 88%, % sand > % gravel, gravel < 15%

Because 50% of the particles are less than 0.075 mm, use flowchart Figure 2.8b. Since LL (oven-dried) to LL (not dried) ratio = 1 (>0.75), the soil is inorganic (flow chart Figure 2.6b).

Following flowchart Figure 2.8b, the Group symbol = ML, group name = silt. You can add a qualifying term "sandy" to indicate that the soil contains some small amount of sand. So the group name can be stated as "sandy silt."

Soil B: % finer than sieve 0.075 mm = 20% (silt), % sand > % gravel, gravel = 16%

Soil B has less than 50% fines; use flowchart Figure 2.8a.

Following flowchart Figure 2.8a, Group symbol = SM, group name = silty sand with gravel

Step 3: Plot the Atterberg limits on the plasticity chart.

Soil A: $PI = 26 - 18 = 8\%$

The point (26, 8) falls above the A-line; the soil is inorganic.

Soil B: Nonplastic and inorganic

Step 4: Use Table 2.8 to make a preliminary assessment.

Soil B, with a rating of 5, is better than soil B, with a rating of 6, for the dam core.

2.8 SUMMARY

We have dealt with a large body of basic information on the physical parameters of soils. Soils are conveniently idealized as three-phase materials: solids, water, and air. The physical parameters of a soil depend on the relative proportion of these constituents in a given mass. Soils are classified into groups by their particle sizes and Atterberg limits. Soils within the same group are likely to have similar mechanical behavior and construction use. Some of the main physical parameters for soils are the particle sizes, void ratio, liquid limit, plastic limit, shrinkage limit, and plasticity and liquidity indexes. Water can significantly change the characteristics of soils.

2.8.1 Practical examples

EXAMPLE 2.14 *Calculating Soil Quantities for a Highway Embankment*

A 2-lane highway is required to be constructed in an area that is subjected to frequent flooding. The proposed width of the 2 lanes is 7.5 m with outside shoulders of 3 m width each with a slope of 1 (*V*) to 2 (*H*) (see Figure E2.14). An embankment 1.25 m high is to be constructed using a soil from a borrow pit. The water content of this soil in the borrow pit is 8% and the degree of saturation is 41%. The swell index is 1.1. The specification requires the embankment to be compacted to a dry unit weight of 18 kN/m³ at a water content of 10 ± 0.5%. Determine, for 1 kilometer length of embankment, the following:

(a) The weight of soil from the borrow pit required to construct the embankment.

(b) The number of truckloads of soil required for the construction. The contractor proposed to use transfer dump trucks with 220 kN payload capacity. Local government regulations require a maximum loaded capacity of 90% of payload.

(c) The minimum cost per kilometer of compacted embankment given the following estimated costs:

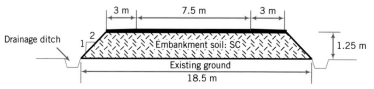

Figure E2.14

Truck rental = $105 per hour.

A load and unload round trip = 1.5 hours.

To place and compact soil = $15 per m of loose soil.

Strategy The strategy is similar to that adopted in Example 2.7.

Solution 2.14

Step 1: Calculate γ_d for the borrow pit material.

$$e = \frac{wG_s}{S} = \frac{0.08 \times 2.7}{0.41} = 0.527$$

$$\gamma_d = \frac{G_s \gamma_w}{1+e} = \frac{2.7 \times 9.81}{1+0.527} = 17.35 \text{ kN/m}^3$$

Step 2: Determine the volume of borrow pit soil required.

Total top width = $7.5 + 2 \times 3 = 13.5$ m; total bottom width = $13.5 + 2(2 \times 1.25) = 18.5$ m (see Figure E2.14).

Volume of finished embankment: $V = \frac{1}{2}(13.5 + 18.5) \times 1.25 \times 1000 = 20{,}000 \text{ m}^3$

Volume of borrow bit soil required: $\frac{(\gamma_d)_{embankment}}{(\gamma_d)_{borrow\ pit}} \times V = \frac{18}{17.35} \times 20{,}000 = 20{,}749 \text{ m}^3$

Step 3: Determine the number of trucks required.

Weight of soil required to be transported = swell factor × volume required × bulk unit weight; that is,

$1.1 \times 20{,}749 \times 17.35(1 + 0.08) = 427{,}674$ kN

Note: 17.25(1 + 0.08) is the bulk unit weight (=dry unit weight × {1 + water content}).

Allowable capacity = $0.9 \times 220 = 198$ kN.

$$\text{Number of trucks} = \frac{427{,}674}{198} = 2160$$

Step 4: Determine cost.

Truck rental = $105 (dollars per hour) × 1.5 (number of hours per round trip) × 2160 (number of truck loads) = $340,200.

Volume of borrow pit material to be transported = $20{,}749 \text{ m}^3$ × swell index = $20{,}749 \times 1.1 = 22{,}824 \text{ m}^3$. Place and compact cost = volume of loose soil × place and compact cost per cubic meter = $22{,}824 \times \$15 = \$342{,}360$.

Total cost = $340,200 + $342,260 = $682,460.

EXAMPLE 2.15 *Estimating Soil Profile Based on Soil Classification*

Three boreholes (BH) along a proposed road intersection are shown in Figure E2.15a, b. The soils in each borehole were classified using USCS. Sketch a soil profile along the center line.

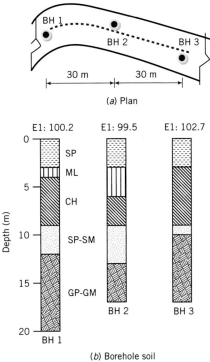

(b) Borehole soil

Figure E2.15a, b

Strategy Assume that the boreholes are all along the center line. You should align the boreholes relative to a single elevation and then sketch the soil profile using USCS.

Solution 2.15

Step 1: Align the boreholes relative to a single elevation.

See Figure E2.15c.

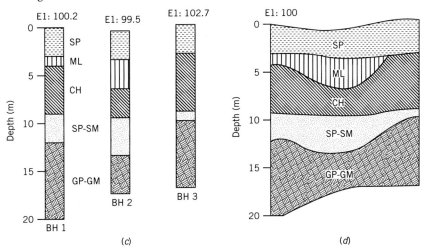

Figure E2.15c, d

Step 2: Sketch soil profile.

Draw smooth curves around each corresponding soil type in the boreholes, as shown in Figure E2.15d. These curves are estimates. The low plasticity silt (ML) is only present in boreholes 1 and 2. You have to make an estimate as to where this layer of silt ends before BH 3.

EXERCISES

Assume $G_s = 2.7$, where necessary, for solving the following problems.

Concept understanding

2.1 A container contains a dry powdered clay. A similar container contains sand filled to the same volume. If water is added to saturate both the clay and the sand, which is likely to have a higher water content and why?

2.2 Which soil type, sand or clay, is likely to have a greater void ratio and why?

2.3 Can the liquid limit of a soil be greater than 100%? Why?

2.4 What are the purposes of a soil classification system?

Problem-solving skills

2.5 The following data are available for the determination of the specific gravity of a sand using a cylinder:

Volume of cylinder filled with water $= 0.0005 \, m^3$.

Weight of sand $= 1.28 \, N$.

Volume of water and sand in cylinder after all air bubbles were removed $= 0.00055 \, m^3$.
Calculate the specific gravity.

2.6 A specific gravity test was done on a silty clay. The data are as shown below. Calculate the specific gravity.

Mass of oven dried silty clay	= 102.3 grams.
Mass of pycnometer, dry soil, and water	= 723.1 grams.
Mass of pycnometer and water	= 661.0 grams.

2.7 A cube of wet soil, $0.15 \, m \times 0.15 \, m \times 0.15 \, m$, weighs $66.5 \, N$. Calculate (a) the bulk unit weight, and (b) the dry unit weight if the water content is 20%.

2.8 Determine (a) water content, (b) void ratio, and (c) porosity given the following:

Wet weight of saturated clay $= 5.69 \, N$.

After oven drying at 110°C, the dry weight $= 3.91 \, N$.

2.9 Determine the saturated unit weight given the following:

Wet weight of a sample of a saturated soil $= 5.34 \, N$.

After oven drying at 110°C, the dry weight $= 4.09 \, N$.

2.10 A soil sample has a bulk unit weight of $17.5 \, kN/m^3$ at a water content of 11%. Determine (a) the void ratio, (b) the percentage air in the voids (air voids), and (c) the degree of saturation of this sample.

2.11 Calculate (a) the water content, (b) void ratio, (c) porosity, and (d) degree of saturation given the following data:

Volume of wet sand sample $= 0.0005\,m^3$.

Weight of wet sand sample $= 7.78\,N$.

After oven drying at 110°C, the dry weight $= 7.34\,N$.

2.12 The porosity of a soil is 50% and its water content is 10%. Determine (a) the degree of saturation, (b) the bulk unit weight, and (c) saturated unit weight.

2.13 A soil sample of diameter 38 mm and length 76 mm has a wet weight of 1.6 N and a dry weight of 1.2 N. Determine (a) the water content, (b) the degree of saturation, (c) the porosity, (d) the bulk unit weight, and (e) the dry unit weight.

2.14 The weight of a wet sample of soil of volume $0.00014\,m^3$ and its container is 3.25 N. The dry weight of the soil and its container is 2.8 N. The weight of the container is 0.58 N. Determine the following:

(a) The bulk, dry, and saturated unit weights of the soil.

(b) The void ratio and the degree of saturation.

(c) Determine the volume of air voids in the soil.

(d) What is the weight of water required to saturate $1\,m^3$ of this soil?

2.15 A sand has a natural water content of 5% and bulk unit weight of $16.5\,kN/m^3$. The void ratios corresponding to the densest and loosest state of this soil are 0.51 and 0.87. Find the relative density, density index, and degree of saturation.

2.16 The void ratio of a soil is 1.2. Determine the bulk and effective unit weights for the following degrees of saturation: (a) 75%, (b) 95%, and (c) 100%. What is the percentage error in the bulk unit weight if the soil were 95% saturated but assumed to be 100% saturated?

2.17 The following results were obtained from a liquid limit test on a clay using Casagrande's cup device:

Number of blows	6	12	20	28	32
Water content (%)	52.5	47.1	43.2	38.6	37.0

(a) Determine the liquid limit of this clay.

(b) If the natural water content is 38% and the plastic limit is 23%, calculate the liquidity index and the consistency index.

(c) Describe the soil based on the consistency index.

(d) Do you expect a brittle type of failure for this soil? Why?

2.18 The following data were recorded from a liquid limit test on a clay using Casagrande's cup device. Determine the liquid limit.

Test number	Mass of container (grams) M_c	Mass of container and wet soil (grams) M_w	Mass of container and dry soil (grams) M_d	Blow count
1	45.3	57.1	52.4	28
2	43.0	59.8	56.0	31
3	45.2	61.7	57.9	22
4	45.6	58.4	55.3	18

2.19 A fall cone test was carried out on a soil to determine its liquid and plastic limits using a cone of mass 80 grams. The following results were obtained:

80-gram cone				
Penetration (mm)	8	15	19	28
Water content (%)	43.1	52.0	56.1	62.9

Determine (a) the liquid and plastic limits and (b) the plasticity index. If the soil contains 45% clay, calculate the activity.

2.20 The following results were recorded in a shrinkage limit test using mercury:

Mass of container	17.0 grams
Mass of wet soil and container	72.3 grams
Mass of dish	132.40 grams
Mass of dish and displaced mercury	486.1 grams
Mass of dry soil and container	58.2 grams

Determine the shrinkage limit if the volume of the container is $32.4\,cm^3$.

2.21 The results of a particle size analysis of a soil are given in the following table. No Atterberg limits tests were conducted.

Sieve opening	9.53	4.75	2.00	0.85	0.425	0.15	0.075
% Finer	100	89.8	70.2	62.5	49.8	28.6	2.1

(a) Would you have conducted Atterberg limit tests on this soil? Justify your answer.

(b) Classify the soil according to USCS.

(c) Is this soil a good foundation material? Justify your answer.

2.22 The results of a particle size analysis of a soil are given in the following table. Atterberg limits tests gave $LL = 58\%$ and $PL = 32\%$. The clay content is 31%.

Sieve opening	9.53	4.75	2.00	0.85	0.425	0.15	0.075
% Finer	100	90.8	82.4	77.5	71.8	65.6	62.8

(a) Classify the soil according to USCS.

(b) Rate this soil as a subgrade for a highway.

Critical thinking and decision making

2.23 A fine-grained soil has a liquid limit of 200% and a plastic limit of 45%. The natural water content of the soil in the field is 60% and the clay content is 63%.

(a) Calculate the plasticity index, the liquidity index and the activity.

(b) What is the soil state (e.g., liquid) in the field?

(c) What is the predominant mineral in this soil?

(d) This soil is under a rectangular concrete slab, 15 m × 45 m used as a foundation for a building. A water pipe, 100 mm diameter, is located in a trench 450 mm below the center of the slab. The trench, 300 mm wide and 450 mm deep, running along the length of the slab, was backfilled with the same soil. If this pipe were to leak, what effect would it have on the foundation? Draw a neat sketch of the existing trench and pipe, and show in another sketch how you would mitigate any water issue related to the pipe and the soil. Explain why your mitigation method is better than the existing construction.

2.24 A portion of a highway embankment from Noscut to Windsor Forest is 5 kilometers long. The average cross section of the embankment is shown in Figure P2.24a. The gradation curves for the soils at the two borrow pits are shown in Figure P2.24b. Pit 1 is located 2 kilometers from the start of the embankment while pit 2 is 1.2 kilometers away. Estimated costs for various earthmoving operations are shown in the table below. You are given 10 minutes by the stakeholder's committee to present your recommendations. Prepare your presentation. The available visual aid equipment is an LCD projector.

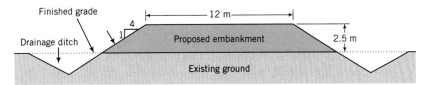

Figure P2.24a

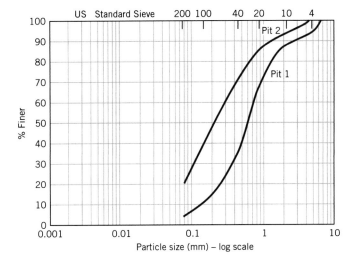

Figure P2.24b

	Cost	
Operation	**Pit 1**	**Pit 2**
Purchase and load borrow pit material at site, haul 2 kilometers round trip, and spread with 200 HP dozer	$50/m³	$30/m³
Extra mileage charge for each kilometer	$0.5/m³	$0.5/m³
Compaction	$15/m³	$16/m³
Miscellaneous	$1/m³	$0.75/m³

2.25 The soil profiles for four boreholes (BH) at a site proposed for an office building are shown in Figure P2.25. The soils in each borehole were classified using USCS. Sketch the soil profiles along a diagonal line linking boreholes 1, 2, and 3 and along a line linking boreholes 3 and 4.

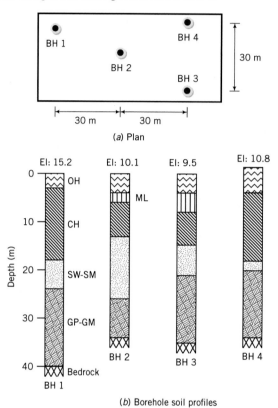

(a) Plan

(b) Borehole soil profiles

Figure P2.25

Chapter 3
Soils Investigation

3.1 INTRODUCTION

Geological forces and processes often result in inhomogeneous and discontinuous formations that significantly influence the stability and costs of civil engineering works. The amount of investigation needed to characterize a site economically, the type and methods of construction, and natural geological hazards such as earthquakes, volcanic activity, and groundwater conditions are important geological factors that must be considered in the practice of geotechnical engineering. Many failures of structures, causing loss of lives and property, have resulted from unrealized geological conditions. Consider the geology at a potential construction site in a county, as shown in Figure 3.1. To map these geological features requires applications of geophysical methods and a series of closely spaced boreholes. The precise size of each geological feature is difficult to ascertain. In building a skyscraper, for example, you must have knowledge of the geological features under and within the vicinity of the building to design a safe and economical foundation.

A soils investigation is an essential part of the design and construction of a proposed structural system (buildings, dams, roads and highways, etc.). Soils are identified, observed, and recovered during a soils investigation of a proposed site. Usually soils investigations are conducted only on a fraction of a proposed site because it would be prohibitively expensive to conduct an extensive investigation of a whole site. The estimates and judgments made based on information from a limited set of observations, and from field and laboratory test data, will have profound effects on the performance and costs of structures constructed at a site.

Learning outcomes

When you complete this chapter, you should be able to do the following:

- Plan a soils investigation.
- Describe soils in the field.
- Appreciate the limitations of a soils investigation.

Soil Mechanics Fundamentals, First Edition. Muni Budhu.
© 2015 John Wiley & Sons, Ltd. Published 2015 by John Wiley & Sons, Ltd.
Companion website: www.wiley.com\go\budhu\soilmechanicsfundamentals

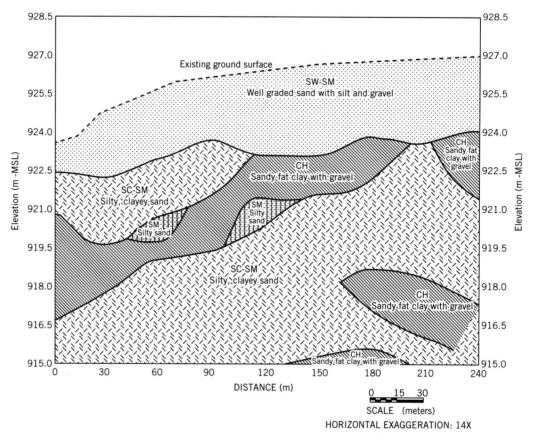

Figure 3.1 Soil profile at a construction site.

3.2 DEFINITIONS OF KEY TERMS

SPT is the standard penetration test.
N is the number of blows for the last 304 mm penetration of an SPT sampler.
Soil sensitivity (S_t) is the ratio of the intact strength to the disturbed strength.
CPT is the cone penetrometer test (sometimes called cone penetration test)
Cone tip resistance (q_c) is the average resistance (average vertical stress) of the cone in a CPT.
Sleeve resistance (f_s) is the average resistance of the sleeve or shaft of the CPT.

3.3 PURPOSES OF A SOILS INVESTIGATION

A soils investigation program is necessary to provide information for design and construction, environmental assessment, and project due diligence (*due diligence* is the process of evaluating a prospective project to facilitate business decisions by the owner). The purposes of a soils investigation are:

1. To evaluate the general suitability of the site for the proposed project.
2. To enable an adequate and economical design to be made.
3. To disclose and make provision for difficulties that may arise during construction due to ground and other local conditions.

3.4 PHASES OF A SOILS INVESTIGATION

The scope of a soils investigation depends on the type, size, and importance of the structure; the client; the engineer's familiarity with the soils at the site; and local building codes. Structures that are sensitive to settlement such as machine foundations and high-use buildings usually require a more thorough soils investigation than a foundation for a house. A client may wish to take a greater risk than normal to save money and set limits on the type and extent of the site investigation. You should be cautious about any attempt to reduce the extent of a soils investigation below a level that is desirable for assuming acceptable risks for similar projects on or within similar ground conditions. If the geotechnical engineer is familiar with a site, he/she may undertake a very simple soils investigation to confirm his/her experience. Some local building codes have provisions that set out the extent of a site investigation. It is mandatory that a visit be made to the proposed site.

A soils investigation has three components. The first component is done prior to design. The second component is done during the design process. The third component is done during construction. The second and third components are needed for contingencies. The first component is generally more extensive and is conducted in phases. These phases are as follows:

Phase I. This phase is sometimes called "desk study." It involves collection of available information such as a site plan; type, size, and importance of the structure; loading conditions; previous geotechnical reports; maps, including topographic maps, aerial photographs, still photographs, satellite imagery, and geologic maps; and newspaper clippings. An assortment of maps giving geology, contours and elevations, climate, land use, aerial photos, regional seismicity, and hydrology are available on the Internet (e.g., http://www.usgs.gov). Geographical information system (GIS)—an integration of software, hardware, and digital data to capture, manage, analyze, and display spatial information—can be used to view, share, understand, question, interpret, and visualize data in ways that reveal relationships, patterns, and trends. GIS data consist of discrete objects such as roads and continuous fields such as elevation. These are stored either as raster or vector objects. Google Earth (http://earth.google.com) can be used to view satellite imagery, maps, terrain, and 3D structures. You can also create project maps using Google Earth.

Phase II. Preliminary reconnaissance or a site visit to provide a general picture of the topography and geology of the site. It is necessary that you take with you on the site visit all the information gathered in Phase I to compare with the current conditions of the site. Your site visit notes should include:

- Photographs of the site and its neighborhood.
- Access to the site for workers and equipment.
- Sketches of all fences, utility posts, driveways, walkways, drainage systems, and so on.
- Utility services that are available, such as water and electricity.
- Sketches of topography including all existing structures, cuts, fills, ground depression, ponds, and so on.

- State of any existing building at the site or nearby. Your notes should include exterior and interior cracks, any noticeable tilt, type of construction (e.g., brick or framed stucco building), evidence of frost damage, molds, and any exceptional features.
- Geological features from any exposed area such as a road cut.
- Occasionally, a few boreholes, trenches, and trial pits may be dug to explore the site.

Phase III. Detailed soils exploration. The objectives of a detailed soils exploration are:

- To determine the geological structure, which should include the thickness, sequence, and extent of the soil strata.
- To determine the groundwater conditions.
- To obtain disturbed and undisturbed samples for laboratory tests.
- To conduct in situ tests.

Phase IV. Laboratory testing. The objectives of laboratory tests are:

- To classify the soils.
- To determine soil strength, failure stresses and strains, stress–strain response, permeability, compactibility, and settlement parameters. Not all of these may be required for a project.

Phase V. Write a report. The report must contain a clear description of the soils at the site, methods of exploration, soil stratigraphy, in situ and laboratory test methods and results, and the location of the groundwater. You should include information on and/or explanations of any unusual soil, water-bearing stratum, and any soil and groundwater conditions such as frost susceptibility or waterlogged areas that may be troublesome during construction.

Key points

1. A soils investigation is necessary to determine the suitability of a site for its intended purpose.
2. A soils investigation is conducted in phases. Each phase affects the extent of the next phase.
3. A clear, concise report describing the conditions of the ground, soil stratigraphy, soil parameters, and any potential construction problems must be prepared for the client.

What's next … In the next section, we will study how a soils exploration program (Phase III) is normally conducted.

3.5 SOILS EXPLORATION PROGRAM

A soils exploration program usually involves test pits and/or soil borings (boreholes). During the site visit (Phase II), you should work out most of the soils exploration program. A detailed soils exploration consists of:

1. Determining the need for and extent of geophysical exploration.
2. Preliminary location of each borehole and/or test pit.
3. Numbering of the boreholes or test pits.

4. Planned depth of each borehole or test pit.
5. Methods and procedures for advancing the boreholes.
6. Sampling instructions for at least the first borehole. The sampling instructions must include the number of samples and possible locations. Changes in the sampling instructions often occur after the first borehole.
7. Determining the need for and types of in situ tests.
8. Requirements for groundwater observations.

3.5.1 Soils exploration methods

The soils at a site can be explored using one or more of the following methods.

3.5.1.1 Geophysical methods

Nondestructive techniques used to provide spatial information on soils, rocks, and hydrological and environmental conditions. Popular methods are:

- *Ground-penetrating radar (GPR):* Also called *georadar*, GPR is a high-resolution, high-frequency (10 MHz to 1000 MHz) electromagnetic wave technique for imaging soils and ground structures. An antenna is used to transmit and recover radar pulses generated by a pulse generator (Figure 3.2). The returned pulse is then processed to produce images of the soil profile. The key geotechnical uses are soil profile imaging and location of buried objects. GPR produces continuous-resolution images of the soil profile with very little soil disturbance. GPR is not suitable for highly conductive (>15 milliohms/m) wet clays and silts. GPR resolution decreases with depth.
- *Seismic surveys:* Seismic investigations utilize the fact that surface waves travel with different velocities through different materials. The subsurface interfaces are determined by recording the magnitude and travel time of the seismic waves, essentially compression waves (P waves), at a point some distance from the source of the wave. The velocity of propagation is the most important parameter in the application of seismic methods. The densities and elastic properties of the geological materials control the velocity of propagation. When a seismic wave encounters a boundary between two elastic media, the wave

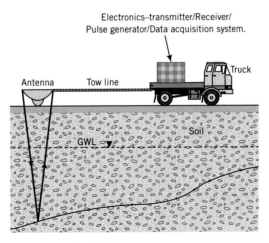

Figure 3.2 Ground-penetrating radar (GPR).

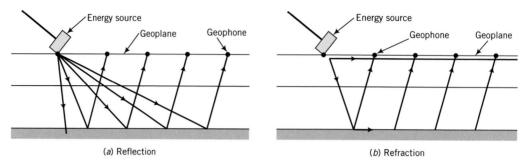

Figure 3.3 Wave transmission in seismic reflection (a) and refraction (b) tests.

energy is transmitted by *reflection, refraction,* and *diffraction. Seismic reflection* and refraction are used in geotechnical site characterization.

In seismic reflection tests, the travel times of waves reflected from subsurface interfaces are measured by geophones (Figure 3.3a). Geophones are motion-sensitive transducers that convert ground motion to electric signals. The travel times are correlated to depth, size, and shape of the interfaces. The angle of reflection of the waves is a function of the material density contrast. Seismic reflection is used when high resolution of soil profile is required, especially at large depths (>50 m).

Seismic refraction surveys are very similar to seismic reflection surveys except that refraction waves are measured and the source geophone is placed at a greater distance (Figure 3.3b). The latter enables the recording of seismic waves that are primarily horizontal rather than vertical. In most refraction surveys, only the initial P waves are recorded. The seismic refraction method is used to determine the depth and thickness of the soil profile and the existence of buried structures.

For shallow depths of investigation, the ground surface is pounded by a sledgehammer to generate the seismic waves; for large depths, a small explosive charge is used. Seismic methods are sensitive to noise and vibration. Various filtering techniques are used to reduce background noise and vibration. Multichannel analysis of surface waves (MASW) is used to map spatial changes in low-velocity materials. A soil profile interpreted from MASW is shown in Figure 3.4.

To get information on the stiffnesses of soil layers, crosshole seismic tests are used. The seismic source is located in one borehole and the geophone is located in an adjacent borehole (Figure 3.5). The P and S (shear) wave velocities are calculated from the arrival times and the geophone distances. These are then used to calculate the soil stiffnesses.

Downhole seismic tests are used to detect layering and the strength of the layers. The seismic source is located on the surface and geophones are located in a borehole (Figure 3.6).

- *Electrical resistivity:* Electrical resistivity measurements can be used for identification and quantification of depth of groundwater, detection of clays, and measurement of groundwater conductivity. Soil resistivity, measured in ohm-centimeters (ohm-cm), varies with moisture content and temperature changes. In general, an increase in soil moisture results in a reduction in soil resistivity. The pore fluid provides the only electrical path in sands, while both the pore fluid and the surface-charged particles provide electrical paths in clays. Resistivities of wet fine-grained soils are generally much lower than those of wet coarse-grained soils. The difference in resistivity between a soil in a dry and in a saturated condition may be several orders of magnitude.

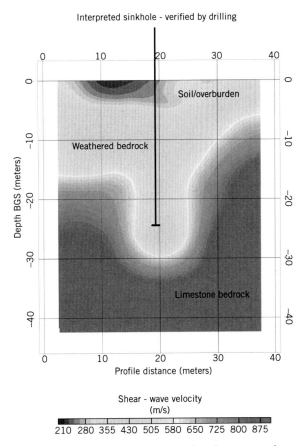

Figure 3.4 Soil profile from a multichannel analysis of surface waves from seismic tests. (Courtesy of Zonge International.)

The method of measuring subsurface resistivity involves placing four electrodes in the ground in a line at equal spacing, applying a measured AC current to the outer two electrodes, and measuring the AC voltage between the inner two electrodes. A measured resistance is calculated by dividing the measured voltage by the measured current. This resistance is then multiplied by a geometric factor that includes the spacing between each electrode to determine the apparent resistivity.

Electrode spacings of 0.75, 1.5, 3.0, 6.0, and 12.0 m are typically used for shallow depths (<10 m) of investigations. Greater electrode spacings of 1.5, 3.0, 6.0, 15.0, 30.0, 100.0, and 150.0 m are typically used for deeper investigations. The depth of investigation is typically less than the maximum electrode spacing. Water is introduced to the electrode holes as the electrodes are driven into the ground to improve electrical contact. A subsurface resistivity profile is typically performed at one location by making successive measurements at several electrode spacings. A soil profile from resistivity measurements is shown in Figure 3.7.

- *Other geophysical methods of geotechnical engineering interest:*
  1. Gamma density, or gamma-gamma, measures electron density and can be used to estimate the total soil density or porosity.

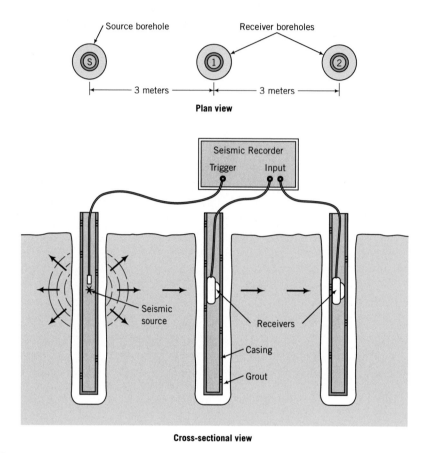

Figure 3.5 Typical setup for a crosshole seismic survey. (Source: ASTM D 4428.)

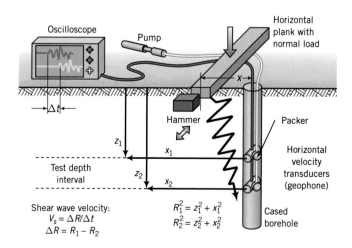

Figure 3.6 Downhole seismic survey. (Courtesy of Professor Paul Mayne, Georgia Tech.)

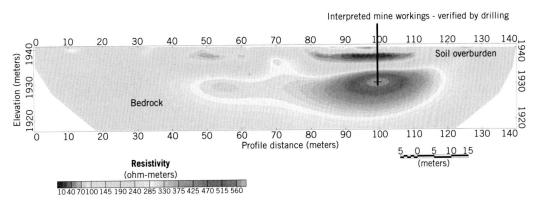

Figure 3.7 Soil profile from electrical resistivity tests. (Courtesy of Zonge International.)

2. Neutron porosity measures hydrogen density. It is used for porosity estimation below the groundwater level.
3. Sonic-VDL measures the seismic velocity. It is useful to measure soil stiffnesses and to detect bedrock elevation.
4. Microgravity is used to detect changes in subsurface densities and is particularly good at detecting cavities. A gravimeter is used at discrete points on the earth's surface to detect small changes in gravity. These changes are called gravity anomalies and are related to density changes.
5. Frequency domain electromagnetics (FDE) is a continuous-wave field method that is primarily used to map lateral variations of a structure in the subsurface. A fixed transmitter is used on a single frequency. The electrical conductivity of the soil is determined by measuring the magnitude and phase of the electromagnetic current.
6. Very low frequency electromagnetics (VLFE) is also a continuous-wave field method that detects increases in electrical conductivity by measuring the distortion of the very low frequency waves. VLFE is very useful in detecting fractures.
7. Time domain electromagnetics (TDE) measures the rate of decay of pulsating currents over time. TDE is useful in determining the variation of conductivity of soils with depth. You can, for example, obtain variations of soil saturation with depth using TDE.

3.5.1.2 Destructive methods

Destructive methods require disturbing the soil mass, often to great depths. These methods allow for the inspection of the soil mass, recovery of soil samples, and in situ tests (destructive tests). The common methods are:

■ *Trial pits or test pits*: A pit is dug by hand using shovels or with a machine such as a backhoe. This method can provide excellent shallow-depth soil stratigraphy (Figure 3.8).
■ *Hand or power augers*: These are tools (Figure 3.9) used to quickly create a hole about 100 mm to 250 mm in diameter in the ground. You can inspect the soil and take undisturbed (some disturbance is inevitable) samples for lab tests.
■ *Wash boring*: Water is pumped though a hollow rod that may or may not be equipped with a drill bit to remove soil from a borehole. The washings can be used to estimate the soil types.
■ *Rotary rigs*: These are mechanical devices used to drill boreholes, extract soil samples, and facilitate in situ tests (Figure 3.10).

Figure 3.8 A test pit.

Figure 3.9 A hand-powered auger.

- *Direct-push:* A split-spoon (see Figure 3.11b) type tube (mandrel) of diameter 30–50 mm and length 1.5–5 m with a plastic insert is pushed by a hydraulic jack to continuously sample the soil.
- *Sonic drilling:* Mechanical vibrations at frequencies between 50 Hz and 150 Hz are used to cut soil or rock cores of diameter 100–200 mm.

The advantages and disadvantages of each of these methods are shown in Table 3.1.

3.5.2 *Soil identification in the field*

In the field, the predominant soil types based on texture are identified by inspection. Gravels and sands are gritty and the individual particles are visible. Silts easily crumble, and water

Figure 3.10 A rotary drill.

Table 3.1 Advantages and disadvantages of soil exploration methods.

Method	Advantages	Disadvantages
Geophysical methods Ground penetration radar Seismic surveys Electrical resistivity Microgravity Electromagnetic survey (e.g., FDE, VLFE, TDE)	▪ Nondestructive ▪ Quick ▪ Provide stratigraphy, groundwater, and relative wetness ▪ Relatively inexpensive ▪ Provide subsurface geologic information with which to plan detailed soils investigations	▪ No soil samples ▪ Limited design parameters ▪ Site may not have enough real estate to conduct tests adequately ▪ Much of the information is qualitative
Test pits A pit is dug either by hand or by a backhoe.	▪ Cost-effective ▪ Provide detailed information on stratigraphy ▪ Large quantities of disturbed soils are available for testing ▪ Large blocks of undisturbed samples can be carved out from the pits ▪ Field tests can be conducted at the bottom of the pit	▪ Depth limited to about 6 m ▪ Deep pits uneconomical ▪ Excavation below groundwater and into rock difficult and costly ▪ Too many pits may scar site and require backfill soils
Hand augers The auger is rotated by turning and pushing down on the handlebar.	▪ Cost-effective ▪ Not dependent on terrain ▪ Portable ▪ Low headroom required ▪ Used in uncased holes ▪ Groundwater location can easily be identified and measured	▪ Depth limited to about 6 m ▪ Labor-intensive ▪ Undisturbed samples can be taken only for soft clay deposits ▪ Cannot be used in rock, stiff clays, dry sand, or caliche soils

Table 3.1 *Continued*

Method	Advantages	Disadvantages
Power augers Truck mounted and equipped with continuous-flight augers that bore a hole 100 mm to 250 mm in diameter. Augers can have a solid or hollow stem.	▪ Quick ▪ Used in uncased holes ▪ Undisturbed samples can be obtained quite easily ▪ Drilling mud not used ▪ Groundwater location can easily be identified	▪ Depth limited to about 15 m; at greater depth drilling becomes difficult and expensive ▪ Site must be accessible to motorized vehicle
Wash boring Water is pumped to bottom of borehole and soil washings are returned to surface. A drill bit is rotated and dropped to produce a chopping action.	▪ Can be used in difficult terrain ▪ Low equipment costs ▪ Used in uncased holes	▪ Depth limited to about 30 m ▪ Slow drilling through stiff clays and gravels ▪ Difficulty in obtaining accurate location of groundwater level ▪ Undisturbed soil samples cannot be obtained
Rotary drills A drill bit is pushed by the weight of the drilling equipment and rotated by a motor.	▪ Quick ▪ Can drill through any type of soil or rock ▪ Can drill to depths of 7500 m ▪ Undisturbed samples can be recovered	▪ Expensive equipment ▪ Quickness is affected by the sequence of augering, drilling, and sampling ▪ The soil mass is not sampled continuously, so some soil stratifications can be excluded ▪ Terrain must be accessible to motorized vehicle ▪ Difficulty in obtaining location of groundwater level ▪ Additional time required for setup and cleanup
Direct push A 30 mm to 50 mm diameter split-spoon type tube (mandrel) with a plastic insert is pushed by a hydraulic jack to continuously sample the soil.	▪ Very quick (compared with rotary drilling) ▪ Continuous sampling ▪ Continuous stratification information ▪ Undisturbed samples can be recovered ▪ Minimal drilling waste	▪ Stiff soil or rock layers or strongly cemented soils require hammering ▪ Terrain must be accessible to motorized vehicle
Sonic drilling Mechanical vibrations at frequencies between 50 HZ and 150 Hz are used to cut soil or rock cores of diameter 100 mm to 200 mm and length of about 3 m.	▪ Very quick (compared with rotary drilling) ▪ Continuous sampling ▪ Continuous stratification information ▪ Can penetrate and sample relatively hard material including gravel, and cobbles ▪ Minimal drilling waste	▪ Terrain must be accessible to motorized vehicle ▪ Undisturbed samples cannot be recovered

migrates to the surface on application of pressure. Clays fail this water migration test since water flows very slowly through clays. Clays feel smooth, greasy, and sticky to the touch when wet but are very hard and strong when dry.

Common descriptive terms and methods of identification are as follows.

- *Color:* Color is not directly related to engineering properties of soils, but is related to soil mineralogy and texture.
  Gray and bluish: unoxidized soils
  White and cream: calcareous soils
  Red and yellow: oxidized soils
  Black and dark brown: soils containing organic matter
- *Moisture:* Appearance due to water is described as wet, dry, or moist.
- *Structure:*
  Homogeneous: Color and texture feel the same throughout.
  Nonhomogeneous: Color and texture vary.
- *Shape:* Angular, subangular, subrounded, rounded, flaky.
- *Weathering:* Fresh, decomposed, weathered.
- *Carbonate:* Effervesces with acid. Add a small amount of hydrochloric acid and check if soil effervesces. If it does, it contains carbonate.
- *Smell:* Organic soils give off a strong odor that intensifies with heat. Nonorganic soils have a subtle odor with the addition of water.
- *Feel:* Use feel (use your fingers) to distinguish between sand, silts, and clays.
  Sand has a gritty feel.
  Silt has a rough feel similar to fine sandpaper.
  Clay feels smooth and greasy. It sticks to fingers and is powdery when dry.
- *Consistency:* Very stiff: Finger pressure barely dents soil, but it cracks under significant pressure.
  Stiff: Finger pressure dents soil.
  Firm: Soil can be molded using strong finger pressure.
  Soft: Easily molded by finger.
  Very soft: Soil flows between fingers when fist is closed.
- *Dilatancy:* Place a small amount of the soil in your palm and shake horizontally. Then strike it with the other hand. If the surface is slurry and water appears, the soil probably has a large amount of silt. This is not the same as dilatancy due to shearing (Chapter 8).
- *Other features:* Roots, human-made residues
- *Packing:* Coarse-grained soils are described as:
  Very loose: collapses with slight disturbance; open structure
  Loose: collapses upon disturbance; open structure
  Medium dense: indents when pushed firmly
  Dense: barely deforms when pushed by feet or by stomping
  Very dense: impossible to depress with stomping

3.5.3 Number and depths of boreholes

It is practically impossible and economically infeasible to completely explore the whole project site. You have to make judgments on the number, location, and depths of borings to provide sufficient information for design and construction. The number and depths of borings should cover the zone of soil that would be affected by the structural loads. There

Table 3.2 Guidelines for the minimum number of boreholes for buildings and subdivisions based on area.

Buildings		Subdivisions	
Area (m²)	No. of boreholes (min.)	Area (m²)	No. of boreholes (min.)
100	2	4000	2
250	3	8000	3
500	4	20,000	4
1000	5	40,000	5
2000	6	80,000	7
5000	7	400,000	15
6000	8		
8000	9		
10,000	10		

is no fixed rule to follow. In most cases, the number and depths of borings are governed by experience based on the geological character of the ground, the importance of the structure, the structural loads, and the availability of equipment. Building codes and regulatory bodies provide guidelines on the minimum number and depths of borings.

The number of boreholes should be adequate to detect variations of the soils at the site. If the locations of the loads on the footprint of the structure are known (this is often not the case), you should consider drilling at least one borehole at the location of the heaviest load. As a guide, a minimum of three boreholes should be drilled for a building area of about 250 m² and about five for a building area of about 1000 m². Some guidelines on the minimum number of boreholes for buildings and for due diligence in subdivisions are given in Table 3.2. Some general guidance on the depth of boreholes is provided in the following:

- In compressible soils such as clays, the borings should penetrate to at least between 1 and 3 times the width of the proposed foundation below the depth of embedment or until the stress increment due to the heaviest foundation load is less than 10%, whichever is greater.
- In very stiff clays and dense, coarse-grained soils, borings should penetrate 5–6 m to prove that the thickness of the stratum is adequate.
- Borings should penetrate at least 3 m into rock.
- Borings must penetrate below any fills or very soft deposits below the proposed structure.
- The minimum depth of boreholes should be 6 m unless bedrock or very dense material is encountered.

General guidelines for the minimum number or frequency of boreholes and their minimum depths for common geotechnical structures are shown in Table 3.3.

3.5.4 Soil sampling

The objective of soil sampling is to obtain soils of satisfactory size with minimum disturbance for observations and laboratory tests. Soil samples are usually obtained by attaching an open-ended, thin-walled tube—called a Shelby tube or, simply, a sampling tube—to drill rods and forcing it down into the soil.

Table 3.3 Guidelines for the minimum number or frequency and depths of boreholes for common geostructures.

Geostructure	Minimum number of boreholes	Minimum depth
Shallow foundation for buildings	1, but generally boreholes are placed at node points along grids of sizes varying from 15 m × 15 m to 40 m × 40 m	5 m or $1B$ to $3B$, where B is the foundation width
Deep (pile) foundation for buildings	Same as shallow foundations	25–30 m; if bedrock is encountered, drill 3 m into it
Bridge	Abutments: 2 Piers: 2	25–30 m; if bedrock is encountered, drill 3 m into it
Retaining walls	length < 30 m: 1 length > 30 m: 1 every 30 m, or 1 to 2 times the height of the wall	1 to 2 times the wall height Walls located on bedrock: 3 m into bedrock
Cut slopes	Along length of slope: 1 every 60 m; if the soil does not vary significantly, 1 every 120 m On slope: 3	6 m below the bottom of the cut slope
Embankments including roadway (highway, motorway)	1 every 60 m; if the soil does not vary significantly, 1 every 120 m	The greater of 2× height or 6 m

The tube is carefully withdrawn, hopefully with the soil inside it. Soil disturbances occur from several sources during sampling, such as friction between the soil and the sampling tube, the wall thickness of the sampling tube, the sharpness of the cutting edge, and the care and handling of the sample tube during transportation. To minimize friction, the sampling tube should be pushed instead of driven into the ground.

Sampling tubes that are in common use have been designed to minimize sampling disturbances. One measure of the effects of sampler wall thickness is the recovery ratio defined as L/z, where L is the length of the sample and z is the distance that the sampler was pushed. Higher wall thickness leads to a greater recovery ratio and greater sampling disturbance.

One common type of soil sampler is the Piston sampler, which is a thin-walled, seamless steel tube of diameter ranging from 50 mm to 100 mm, the latter most popular, wall thickness about 1.75 mm, and length of 600–1000 mm, the latter most popular (Figure 3.11a) connected to a hydraulic cylinder. This sampler is particularly useful fine-grained soils especially soft clays. Another popular sampler is the "standard" sampler, also known as the split-spoon sampler (split-barrel sampler), which has an inside diameter of about 35 mm and an outside diameter of about 50 mm (Figure 3.11b). The sampler has a split barrel that is held together using a screw-on driving shoe at the bottom end and a cap at the upper end. In some countries, a steel liner is used inside the sampler, but in the United States, it is standard practice not to use this liner.

Consequently, the soil sample has a greater diameter. The results of SPT (see in situ tests later in this chapter) are different for lined and unlined samplers. The thicker wall of the standard sampler permits higher driving stresses than the Shelby tube, but does so at the

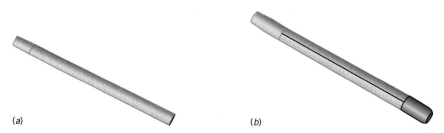

Figure 3.11 (a) A thin-walled tube and (b) a split-tube sampler.

expense of higher levels of soil disturbances. Split-spoon samples are disturbed. They are used for visual examination and for classification tests.

3.5.5 Groundwater conditions

If you dig a hole into a soil mass that has all the voids filled with water (fully saturated), you will observe water in the hole up to a certain level. This water level is called groundwater level or groundwater table.

The top of the groundwater level is under atmospheric pressure and is sometimes called the free surface. We will denote groundwater level by the symbol ▼ or ∇. Many construction failures, court battles, and construction cost overruns are due to the nonidentification or nondisclosure of groundwater conditions at a site. The water table invariably fluctuates depending on environmental conditions (e.g., rainfall patterns, winter rains, monsoons, drought), human activities (e.g., pumping groundwater from wells and drawdown during construction), and geological conditions.

The water-bearing soils below the groundwater level are called an aquifer. Aquifers can be unconfined or confined or semiconfined (Figure 3.12). In an unconfined aquifer the groundwater level is free to fluctuate up and down, depending on the availability of water. During winter and spring, the groundwater level usually rises. In drier months, the groundwater level drops.

Aquifers are sometimes separated by geological formations that may restrict groundwater flow. If the formations are impermeable, such as fine-grained soils (e.g., clay) and/or nonporous rock, they are called aquicludes. If the formations are semi-impermeable, they are called aquitards.

A confined aquifer is a water-bearing stratum that is confined by aquicludes (impermeable geological formations) above and below it. The water held in an unconfined aquifer is under pressure because of the confinement. If one of the impermeable formations, usually the top, is penetrated, water can rise above the ground surface. In some unconfined aquifers, water has risen more than 50 m above the aquifer surface during well drilling. Confined aquifers (also called artesian aquifers) are not directly affected by seasonal climatic changes. There is really no true confined aquifer, as some infiltration does occur from the overlying soil or rock. The geological formations are rarely continuous, especially in alluvial aquifers. Often the aquifer consists of fingerings or zones or lenses of impermeable and semi-impermeable materials. Sometimes a zone or zones of water is/are collected within the unsaturated geological formation. These are known as perched aquifers, and the groundwater levels within them are called perched water tables. If these perched aquifers are not identified and

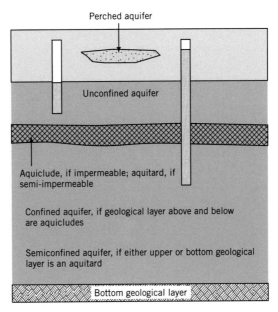

Figure 3.12 Unconfined, confined, and perched aquifers, aquiclude, and aquitard.

reported in the soils report, they may cause instability, construction problems, and litigation.

You should identify not only the groundwater level and any special conditions (e.g., artesian condition) but also the possible range of groundwater level fluctuations.

3.5.6 *Types of in situ or field tests*

Over the years, several in situ testing devices have emerged to characterize the soil and to estimate strength and deformation properties. The most popular devices are:

- Vane shear test (VST)
- Standard penetration test (SPT)
- Cone penetrometer test (CPT)
- Pressuremeter test (PMT)
- Flat plate dilatometer (DMT)

The advantages of in situ testing especially using SPT, VST, and CPT are:

1. Continuous or semicontinuous soil profiling.
2. Soil response provided in its natural state based on the type of in situ test used.
3. Tests less costly than laboratory tests.

The disadvantages are:

1. Stress and strain states, and rate of loading imposed by the particular in situ test, may not be representative of the stress/strain states and loading rate imposed by the structure.
2. Drainage and boundary conditions cannot be controlled.

3. Design soil parameters are not obtained directly (notable exception is the pressure-meter). Empirical correlations have to be used or be developed using carefully controlled laboratory tests. Often the correlations are weak (low value of regression coefficient or *R*-squared)

The interpretation of the test results from some of these tests is presented in Chapter 8.

3.5.6.1   Vane shear test (VST)

The shear vane device consists of four thin metal blades welded orthogonally (90°) to a rod (Figure 3.13). The vane is pushed, usually from the bottom of a borehole, to the desired depth. A torque is applied at a rate of about 6° per minute by a torque head device located above the soil surface and attached to the shear vane rod.

After the maximum torque (T_{max}) is obtained, the shear vane is rotated an additional 8 to 10 revolutions to measure the residual torque, T_{res}. The ratio of the maximum torque to the residual torque is the soil sensitivity, S_t, where

$$S_t = \frac{T_{max}}{T_{res}}$$
(3.1)

Sensitivity is a measure of the reduction of undrained shear strength (see Chapter 8) due to soil disturbance. The results of a vane shear test are displayed as undrained or vane shear strength versus depth. An example of a VST result is shown in Figure 3.14.

The VST is simple, inexpensive, and quick to perform, and the equipment is widely available. The insertion of the vane causes soil remolding. Higher blade thickness results in greater

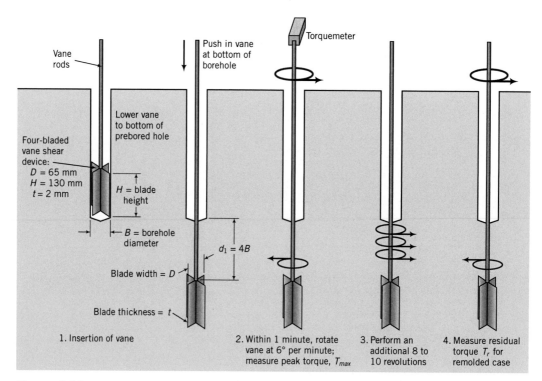

Figure 3.13   Vane shear tester. (Courtesy of Professor Paul Mayne, Georgia Tech.)

FIELD VANE SHEAR TEST RECORD

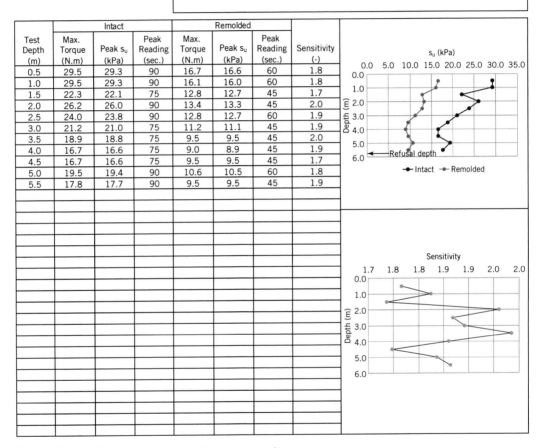

	FV No:	FV-16
	Phase	
	Date Started:	
	Date Finished:	
	Time Started:	
	Time Finished:	
	Contractor:	

Vane Information

Type of Machine:	GEONOR-H10
Type:	Rectangular
Vane Dia., mm:	65
Vane Height, mm:	130
Loading, deg./sec.:	0.2

Field Vane Test Information

No. of Test, #:	11
Test Interval, #/m:	0.5
Final Test Depth, m:	5.5
Refusal Depth, m:	5.7

FV Test Coord. & Elev.

East, m	North, m	Elev.,m
600018.5	1179098.7	−1.25

Notes:
1. Casing, vane and vane shoe equipment coupled with torque rods were used to perform vane shear test.
2. Peak undrained shear strength (s_u) herein refers to "raw field" shear strength.

Test Depth (m)	Intact			Remolded			Sensitivity (-)
	Max. Torque (N.m)	Peak s_u (kPa)	Peak Reading (sec.)	Max. Torque (N.m)	Peak s_u (kPa)	Peak Reading (sec.)	
0.5	29.5	29.3	90	16.7	16.6	60	1.8
1.0	29.5	29.3	90	16.1	16.0	60	1.8
1.5	22.3	22.1	75	12.8	12.7	45	1.7
2.0	26.2	26.0	90	13.4	13.3	45	2.0
2.5	24.0	23.8	90	12.8	12.7	60	1.9
3.0	21.2	21.0	75	11.2	11.1	45	1.9
3.5	18.9	18.8	75	9.5	9.5	45	2.0
4.0	16.7	16.6	75	9.0	8.9	45	1.9
4.5	16.7	16.6	75	9.5	9.5	45	1.7
5.0	19.5	19.4	90	10.6	10.5	60	1.8
5.5	17.8	17.7	90	9.5	9.5	45	1.9

Figure 3.14 Example of a vane shear test result.

remolding and lower soil strengths. The blade thickness should not exceed 5% of the vane diameter. The soil resistance usually changes with rate of rotation. Errors in the measurements of the torque include excessive friction, variable rotation, and calibration. The VST cannot be used for coarse-grained soils and very stiff clays.

3.5.6.2 Standard penetration test (SPT)

The standard penetration test (SPT) was developed circa 1927 and it is perhaps the most popular field test. The SPT is performed by driving a standard split-spoon sampler into the ground by blows from a drop hammer of mass 63.5 kg falling 760 mm (Figure 3.15). The

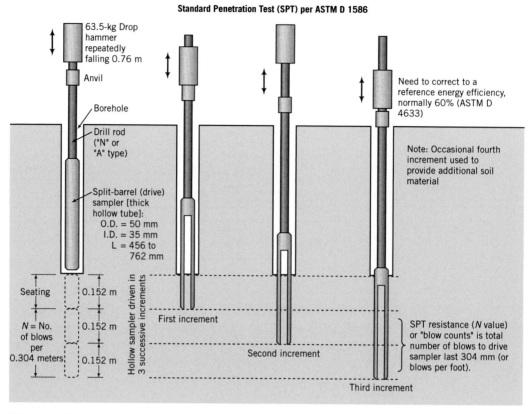

Figure 3.15 Driving sequence in an SPT. (Courtesy of Professor Paul Mayne, Georgia Tech.)

sampler is driven 152 mm into the soil at the bottom of a borehole, and the number of blows (N) required to drive it an additional 304 mm is counted. The number of blows (N) is called the standard penetration number. For very dense coarse-grained soils or when an obstacle is encountered, the number of blows may exceed 50 per 25 mm penetration. When this occurs, the SPT record would read "refusal." In practice, SPT are usually conducted at 0.6-m intervals. An example of SPT record is shown in Figure 3.16.

Various corrections are applied to the N values to account for energy losses, overburden pressure, rod length, and so on. It is customary to correct the N values to a rod energy ratio of 60%. The rod energy ratio is the ratio of the energy delivered to the split spoon sampler to the free-falling energy of the hammer. The corrected N values are denoted as N_{60} and given as

$$N_{60} = N\left(\frac{ER_r}{60}\right) = NC_E \tag{3.2}$$

where ER_r is the energy ratio and C_E is the 60% rod energy ratio correction factor. Correction factors for rod lengths, sampler type, borehole diameter, and equipment (60% rod energy ratio correction) are given in Table 3.4.

We can write a composite correction factor, C_{RSBEN}, for the correction factors given in Table 3.4 as

$$C_{RSBEN} = C_R C_S C_B C_E C_N \tag{3.3}$$

							BORING LOG:							
							SHEET 1 of 3							
							NCS PROJECT #:							

PROJECT:
CLIENT:
CLIENT PROJECT:

CONTRACTOR:
DRILLER:
INSPECTOR:

COORDINATES N: **REF. ALIGNMENT:**
      E: **STATION:**
LOCATION: Tucson, Arizona **OFFSET:**

RIG TYPE: 200 mm O.D. HSA
DRILLING METHOD: 200 mm O.D. HSA
HAMMER TYPE: Auto Hammer

COMMENTS:

SURFACE ELEV.: 732
TOTAL DEPTH: 25.5 m

START DATE: 10/26/2005 **TIME:** 09:20 AM
FINISH DATE: 10/26/2005 **TIME:** 01:45 PM

Type/Symbol	Casing	Split Spoon S	Ring Sample R	Split Spoon S2	Cuttings CU	Core Barrel C	GROUNDWATER DATA					
							Date	Time	Water Depth (m)	Casing Depth (m)	Hole Depth (m)	Symbol
I.D.		35 mm	63.5 mm	22 mm	–		10/26/05	06:00 AM	5	8	8	▽
O.D.		50 mm	75 mm	38 mm	–							
Length		456 mm	456 mm	46 mm	–							
Hammer WT.	63.5 kg		Drill Rod Size I.D. (O.D.)									
Hammer Fall	760 mm											

DEPTH BELOW SURFACE (m)	ELEVATION (m)	GRAPHIC	SOIL SAMPLE			DEPTH (m)		BLOWS				RECOVERY (mm)	VISUAL MATERIAL CLASSIFICATION AND REMARKS	MOISTURE, %	DRY UNIT WEIGHT (kN/m³)	SAMPLES SENT TO LAB
			TYPE	NUMBER	SYMBOL	FROM	TO	0–152 mm	152–304 mm	304–456 mm	N-VALUE					
	731.0												POORLY GRADED SAND WITH SILT (fill), medium dense, moist, brown, fine SAND, few non plastic fines, drywall, brown glass, brick and other debris present, weak cementation, strong reaction with HCl. (SP-SM)			
1.5			R	1	▮	1.5	1.82	6	7		13					
	729.5															
3.0	728.0		S	2	☒	3	3.46	2	2	3	5	356	POORLY GRADED SAND (fill), loose, moist, brown, fine SAND, trace non plastic fines, weak cementation, no reaction with HCl. (SP)			
4.5	726.5		S	3	☒	4.5	4.96	3	4	6	10	152	CLAYEY SAND (native), medium dense, brown to dark brown, fine SAND, some medium plastic fines, weak cementation, no reaction with HCl. (SC)			
6.0	725.0		S	4	☒	6	6.46	23	34	33	67	380	WELL-GRADED GRAVEL WITH SAND, very dense, moist, brown, fine to coarse GRAVEL, some fine to coarse sand, trace non plastic fines, sub-angular particles, no cementation, no reaction to HCl. (GW)			
													WELL-GRADED SAND WITH CLAY, medium dense, wet, brown, fine to coarse			

Figure 3.16 Example of an SPT record.

Table 3.4 Correction factors for rod length, sampler type, and borehole size.

Correction factor	Item	Correction factor
C_R	Rod length (below anvil)	$C_R = 0.8$; $L \leq 4\,m$ $C_R = 0.05L + 0.61$; $4\,m < L \leq 6\,m$ $C_R = -0.0004L^2 + 0.017L + 0.83$; $6\,m < L < 20\,m$, $C_R = 1$; $L \geq 20\,m$ $L =$ rod length
C_S	Standard sampler US sampler without liners	$C_S = 1.0$ $C_S = 1.2$
C_B	Borehole diameter: 65 mm to 115 mm 152 mm 200 mm	 $C_B = 1.0$ $C_B = 1.05$ $C_B = 1.15$
C_E	Equipment: Safety hammer (rope, without Japanese "throw" release) Donut hammer (rope, without Japanese "throw" release) Donut hammer (rope, with Japanese "throw" release) Automatic trip hammer (donut or safety type)	 $C_E = 0.7$ to 1.2 $C_E = 0.5$ to 1.0 $C_E = 1.1$ to 1.4 $C_E = 0.8$ to 1.4

Source: Youd et al. (2001).

where C_R, C_S, C_B, C_N, and C_E are correction factors for rod length, sampler type, borehole diameter, overburden pressure (the effective soil pressure, which is the unit weight of the soil multiplied by the depth minus the groundwater pressure at that depth (see Chapter 6) and rod energy correction, respectively. The correction factor, C_N, is

$$C_N = \sqrt{\frac{98.5}{\sigma'_{zo}}}; \quad C_N \leq 2; \text{ the unit for } \sigma'_{zo} \text{ is kPa} \tag{3.4}$$

The corrected N value, often written as $N_{1,60}$, is

$$N_{1,60} = C_{RSBEN}N \tag{3.5}$$

All corrected N values should be reported in whole numbers.

The SPT is very useful for determining changes in stratigraphy and locating bedrock. Also, you can inspect the soil in the split spoon sampler to describe the soil profile and extract disturbed samples for laboratory tests.

The SPT is simple and quick to perform. The equipment is widely available and can penetrate dense materials. SPT results have been correlated with engineering properties of soils, bearing capacity, and settlement of foundations. Most of these correlations are, however, weak (regression coefficient, R-squared value < 60%). They pertain only to soils under the conditions of the tests for which the correlations were developed. There are multiple sources of errors including test performance and the use of nonstandard equipment. Test performance errors include faulty methods of lifting and dropping the hammer, improper cleaning

Table 3.5 Compactness of coarse-grained soils based on N values.

N	γ (kN/m³)	D_r (%)	Compactness
0–4	11–13	0–20	Very loose
4–10	13–16	20–40	Loose
10–30	16–19	40–70	Medium
30–50	19–21	70–85	Dense
>50	>21	>85	Very dense

Source: Terzaghi and Peck (1948).

Table 3.6 Consistency of saturated fine-grained soils based on SPT.

Description	N
Very soft	0–2
Soft	2–4
Medium stiff	4–8
Stiff	8–15
Very stiff	15–30
Hard	>30

Source: Terzaghi and Peck (1948).

of the bottom of the borehole before the test commences, and not maintaining the ground-water level, if one is encountered. These errors lead to N values that are not representative of the soil. SPT tests are unreliable for coarse gravel, boulders, soft clays, silts, and mixed soils containing boulders, cobbles, clays, and silts. During an SPT, dynamic stresses are applied to the surrounding soil. For non–free-draining soils such as clays and mixed soils, excess pressures of unknown amount in the water in the void spaces are created. Because water is incompressible, it behaves as a rigid material (like concrete) when the sampler of the SPT is driven into non–free-draining soils. Consequently, the recorded N value for these soils cannot be related to soil strength or any other soil parameter. Since SPTs are not performed continuously, the test may fail to capture important soil layers, especially weak, porous soils. SPT must be conducted in conjunction with other testing methods such as the cone penetration test (see the next section).

Compactness of coarse-grained soils based on N values is given in Table 3.5. The consistency of fine-grained soils based on SPT is given in Table 3.6. The values in these tables are estimates to provide a preliminary assessment of the character of the ground.

EXAMPLE 3.1 *Correcting SPT Values*

The blow counts for an SPT test at a depth of 6 m in a coarse-grained soil at every 152 mm are 8, 12, and 15. A donut automatic trip hammer and a standard sampler were used in a borehole 152 mm in diameter. The effective overburden pressure was 90 kPa.

(a) Determine the N value.

(b) Correct the N value to N_{60} value.

(c) Correct the N value to $N_{1,60}$ value.

(d) Make a preliminary description of the compactness of the soil.

(e) Estimate the bulk unit weight.

Strategy The N value is the sum of the blow counts for the last 2 to 6 in penetration. Just add the last two blow counts.

Solution 3.1

Step 1: Determine N by adding the last two blow counts.

$$N = 12 + 15 = 27$$

Step 2: Correct N to N_{60}.

$$N_{60} = N\left(\frac{ER_r}{60}\right) = NC_E$$

From Table 3.4: Donut automatic trip hammer, $C_E = 0.8$ to 1.4; use $C_E = 1$.

$$N_{60} = NC_E = 27 \times 1 = 27$$

Step 3: Correct N to $N_{1,60}$.

From Table 3.4, $C_R = 0.05L + 0.61$; $\leq 20\,\text{ft} = 0.05 \times 6 + 0.61 = 0.91$

Standard sampler, $C_S = 1.0$; borehole of diameter $= 152\,\text{mm}$ $C_B = 1.05$

$$C_N = \sqrt{\frac{98.5}{\sigma'_{zo}}} = \sqrt{\frac{98.5}{90}} = 1.05 < 2; \quad \text{use } C_N = 1.05$$

$$N_{1,60} = C_{\text{RSBEN}}N = 0.91 \times 1.0 \times 1.05 \times 1 \times 1.05 \times 27 = 27$$

Step 4: Use Table 3.5 to describe the compactness and bulk unit weight.

For $N = 27$, the soil is medium dense.

By interpolation,

$$\gamma = 16 + \frac{(27-10)}{(30-10)} \times (19 - 16) = 18.6\ \text{kN/m}^3$$

EXAMPLE 3.2 *Estimating Soil Compactness from SPT Values*

Delineate and estimate the compactness of the soil based on the SPT record shown in Figure 3.16.

Strategy Make a plot of N values versus depth and then use Table 3.5 to estimate the compactness.

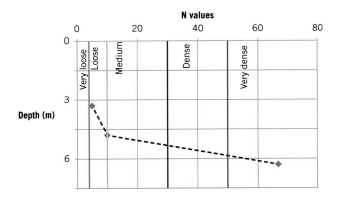

Figure E3.2

Solution 3.2

Step 1: Plot N values versus depth.

(a) Make a table of the values of average depth over the last 0.304 m and N.

Average depth (m)	N
3.23	5
4.73	10
6.23	67

(b) Plot the values (see Figure E3.2).

The linking of the points by straight lines is a gross approximation. Soils are notorious in changing types even over small distances.

Use Table 3.5 to estimate the compactness as illustrated in Figure E3.2.

For example, the soil from about 3 m to 5 m is loose.

3.5.6.3 Cone penetrometer test (CPT)

The cone penetrometer is a cone about 36 mm in diameter with a base area of about $10\,cm^2$ and a cone angle of $60°$ (Figure 3.17a) that is attached to a rod. An outer sleeve of surface area about $150\,cm^2$ encloses the rod just above the cone base. The thrusts required to drive the cone and the sleeve into the ground at a rate of 2 cm/s are measured independently so that the end resistance or cone tip resistance, q_c, and side friction or sleeve resistance, f_s, may be estimated separately. The friction ratio, $R_f = 100\ f_s/q_c$ is often used to present the sleeve resistance. Although originally developed for the design of piles, the cone penetrometer has also been used to estimate the bearing capacity and settlement of all types of foundations. There is no internationally accepted cone geometry. So different geometry of cones are in use. The results from these different cones will be different. The geometry of the cones shown and the method of cone penetrometer testing in this textbook are typically used in the United States.

The piezocone (uCPT or CPTu) is an electric cone penetrometer (Figure 3.17b) that has porous elements inserted into the cone or sleeve to allow for porewater pressure measurements (Figure 3.17c). The measured porewater pressure depends on the location of the

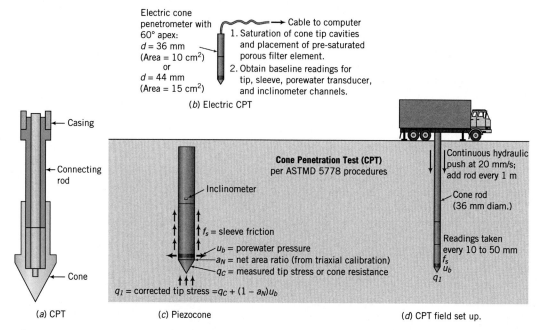

Figure 3.17 (a) CPT, (b) electric CPT, (c) piezocone, (d) CPT field set up. (Courtesy of Professor Paul Mayne, Georgia Tech.)

porous elements. A load cell is often used to measure the force of penetration. The piezocone is a very useful tool for soil profiling. Researchers have claimed that the piezocone provides useful data to estimate the shear strength, bearing capacity, and consolidation characteristics of soils. The CPT field set up is shown in Figure 3.17d. Typical results from a piezocone are shown in Figure 3.18.

Other CPT variants include the seismic cone (SCPT) and the vision cone (VisCPT or VisCPTu). In the SCPT, geophones are installed inside the cone. Hammers on the surface are used to produce surface disturbances, and the resulting seismic waves are recorded by the geophones (usually three). The recorded data are then analyzed to give damping characteristics and soil strength parameters.

The VisCPT and VisCPTu have miniature cameras installed in the CPT probe that provide continuous images of the soil adjacent to the cone. Through image processing, the soil texture can be inferred. The VisCPTu can also be used to detect liquefiable soils.

In most cases, CPT and CPTu are used so that the soil stratigraphy has to be inferred from measured cone resistance values. Generally, sands have $q_c > 4500$ kPa and fine-grained soils have $q_c < 2000$ kPa. One popular classification chart proposed is shown in Figure 3.19. This chart is based on CPT data for depths less than 30 m. It is generally observed that q_c and f_s increase with depth so that you need to be extra cautious in using soil classification based on Figure 3.19 for depths in excess of 30 m.

Regardless of which CPT probe is used, the results are average values of the soil resistance over a length of about 10 cone diameters—about 5 diameters above the tip plus about 5 diameters below the tip. In layered soils, the soil resistances measured by the cone may not represent individual layers, especially thin layers (<5 cone diameters).

The cone resistance is influenced by several soil variables such as stress level, soil density, stratigraphy, soil mineralogy, soil type, and soil fabric. Results of CPT have been correlated

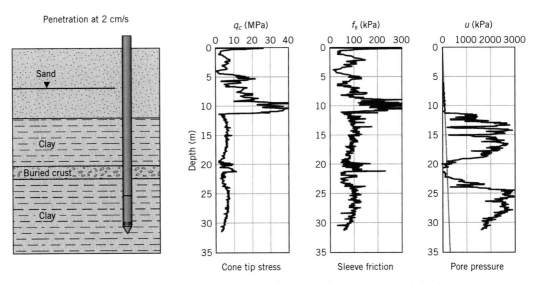

Figure 3.18 Piezocone results. (Courtesy of Professor Paul Mayne, Georgia Tech.)

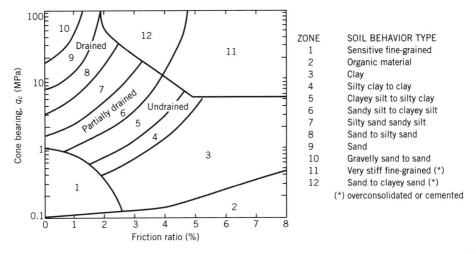

Figure 3.19 Soil classification based on cone tip resistance and friction ratio. (Source: Modified from Robertson, 1990.)

with laboratory tests to build empirical relationships for strength and deformation parameters. Investigators have also related CPT results to other field tests, particularly SPT. In practice, the cone data are inspected and large values are eliminated based on experience. A composite distribution of cone resistance (tip resistance) is drawn, as illustrated in Figure 3.20. In this case, three zones or layers of soils are identified; each assumed to have a uniform cone resistance. For example, the cone resistance for zone 2 is about 1900 kPa.

CPT is quick to perform, with fewer performance errors compared with SPT. It can provide continuous records of soil conditions. CPT cannot be used in dense, coarse-grained soils (e.g., coarse gravel, boulders) and mixed soils containing boulders, cobbles, clays, and silts.

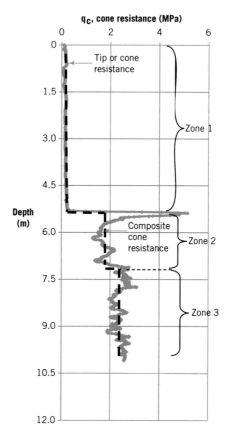

Figure 3.20 Estimation of composite cone tip resistance.

The cone tip is prone to damage from contact with dense objects. The more sophisticated uCPT, SCPT, and VisCPT usually require specialists to perform and to interpret the results.

3.5.6.4 Pressuremeter

The Menard pressuremeter (Figure 3.21) is a probe that is placed at the desired depth in an unlined borehole, and pressure is applied to a measuring cell of the probe. The pressure applied is analogous to the expansion of a cylindrical cavity. The pressure is raised in stages at constant time intervals, and volume changes are recorded at each stage. A pressure–volume change curve is then drawn from which the elastic modulus, shear modulus, horizontal stress, friction angle and undrained shear strength (see Chapter 8) may be estimated. The pressuremeter test is more costly than CPT and the flat plate dilatometer, and is not widely available. The drainage condition is unknown, and this leads to uncertainty in the interpretation of the test data to estimate the shear modulus and shear strength. The dimensions shown in Figure 3.21 vary internationally.

3.5.6.5 Flat plate dilatometer (DMT)

The flat plate dilatometer consists of a tapered blade (Figure 3.22). The dimensions shown in Figure 3.22 vary internationally. On the flat face, the dilatometer is a flexible steel

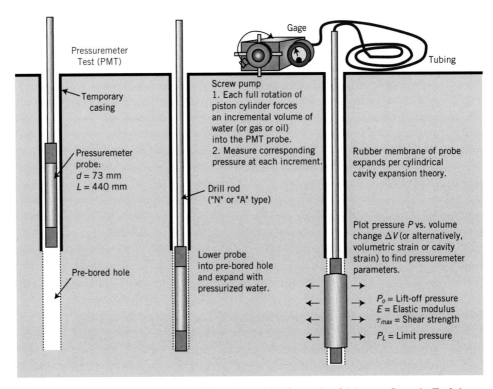

Figure 3.21 Pressuremeter test CPT. (Courtesy of Professor Paul Mayne, Georgia Tech.)

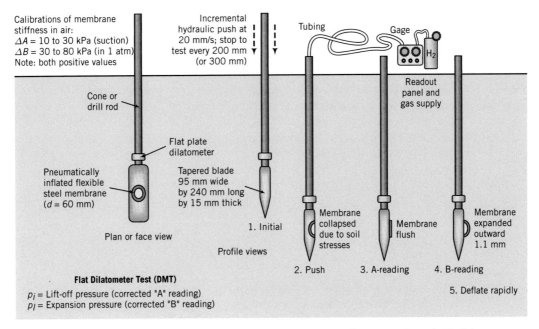

Figure 3.22 Flat plate dilatometer. (Courtesy of Professor Paul Mayne, Georgia Tech.)

membrane that, when inflated, pushes the soil laterally. The blade is attached to drill rods and is pushed into the soil at a rate of about 2 cm/s by a drill rig. Tests are normally conducted every 200 mm. The pneumatic pressures (1) to bring the membrane flush with the soil surface, (2) to push the soil laterally for a distance of 1.1 mm, and (3) at which the membrane returns to its original position are recorded.

Results from dilatometers have been related to undrained shear strength, lateral earth pressures, overconsolidation ratio, and elastic modulus. DMT is simple and quick to conduct. It provides reasonable estimates of horizontal stress and is less costly than the pressuremeter test. Dilatometers cause significant remolding of the soil before the test commences, and the results obtained should be used with caution. The dilatometer test is best suited for clays and sands.

3.5.7 Soils laboratory tests

Samples are normally taken from the field for laboratory tests to characterize the physical and mechanical (strength and deformation) properties. These parameters are used to design foundations and to determine the use of soils as a construction material. Disturbed samples such as from a standard sampler are usually used for visual inspection and for tests to determine the physical properties such as plasticity and grain size and shape. Undisturbed samples such as from a thin-walled sampler are used for both physical and mechanical properties. Test results, especially those that relate to the mechanical properties, are strongly affected by sampling, handling, transportation, and sample preparation disturbances. Care must therefore be exercised to protect the intact condition of the soil samples. Wax is often used to coat the soil samples to prevent moisture losses.

3.5.8 Types of laboratory tests

Laboratory tests are needed to classify soils and to determine strength, settlement, and stiffness parameters for design and construction. They allow for better control of the test conditions applied to the soil than in situ tests. Laboratory test samples are invariably disturbed, and the degree of disturbance can significantly affect the test results. Sufficient care must be taken to reduce testing disturbances. Laboratory tests can be divided into two classes: Class I tests are tests to determine the physical properties; Class II tests are used to determine the mechanical properties. Table 3.7 and Table 3.8 summarize these tests. You will learn about them and the meaning and importance of the soil parameters that they measure in later chapters of this textbook.

Key points

1. A number of tools are available for soil exploration. You need to use judgment as to the type that is appropriate for a given project.
2. Significant care and attention to details are necessary to make the results of a soils investigation meaningful.

Table 3.7 Summary of laboratory tests to determine physical properties.

Physical Properties	Test objective	Parameters determined	Purpose
Specific gravity	To determine the specific gravity of soils	G_s	To calculate soil density (unit weight)
Grain size determination	To determine the grain size distribution	D_{10}, D_{50} $$Cu = \frac{D_{60}}{D_{10}}$$ $$CC = \frac{D_{30}{}^2}{D_{10}D_{60}}$$	Soil classification
Water content	To determine the water content of a soil	w	Qualitative information on strength and deformation
Index test	To determine the water content at which soil changes phases	PL, LL, PI, SL, LI	Soil classification; qualitative information on strength and settlement
Compaction	To determine the maximum dry density and optimum water content	$(\gamma_d)_{max}, w_{opt}$	Specification of compaction in the field
Permeability	To determine the hydraulic conductivity	k	Estimate of flow of water and seepage forces; stability analysis
Maximum and minimum dry Density	To determine the maximum and minimum dry density of coarse-grained soil	e_{max} e_{min}	Soil classification

3.6 SOILS REPORT

A clear, concise, and accurate report of the site investigation must be prepared. The report should contain at a minimum the following:

1. A document (often a letter) authorizing the investigation.
2. A summary of the work done and recommendations (about one page).
3. Scope of work.
4. Description of the site.
5. Details of the types of investigation conducted, soil and groundwater information including lab and field test results, assumptions and limitations of the investigation, and possible construction difficulties. Soil boring logs (a typical one is shown in Figure 3.23) are normally used to summarize the soil data. Typically, the boring log should contain the following:
   (a) Name of project and location, including street name.
   (b) Location of boring: station and offset.
   (c) Date boring was performed.

      (d) Surface elevation.

      (e) Depth and thickness of each stratum, with fill pattern to quickly identify different soil types. A legend of the fill pattern must be included in the soils report.

      (f) Depths at which samples or in situ tests were conducted, with sample or test numbers.

      (g) Soil classification of each stratum.

      (h) Depth to water (if encountered).

  6. Analysis and interpretation of the data collected.

  7. Recommendations for design and construction, with discussions of any special provisions.

Table 3.8 Summary of laboratory mechanical tests.

Test	Stress condition	Drainage condition	Soil type	Parameters	Advantages	Disadvantages
Direct shear (DS)	Plane strain: stress or strain control	Drained	Coarse-grained	$\phi'_{cs}, \phi'_{p}, \alpha_p$	▪ Simple ▪ Quick ▪ Commonly available	▪ Soil fails on predetermined failure plane ▪ Nonuniform stress distribution ▪ Strains cannot be determined ▪ Cannot determine stress state
Triaxial (T)	Axisymmetric stress or strain control	Drained or undrained	All	$\phi'_{cs}, \phi'_{p}, s_u, E,$ and G	▪ Versatile: two stresses (axial and radial stresses) can be controlled independently ▪ External drainage can be controlled ▪ Commonly available	▪ Principal axes rotate only by 90° instantaneously ▪ Nonuniform stress distribution: reduced by lubricating platen
One-dimensional consolidation	Axisymmetric	drained	Fine-grained	$C_c, C_r, C_\alpha, C_v,$ σ'_{zc}, m_v	▪ Simple ▪ Readily available	▪ One-dimensional
Direct simple shear (DSS)	Plane strain	Drained (constant load) or undrained (constant Volume)	All	$\phi'_{cs}, \phi'_{p}, s_u, G$	▪ Principal axes rotate during test ▪ Closely approximates many field conditions	▪ Nonuniform stress and strain distributions ▪ Not readily available

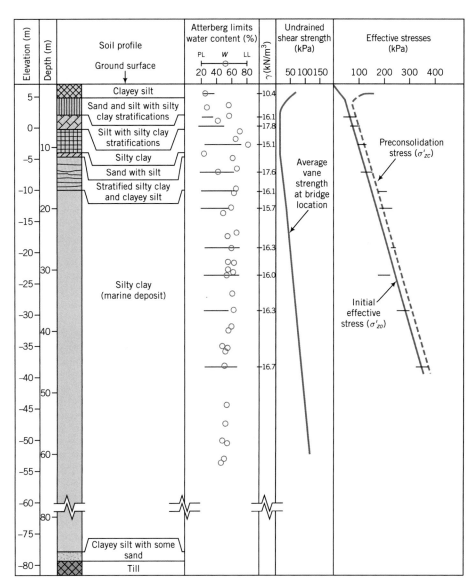

Figure 3.23 A borehole log. (Source: Modified from Blanchet et al., 1980.)

3.7 SUMMARY

At a project site, the soils must be identified and characterized through a soils investigation. Such an investigation is done in phases and may include geophysical investigations, boreholes, and field and laboratory tests. At the completion of a soils investigation, the client normally requires a carefully written report.

EXERCISES

Critical thinking and decision making

3.1 In your area, choose a project under construction or a recently constructed project such as a road or a building. Obtain the soils (geotechnical) report and review it.

3.2 Obtain borehole logs from a building site in your area. Describe the geology, the methods used in the soils exploration, and the type of field tests used, if any.

3.3 On Google Earth (earth.google.com), locate where you live. Conduct a Phase I (desk study) investigation, assuming that a five-story office building (50 m wide × 75 m long × 20 m high) is planned for that location.

3.4 A property developer wants to build a subdivision consisting of 500 residences, a shopping mall, and five office buildings near your college. Assume that the total area is 50 hectacres. The developer hires you to conduct a soils investigation as part of the due diligence process. Describe how you would conduct this investigation.

3.5 A borehole log from 8.5 m to 20.5 m is shown in Figure P3.5. (a) Determine the N values at the depths at which SPT were conducted. (b) Make a plot of N values versus depth. (c) Delineate and estimate the degree of compactness of the soils with depth. (d) Based on the degree of compactness in (c), estimate the unit weights for each soil zone (a zone is a soil layer with a given degree of compactness; e.g., if a soil between depth A and depth B is characterized as loose, you can describe this as a zone) (e) Calculate $N_{1,60}$ for each zone assuming the length of rod above the ground level is 1 m for each test, the use of a standard sample in a borehole of diameter of 200 mm and an automatic trip hammer, $C_E = 1$. Note that "R" denotes refusal.

3.6 The cone resistance (tip resistance) for a CPT test conducted on an alluvial deposit of soil is shown in Figure P3.6. (a) Draw the composite cone resistance with depth. (b) How many soil zones or layers can you identify? Show them in your composite cone resistance with depth diagram. (c) Describe the soil type in each layer.

BORING LOG:	F4
SHEET 2 of 3 **NCS PROJECT #:**	

PROJECT: **CLIENT:** **CLIENT PROJECT:**	**CONTRACTOR:** **DRILLER:** **INSPECTOR:**

DEPTH BELOW SURFACE (m)	ELEVATION (m)	GRAPHIC	SOIL SAMPLE					BLOWS			N-VALUE	RECOVERY (mm)	VISUAL MATERIAL CLASSIFICATION AND REMARKS	MOISTURE, %	DRY UNIT WEIGHT (kN/m³)	SAMPLES SENT TOLAB
			TYPE	NUMBER	SYMBOL	DEPTH (m) FROM	TO	0–152 mm	152–304 mm	304–456 mm						
	723.0		S	5	⊠	8.5	8.946	11	13	17		406	SAND, few low plastic fines, no reaction with HCI, (SW-SC)			
10.0	721.5		S	6	⊠	10.0	10.456	5	5	4		76	CLAYEY SAND, loose, moist, brown, fine to medium SAND, some low to medium plastic fines, no reaction wiht HCI. (SC)			
11.5	720.0		R	7	■	11.5	11.946						LEAN CLAY, very stiff, moist, brown, medium plastic CLAY, few fine sand, trace fine gravel, no reaction with HCI, max, particle size 75 mm. (CL)			
13.0	718.5		S	8	⊠	13.0	13.456	6	11	18		432				x
14.5	717.0		S	9	⊠	14.5	14.946	8	24	50/127 mm	R‡	381	SILTY SAND, very dense, moist, brown, fine SAND, little non plastic fines, no reaction with HCL (SM)			
16.0	715.5		S	10	⊠	16.0	16.456	8	25	30		356	WELL-GRADED SAND WITH CLAY AND GRAVEL, very dense, moist, brown, fine to coarse SAND, some fine gravel, few medium plastic fines, sub-angular particles, no reation with HCL, max, particle size 75 mm. (SW-SC)			x
17.5	714.0		S	11	⊠	17.5	17.946	32	35	50/127 mm	R‡	406				
19.0	712.5		S	12	⊠	19.0	19.456	25	32	50/127 mm	R‡	356				
20.5													CLAYEY SAND, medium dense, moist, brown, fine to coarse SAND, some low			

‡ R = refusal

Figure P3.5

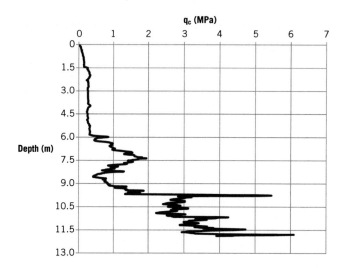

Figure P3.6

Chapter 4
One- and Two-Dimensional Flows of Water through Soils

4.1 INTRODUCTION

In this chapter, we will discuss the flow of water through soils. The flow of water has caused instability and failure of many geotechnical structures (e.g., roads, bridges, dams, and excavations). Therefore, you need to understand how water flows through soil and the stresses it induces.

Learning outcomes

When you complete this chapter, you should be able to do the following:

- Determine the rate of flow of water through soils.
- Determine the hydraulic conductivity of soils.
- Calculate flow under earth structures.
- Calculate seepage stresses, porewater pressure distribution, uplift forces, hydraulic gradients, and the critical hydraulic gradient.

4.2 DEFINITIONS OF KEY TERMS

Groundwater is water under gravity that fills the soil pores.

Head (H) is the mechanical energy per unit weight

Hydraulic conductivity, sometimes called the coefficient of permeability, (k) is a proportionality constant used to determine the flow velocity of water through soils.

Porewater pressure (u) is the average pressure of water within the soil pores.

Equipotential line is a line representing constant head.

Flow line is the flow path of a particle of water.

Flownet is a graphical representation of a flow field.

Soil Mechanics Fundamentals, First Edition. Muni Budhu.
© 2015 John Wiley & Sons, Ltd. Published 2015 by John Wiley & Sons, Ltd.
Companion website: www.wiley.com\go\budhu\soilmechanicsfundamentals

Static liquefaction is the behavior of a soil as a viscous fluid when the porewater pressure reduces the soil's effective stress to zero.

4.3 ONE-DIMENSIONAL FLOW OF WATER THROUGH SATURATED SOILS

If you dig a hole into a soil mass that has all the voids filled with water (fully saturated), you will observe water in the hole up to a certain level. This water level is called groundwater level or groundwater table. Recall from Chapter 3 that we will use the symbol ∇ or $\blacktriangledown$ to denote groundwater level. The pressure of the water below the groundwater level is called the hydrostatic pressure. The depth of the water below the groundwater level is called the hydrostatic head or static pressure head, h_p, or just pressure head. The porewater pressure is the pressure head times the unit weight of water, that is, $u = \gamma_w h_p = 9.81 h_p$ (kPa where the unit of h_p is m).

Porewater pressures are measured by porewater pressure transducers (Figure 4.1) or by piezometers (Figure 4.2). Piezometers are porous tubes that allow the passage of water. In a simple piezometer, you can measure the height of water in the tube from a fixed elevation and then calculate the porewater pressure by multiplying the height of water by the unit weight of water. A borehole cased to a certain depth acts like a piezometer. Modern piezometers are equipped with porewater pressure transducers for electronic reading and data acquisition.

Gravitational flow of water can only occur if there is a gradient, called hydraulic gradient, *i*. The hydraulic gradient is the change in total head divided by the flow distance over which the change occurs. The total head is the sum of the velocity head, the pressure head (h_p) and the elevation head (h_z). The elevation head is the distance above a selected datum or reference elevation taken as positive and as a negative value if below. In problem solving, you should always define your datum for the problem. One of the best selections for datum is

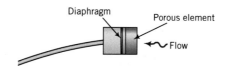

Figure 4.1 Schematic of a porewater pressure transducer.

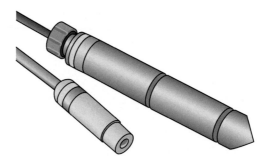

Figure 4.2 Piezometers.

the exit elevation of the flow. The velocity head in soils is small will be neglected. Therefore, for soils, the total head is the sum of the elevation head and the pressure head. The equation describing the hydraulic gradient is

$$i = \frac{\Delta H}{L} \qquad (4.1)$$

where ΔH is change in total head, also called head loss, and L is the distance over which the head change occurs.

Darcy's law (1856) governs the flow of water through saturated soils. He proposed that average flow velocity through soils is proportional to the hydraulic gradient, that is,

$$v_z = k_z i \qquad (4.2)$$

where v_z is the average flow velocity in the vertical direction and k_z is a coefficient of proportionality called the hydraulic conductivity (sometimes called the coefficient of permeability) in the vertical direction. By default, k_z is the hydraulic conductivity for a saturated soil. The application of Darcy's law assumes steady, laminar, inviscid (no change in viscosity), and incompressible (no change in volume) flow. The unit of measurement for k_z is length/ time, that is, cm/s.

The average velocity in the vertical direction is for the total cross-sectional area normal to the direction of flow. Flow through soils, however, occurs only through the interconnected voids. The velocity through the void spaces is called seepage velocity (v_s) and is obtained by dividing the average velocity by the porosity of the soil:

$$v_s = \frac{k_z}{n} i \qquad (4.3)$$

The volume rate of flow in the vertical direction, q_z, or, simply, flow rate is the product of the average velocity and the cross-sectional area:

$$q_z = v_z A = A k_z i \qquad (4.4)$$

The unit of measurement for q_z is cm³/s or m³/yr. It is the total volume of water (Q) divided by the time taken for that volume of water to flow.

In addition to the hydraulic gradient, the hydraulic conductivity depends on:

1. *Soil type:* Coarse-grained soils have higher hydraulic conductivities than fine-grained soils. The water in the double layer in fine-grained soils significantly reduces the flow rate.
2. *Particle size:* Hydraulic conductivity depends on D_{50}^2 (or D_{10}^2) for coarse-grained soils. Small particles, especially clay particles, can significantly reduce the hydraulic conductivity of a soil.
3. *Pore fluid properties, particularly viscosity:* $(k_z)_1 : (k_z)_2 \approx \mu_2 : \mu_1$, where μ is dynamic viscosity (dynamic viscosity of water is 1.12×10^{-3} N.s/m² at 15.6°C) and the subscripts 1 and 2 denote two types of pore fluids in a given soil.
4. *Soil fabric:* The structural arrangement of the soil particles plays a key role in the flow of water through a soil. If the fabric of a soil is such that the void spaces are interconnected uninterruptedly, then the hydraulic conductivity would be larger than the same soil for which the fabric results in discontinuous void spaces. Thus, the greater the

interconnected pore space, the higher the hydraulic conductivity. Two soils with the same void ratio can have different hydraulic conductivities because it is the continuity of the interconnected voids not the value of the void ratio that controls the steady-state flow of water through soils.

5. *Pore size:* Large pores do not indicate high porosity, but interconnected large pores result in high hydraulic conductivity. The flow of water through soils is related to the square of the pore size, and not the total pore volume.

6. *Homogeneity, layering, and fissuring:* Water tends to seep quickly through loose layers, through fissures, and along the interface of layered soils. Catastrophic failures can occur from such seepage.

7. *Degree of saturation:* The flow of water through soils depends on degree of saturation. The hydraulic conductivity of an unsaturated soil is lower than the saturated soil because of capillary action or soil suction. Entrapped gases tend to reduce the hydraulic conductivity. It is often very difficult to get gas-free soils. Even soils that are under groundwater level and are assumed to be saturated may still have some entrapped gases.

8. *Validity of Darcy's law:* Darcy's law is valid only for laminar flow (Reynolds number less than 2100 for pipe flow). Fancher et al. (1933) gave the following criterion for the applicability of Darcy's law for hydraulic conductivity determination:

$$\frac{v D_s \gamma_w}{\mu g} \leq 1 \tag{4.5}$$

where v is velocity, D_s is the diameter of a sphere of equivalent volume to the average soil particles, μ is dynamic viscosity of water, and g is the acceleration due to gravity.

9. *Stress level:* Higher normal stress level tends to reduce the hydraulic connectivity by forcing a tighter configuration (compression) of the soil fabric.

For natural soils, the hydraulic conductivity will vary in different flow directions. In many cases, especially horizontally layered soils, the hydraulic conductivity is larger (could be greater than 10 times larger) in the lateral directions compared to the vertical direction.

Typical ranges of k_z for various soil types are shown in Table 4.1.

Homogeneous clays are practically impervious. Two popular uses of "impervious" clays are in dam construction to curtail the flow of water through the dam and as barriers in

Table 4.1 Hydraulic conductivity for common saturated soil types.

Soil type	k_z (cm/s)	Description	Drainage
Clean gravel (GW, GP)	>1.0	High	Very good
Clean sands, clean sand and gravel mixtures (SW, SP)	1.0 to 10^{-3}	Medium	Good
Fine sands, silts, mixtures comprising sands, silts, and clays (SM-SC) Weathered and fissured clays	10^{-3} to 10^{-5}	Low	Poor
Silt, silty clay (MH, ML)	10^{-5} to 10^{-7}	Very low	Poor
Homogeneous clays (CL, CH)	<10^{-7}	Practically impervious	Very poor

landfills to prevent migration of effluent to the surrounding area. Clean sands and gravels are pervious and can be used as drainage materials or soil filters. The values shown in Table 4.1 are useful only to prepare estimates and in preliminary design for saturated soils.

4.4 FLOW OF WATER THROUGH UNSATURATED SOILS

The hydraulic conductivity is related to the water in the interconnected void space. Thus, the flow of water through unsaturated soils depends on the degree of saturation. The flow of water through an unsaturated soil is initially slow (because of capillarity) and then increases as the soil becomes saturated. The exponent for the order of magnitude for the hydraulic conductivity can vary by as much as 4 between a saturated and an unsaturated soil.

The failure of slopes, soil collapse, soil expansion, and loss of shear strength in unsaturated soils after saturation are important geotechnical issues that require knowledge of flow of water through unsaturated soils. However, such knowledge is beyond the scope of this introductory textbook. Several empirical equations (van Genuchten, 1980; Fredlund and Xing, 1994) are available in the literature to estimate the hydraulic conductivity for unsaturated soils.

4.5 EMPIRICAL RELATIONSHIP FOR k_z

A number of empirical relationships have been proposed linking k_z to void ratio and grain size for coarse-grained soils. Hazen (1892) proposed one of the early relationships as

$$k_z = CD_{10}^2 \text{ (unit: cm/s)} \tag{4.6}$$

where C is a constant varying between 0.4 and 1.4 if the unit of measurement of D_{10} is mm. Typically, $C = 1.0$. Hazen's tests were done on sands with D_{10} ranging from 0.1 mm to 3 mm and Cu < 5. You have to be cautious in using empirical relationships for k_z such as Hazen's relationship because they do not consider many of the effects presented in Section 4.3. Typically, Hazen's equation is used to give a first approximation for coarse-grained soils with particle size within the range 0.1 mm to 3 mm.

Key points

1. The flow of water through soils is governed by Darcy's law, which states that the average flow velocity is proportional to the hydraulic gradient.
2. The proportionality coefficient in Darcy's law is called the hydraulic conductivity, k_z.
3. The value of k_z is influenced by the void ratio, pore size, interconnected pore space, particle size distribution, homogeneity of the soil mass, properties of the pore fluid, stress level, and the amount of undissolved gas in the pore fluid.
4. Homogeneous clays are practically impervious, while sands and gravels are pervious.

EXAMPLE 4.1   *Calculating Flow Parameters*

The flow rate of water from a soil of cross sectional area, $2500\,\mathrm{mm^2}$, is $0.05 \times 10^{-6}\,\mathrm{m^3/s}$. The head difference measured over a length of $300\,\mathrm{mm}$ is $60\,\mathrm{mm}$. Determine the **(a)** hydraulic gradient, **(b)** average velocity, **(c)** seepage velocity if $e = 0.6$, and **(d)** the hydraulic conductivity. Estimate the soil type.

Strategy Use the head difference to calculate the hydraulic gradient and then use Darcy's law to find the hydraulic conductivity.

Solution 4.1

Step 1: Find the hydraulic gradient.

$$L = 300 \text{ mm}, \quad \Delta H = 60 \text{ mm}, \quad i = \frac{\Delta H}{L} = \frac{60}{300} = 0.2$$

Step 2: Determine the average velocity.

$$q_z = Av_z$$

$$A = 2500 \text{ mm}^2; \quad q_z = 0.05 \times 10^{-6} \text{ m}^3/\text{s}$$

$$v_z = \frac{q_z}{A} = \frac{0.05 \times 10^{-6}}{2500 \times 10^{-6}} = 2 \times 10^{-5} \text{ m/s}$$

Step 3: Determine seepage velocity.

$$v_s = \frac{v_z}{n}$$

$$n = \frac{e}{1+e} = \frac{0.6}{1+0.6} = 0.375$$

$$v_s = \frac{2 \times 10^{-5}}{0.375} = 5.3 \times 10^{-5} \text{ m/s}$$

Step 4: Determine the hydraulic conductivity.

From Darcy's law, $v_z = k_z i$.

$$\therefore k_z = \frac{v_z}{i} = \frac{2 \times 10^{-5}}{0.2} = 10 \times 10^{-5} \text{ m/s} = 1 \times 10^{-1} \text{ cm/s}$$

Step 5: Estimate the soil type.

From Table 4.1, with $k_z = 1 \times 10^{-1}\,\mathrm{cm/s}$, the soil type is likely a clean sand or clean sand and gravel mixture comprising sand, silt, and clay (SW, SP).

EXAMPLE 4.2 *Calculating Hydrostatic Pressures*

The groundwater level in a soil mass is 1 m below the existing surface. Plot the variation of hydrostatic pressure with depth up to a depth of 3 m.

Strategy Since the hydrostatic pressure is linearly related to depth, the distribution will be a straight line starting from the groundwater level, not the surface.

Solution 4.2

Step 1: Plot hydrostatic pressure distribution.

$$u = \gamma_w h_p = 9.81 h_p$$

At a depth of 3 m, $h_p = 3 - 1 = 2$ m and $u = 9.81 \times 2 = 19.6$ kPa

The slope of the hydrostatic pressure distribution $= 9.81$ kN/m^3, which is γ_w.

See Figure E4.2.

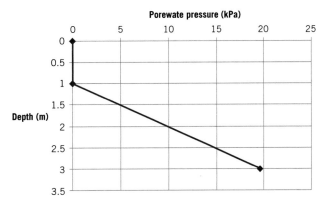

Figure E4.2

> *What's next ...* We have considered flow only through homogeneous soils. In reality, soils are stratified or layered with different soil types. In calculating flow through layered soils, an average or equivalent hydraulic conductivity representing the whole soil mass is determined from the hydraulic conductivity of each layer. Next, we will consider flow of water through layered soil masses: one flow occurs parallel to the layers, and the other flow occurs normal to the layers.

4.6 FLOW PARALLEL TO SOIL LAYERS

When the flow is parallel to the soil layers (Figure 4.3), the hydraulic gradient is the same at all points. The flow through the soil mass as a whole is equal to the sum of the flow through each of the layers. Flow parallel to soil layers is analogous to flow of electricity through resistors in parallel. If we consider a unit width (in the y direction) of flow the equivalent hydraulic conductivity, $k_{x(eq)}$, in the horizontal (x) direction is

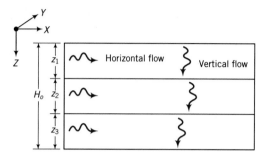

Figure 4.3 Flow through stratified layers.

$$k_{x(eq)} = \frac{1}{H_o}(z_1 k_{x1} + z_2 k_{x2} + \cdots + z_n k_{xn}) \tag{4.7}$$

where H_o is the total thickness of the soil mass, z_1 to z_n are the thicknesses of the first to the nth layer, and k_{x1} to k_{xn} are the horizontal hydraulic conductivities of the first to the nth layer. Equation (4.7) and Equation (4.9) to follow are approximations since no consideration is given to the condition of the interfaces between soil layers. For example, the interface can act as a conduit for the flow.

4.7 FLOW NORMAL TO SOIL LAYERS

For flow normal to the soil layers, the head loss in the soil mass is the sum of the head losses in each layer:

$$\Delta H = \Delta h_1 + \Delta h_2 + \cdots + \Delta h_n \tag{4.8}$$

where ΔH is the total head loss, and Δh_1 to Δh_n are the head losses in each of the n layers. The velocity in each layer is the same. The analogy to electricity is flow of current through resistors in series. The equivalent hydraulic conductivity, $k_{z(eq)}$, in the vertical direction is

$$k_{z(eq)} = \frac{H_o}{(z_1/k_{z1}) + (z_2/k_{z2}) + \cdots + (z_n/k_{zn})} \tag{4.9}$$

where k_{z1} to k_{zn} are the vertical hydraulic conductivities of the first to the nth layer. Values of $k_{z(eq)}$ are generally less than $k_{x(eq)}$—sometimes as much as 10 times less.

4.8 EQUIVALENT HYDRAULIC CONDUCTIVITY

The equivalent hydraulic conductivity for flow parallel and normal to soil layers is

$$k_{eq} = \sqrt{k_{x(eq)} k_{z(eq)}} \tag{4.10}$$

EXAMPLE 4.3 *Vertical and Horizontal Flows in Layered Soils*

A canal is cut into a soil with a stratigraphy shown in Figure E4.3. Assuming that flow takes place laterally and vertically through the sides of the canal and vertically below the canal, determine the equivalent hydraulic conductivity in the horizontal and vertical directions. The vertical and horizontal hydraulic conductivities for each layer are assumed to be the same. Calculate the ratio of the equivalent horizontal hydraulic conductivity to the equivalent vertical hydraulic conductivity for flow through the sides of the canal.

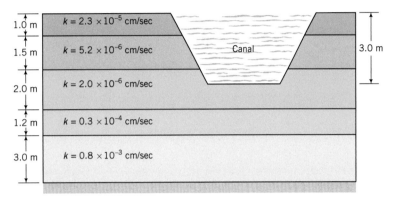

Figure E4.3

Strategy Use Equation (4.7) to find the equivalent horizontal hydraulic conductivity over the depth of the canal (3.0 m) and then use Equation (4.9) to find the equivalent vertical hydraulic conductivity below the canal. To make the calculations easier, convert all exponential quantities to a single exponent.

Solution 4.3

Step 1: Find $k_{x(eq)}$ and $k_{z(eq)}$ for flow through the sides of the canal.

$$H_o = 3 \text{ m}$$

$$
\begin{aligned}
k_{x(eq)} &= \frac{1}{H_o}(z_1 k_{x1} + z_2 k_{x2} + \cdots + z_n k_{xn}) \\
&= \frac{1}{3}(1 \times 0.23 \times 10^{-6} + 1.5 \times 5.2 \times 10^{-6} + 0.5 \times 2 \times 10^{-6}) \\
&= 3 \times 10^{-6} \text{ cm/s}
\end{aligned}
$$

$$
\begin{aligned}
k_{z(eq)} &= \frac{H_o}{(z_1/k_{z1}) + (z_2/k_{z2}) + \cdots + (z_n/k_{zn})} \\
&= \frac{3}{\dfrac{1}{10^{-6}}\left(\dfrac{1}{0.23} + \dfrac{1.5}{5.2} + \dfrac{0.5}{2}\right)} = 0.61 \times 10^{-6} \text{ cm/s}
\end{aligned}
$$

Step 2: Find the $k_{x(eq)}/k_{z(eq)}$ ratio.

$$\frac{k_{x(eq)}}{k_{z(eq)}} = \frac{3 \times 10^{-6}}{0.61 \times 10^{-6}} = 4.9$$

Step 3: Find $k_{z(eq)}$ below the bottom of the canal.

$$H_o = 1.5 + 1.2 + 3.0 = 5.7 \text{ m}$$

$$k_{z(eq)} = \frac{H_o}{\dfrac{z_1}{k_{z1}} + \dfrac{z_2}{k_{z2}} + \cdots + \dfrac{z_n}{k_{zn}}} = \frac{5.7}{\dfrac{1.5}{2 \times 10^{-6}} + \dfrac{1.2}{30 \times 10^{-6}} + \dfrac{3}{800 \times 10^{-6}}}$$
$$= 7.2 \times 10^{-6} \text{ cm/s}$$

What's next ... In order to calculate flow, we need to know the hydraulic conductivity k. We will discuss how this is determined in the laboratory and in the field.

4.9 LABORATORY DETERMINATION OF HYDRAULIC CONDUCTIVITY

4.9.1 *Constant-head test*

The constant-head test is used to determine the vertical hydraulic conductivity of coarse-grained soils. A typical constant-head apparatus is shown in Figure 4.4. Water is allowed to flow through a cylindrical sample of soil under a constant head (h). The outflow (Q) is collected in a graduated cylinder at a convenient duration (t).

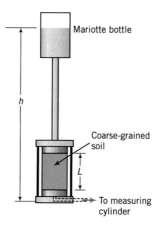

Figure 4.4 A constant-head apparatus.

With reference to Figure 4.4,

$$\Delta H = h \quad \text{and} \quad i = \frac{\Delta H}{L} = \frac{h}{L}$$

The flow rate through the soil is $q_z = Q/t$, where Q is the total quantity of water collected in the measuring cylinder over time t.

From Equation (4.4),

$$k_z = \frac{q_z}{Ai} = \frac{QL}{tAh} \tag{4.11}$$

where k_z is the hydraulic conductivity in the vertical direction and A is the cross-sectional area.

The viscosity of the fluid, which is a function of temperature, influences the value of k_z. The experimental value $(k_{T^\circ C})$ is corrected to a baseline temperature of 20°C using

$$k_{20^\circ C} = k_{T^\circ C} \frac{\mu_{T^\circ C}}{\mu_{20^\circ C}} = k_{T^\circ C} R_T \tag{4.12}$$

where μ is the dynamic viscosity of water, T is the temperature in °C at which the measurement was made, and $R_T = \mu_{T^\circ C}/\mu_{20^\circ C}$ is the temperature correction factor that can be calculated from

$$R_T = 2.42 - 0.475 \ln(T) \tag{4.13}$$

4.9.2 Falling-head test

The falling-head test is used for fine-grained soils because the flow of water through these soils is too slow to get reasonable measurements from the constant-head test. A compacted soil sample or a sample extracted from the field is placed in a metal or acrylic cylinder (Figure 4.5).

Porous stones are positioned at the top and bottom faces of the sample to prevent its disintegration and to allow water to percolate through it. Water flows through the sample

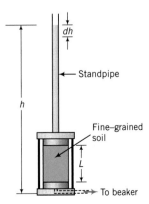

Figure 4.5 A falling-head apparatus.

from a standpipe attached to the top of the cylinder. The head of water (h) changes with time as flow occurs through the soil. At different times, the head of water is recorded. Let dh be the drop in head over a time period dt. The velocity or rate of head loss in the tube is

$$v_z = \frac{dh}{dt}$$

and the inflow of water to the soil is

$$(q_z)_{in} = av_z = -a\frac{dh}{dt}$$

where a is the cross-sectional area of the tube. We now appeal to Darcy's law to get the outflow:

$$(q_z)_{out} = Ak_z i = Ak_z\frac{h}{L}$$

where A is the cross-sectional area, L is the length of the soil sample, and h is the head of water at any time t. The continuity condition requires that $(q_z)_{in} = (q_z)_{out}$. Therefore,

$$-a\frac{dh}{dt} = Ak_z\frac{h}{L}$$

By separating the variables (h and t) and integrating between the appropriate limits, the last equation becomes

$$\frac{Ak_z}{aL}\int_{t_1}^{t_2} dt = -\int_{h_1}^{h_2}\frac{dh}{h}$$

and the solution for k_z is

$$k_z = \frac{aL}{A(t_2 - t_1)}\ln\left(\frac{h_1}{h_2}\right) \tag{4.14}$$

The hydraulic conductivity is corrected using Equation (4.13).

Key points

1. The constant-head test is used to determine the hydraulic conductivity of coarse-grained soils.
2. The falling-head test is used to determine the hydraulic conductivity of fine-grained soils.

EXAMPLE 4.4 *Interpretation of Constant-Head Test Data*

A sample of sand, 5 cm diameter and 15 cm long, was prepared at a porosity of 60% in a constant-head apparatus. The total head was kept constant at $h = 20$ cm and the volume of water collected in 5 seconds was 4 cm³. The test temperature was 20°C. Calculate the hydraulic conductivity and the seepage velocity.

Strategy From the data given, you can readily apply Darcy's law to find k_z.

Solution 4.4

Step 1: Calculate the sample cross-sectional area, hydraulic gradient, and flow.

$$A = \frac{\pi \times D^2}{4} = \frac{\pi \times 5^2}{4} = 19.6 \text{ cm}^2$$

$$\Delta H = h = 20 \text{ cm}$$

$$i = \frac{\Delta H}{L} = \frac{20}{15} = 1.33$$

$$q_z = \frac{Q}{t} = \frac{4}{5} = 0.8 \text{ cm}^3/\text{s}$$

Step 2: Calculate k_z.

$$k_z = \frac{q_z}{Ai} = \frac{0.8}{19.6 \times 1.33} = 3 \times 10^{-2} \text{ cm/s}$$

Step 3: Calculate the seepage velocity.

$$v_s = \frac{k_z i}{n} = \frac{3 \times 10^{-2} \times 1.33}{0.6} = 6.7 \times 10^{-2} \text{ cm/s}$$

EXAMPLE 4.5 *Interpretation of Falling-Head Test Data*

The data from a falling-head test on a silty clay are:

Diameter of soil = 100 mm
Length of soil = 150 mm
Initial head = 890 mm
Final head = 838 mm
Duration of test = 15 minutes
Diameter of tube = 6 mm
Temperature = 22°C

Determine k_z.

Strategy Since this is a falling-head test, you should use Equation (4.14). Make sure you are using consistent units.

Solution 4.5

Step 1: Calculate the parameters required in Equation (4.14).

$$a = \frac{\pi \times 0.6^2}{4} = 0.28 \text{ cm}^2$$

$$A = \frac{\pi \times 10^2}{4} = 78.5 \text{ cm}^2$$

$$t_2 - t_1 = 15 \times 60 = 900 \text{ seconds}$$

Step 2: Calculate k_z.

$$k_z = \frac{aL}{A(t_2 - t_1)} \ln\left(\frac{h_1}{h_2}\right) = \frac{0.28 \times 15}{78.5 \times 900} \ln\left(\frac{89}{83.8}\right) = 3.6 \times 10^{-6} \text{ cm/s}$$

From Equation (4.13), $R_T = 2.42 - 0.475 \ln(T) = 2.42 - 0.475 \ln(22) = 0.95$

$$k_{20°C} = k_z R_T = 3.6 \times 10^{-6} \times 0.95 \approx 3.4 \times 10^{-6} \text{ cm/s}$$

What's next … Flow of water in soils rarely takes place in one direction. Rather, it takes place in three dimensions. In the next section, we consider two-dimensional flow of water through soils, which is satisfactory for flow across long structures such as flow through a dam.

4.10 TWO-DIMENSIONAL FLOW OF WATER THROUGH SOILS

The two-dimensional flow of water through soils is described by Laplace's equation given as

$$k_x \frac{\partial^2 H}{\partial x^2} + k_z \frac{\partial^2 H}{\partial z^2} = 0 \tag{4.15}$$

where H is the total head and k_x and k_z are the hydraulic conductivities in the x and z directions. The assumptions in the derivation of Laplace's equation are that Darcy's law is valid, flow is laminar, irrotational, and inviscid (shear stresses are neglible), the soil and water are incompressible, and the soil is homogeneous and saturated. Laplace's equation expresses the condition that the changes of hydraulic gradient in one direction are balanced by changes in the other directions. For isotropic soils, $k_x = k_z$ and Laplace's equation becomes

$$\frac{\partial^2 H}{\partial x^2} + \frac{\partial^2 H}{\partial z^2} = 0 \tag{4.16}$$

The solution of Laplace's equation requires knowledge of the boundary conditions, which for most "real" structures are complex. We will only consider simple boundary conditions for Equation (4.16) to set the stage for a graphical solution. Computational software programs are available to solve Equation (4.15) and Equation (4.16) for complex boundary conditions. The solution of Equation (4.16) depends only on the values of the total head within the flow field. Let us introduce a velocity potential (ξ), which describes the variation of total head in a soil mass as

$$\xi = kH \tag{4.17}$$

where k is a generic hydraulic conductivity. The velocities of flow in the x (lateral) and z (vertical) directions are

$$v_x = k_x \frac{\partial H}{\partial x} = \frac{\partial \xi}{\partial x} \tag{4.18}$$

$$v_z = k_z \frac{\partial H}{\partial z} = \frac{\partial \xi}{\partial z} \tag{4.19}$$

We can infer from Equations (4.18) and (4.19) that the velocity (v) of the flow is normal to lines of constant total head (equipotential lines), as illustrated in Figure 4.6. The direction of v is in the direction of decreasing total head. The head difference between two equipotential lines is called a potential drop or head loss.

If lines are drawn that are tangents to the velocity of flow at every point in the flow field in the xz plane, we will get a series of lines that are normal to the equipotential lines. These tangential lines are called streamlines or flow lines (Figure 4.6). A flow line represents the

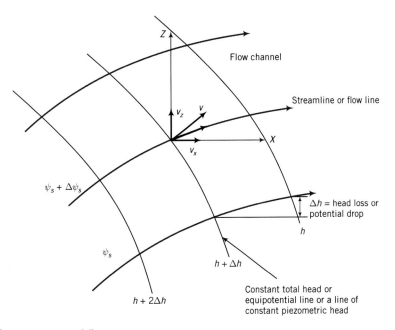

Figure 4.6 Illustration of flow terms.

path that a particle of water is expected to take in steady-state flow. A family of streamlines is represented by a stream function, $\psi_s(x, z)$.

The components of velocity in the x and z directions in terms of the stream function are

$$v_x = \frac{\partial \psi_s}{\partial z} \tag{4.20}$$

$$v_z = \frac{\partial \psi_s}{\partial x} \tag{4.21}$$

Since flow lines are normal to equipotential lines, there can be no flow across flow lines. The rate of flow between any two flow lines is constant. The area between two flow lines is called a flow channel (Figure 4.6). Therefore, the rate of flow is constant in a flow channel.

Key points

1. Laplace's equation describes the two-dimensional flow of water through soils.
2. Key assumptions for the derivation of Laplace's equation are Darcy's law is valid, flow is irrotational and inviscid, the soil and water are incompressible, and the soil is homogeneous and saturated.
3. Flow lines represent flow paths of particles of water.
4. The area between two flow lines is called a flow channel.
5. The rate of flow in a flow channel is constant.
6. Flow cannot occur across flow lines.
7. The velocity of flow is normal to the equipotential line.
8. Flow lines and equipotential lines are perpendicular to each other.

What's next ... In the next section, we will describe flownet sketching and provide guidance in interpreting a flownet to determine flow through soils, the distribution of porewater pressures, and the hydraulic gradients.

4.11 FLOWNET SKETCHING

Computer program utility

Access www.wiley.com\go\budhu\soilmechanicsfundamentals, Chapter 4, to learn about sketching flownets; calculate porewater pressure and seepage forces. Click on "2D Flow" to download an application program that interactively plots flownets for retaining walls and dams. Input different boundary conditions and geometry to explore changes in the flownet. For example, you can drag a sheet pile up or down or position it at different points below a dam and explore how the flownet changes.

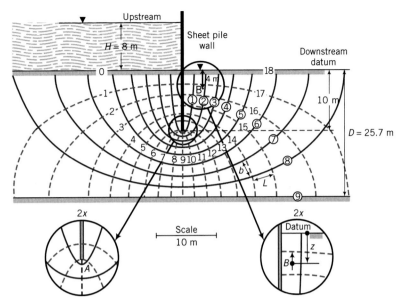

Figure 4.7 Flownet for a sheet pile.

One approximate method for the solution of Laplace's equation often used in practice is called flownet sketching. The flownet sketching technique is simple and flexible, and it conveys a picture of the flow regime. A flownet comprises flow lines and equipotential lines (Figure 4.7). Flow lines are traces of the flow path while equipotential lines are traces of equal heads. Flow lines are normal to equipotential lines. No flow can occur across flow lines. The area between two flow lines is called a flow channel. The rate of flow in a flow channel is constant. The distance between two equipotential lines is the head drop or head loss or potential drop. The velocity of flow is normal to an equipotential line.

4.11.1 Criteria for sketching flownets

A flownet must meet the following criteria:

1. The boundary conditions must be satisfied.
2. Flow lines must intersect equipotential lines at right angles.
3. The area between flow lines and equipotential lines must be curvilinear squares. A curvilinear square has the property that an inscribed circle can be drawn to touch each side of the square and continuous bisection results, in the limit, in a point.
4. The quantity of flow through each flow channel is constant.
5. The head loss between each consecutive equipotential line is constant.
6. A flow line cannot intersect another flow line.
7. An equipotential line cannot intersect another equipotential line.
8. An infinite number of flow lines and equipotential lines can be drawn to satisfy Laplace's equation. However, only a few are required to obtain an accurate solution. The procedure for constructing a flownet is described next.

4.11.2 Flownet for isotropic soils

1. Draw the structure and soil mass to a suitable scale.
2. Identify impermeable and permeable boundaries. The soil–impermeable boundary interfaces are flow lines because water can flow along these interfaces. The soil–permeable boundary interfaces are equipotential lines because the total head is constant along these interfaces.
3. Sketch a series of flow lines (four or five) and then sketch an appropriate number of equipotential lines such that the area between a pair of flow lines and a pair of equipotential lines (cell) is approximately a curvilinear square. You would have to adjust the flow lines and equipotential lines to make curvilinear squares. You should check that the average width and the average length of a cell are approximately equal by drawing an inscribed circle. You should also sketch the entire flownet before making adjustments.

The flownet in confined areas between parallel boundaries usually consists of flow lines and equipotential lines that are elliptical in shape and symmetrical (Figure 4.7). Try to avoid making sharp transitions between straight and curved sections of flow and equipotential lines. Transitions should be gradual and smooth. For some problems, portions of the flownet are enlarged and are not curvilinear squares, and they do not satisfy Laplace's equation. For example, the portion of the flownet below the bottom of the sheet pile in Figure 4.7 does not consist of curvilinear squares. For an accurate flownet, you should check these portions to ensure that repeated bisection results in a point. Figure 4.7 shows a flownet for a sheet pile wall, and Figure 4.8 shows a flownet beneath a dam.

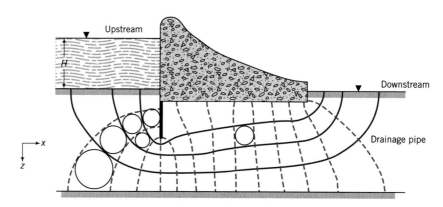

Figure 4.8 Flownet under a dam with a cutoff curtain (sheet pile) on the upstream end.

4.12 INTERPRETATION OF FLOWNET

4.12.1 Flow rate

Let the total head loss across the flow domain be ΔH, that is, the difference between upstream and downstream water level elevation. Then the head loss (Δh) between each consecutive pair of equipotential lines is

$$\Delta h = \frac{\Delta H}{N_d} \qquad (4.22)$$

where N_d is the number of equipotential drops, that is, the number of equipotential lines minus one. In Figure 4.7, $\Delta H = H = 8\,\text{m}$ and $N_d = 18$. Therefore, $\Delta h = \Delta H / N_d = 8/18 = 0.444\,\text{m}$. From Darcy's law, the flow rate is

$$q = k\Delta H \frac{N_f}{N_d} \qquad (4.23)$$

where N_f is the number of flow channels (number of flow lines minus one). In Figure 4.7, $N_f = 9$. The ratio N_f/N_d is called the shape factor. Finer discretization of the flownet by drawing more flow lines and equipotential lines does not significantly change the shape factor.

4.12.2 Hydraulic gradient

You can find the hydraulic gradient over each curvilinear square by dividing the head loss by the length, L; that is,

$$i = \frac{\Delta h}{L} \qquad (4.24)$$

You should notice from Figure 4.7 that L is not constant. Therefore, the hydraulic gradient is not constant. The maximum hydraulic gradient occurs where L is a minimum; that is,

$$i_{max} = \frac{\Delta h}{L_{min}} \qquad (4.25)$$

where L_{min} is the minimum length of the cells within the flow domain. Usually, L_{min} occurs at exit points or around corners (e.g., point A in Figure 4.7), and it is at these points that we usually get the maximum hydraulic gradient.

4.12.3 Critical hydraulic gradient

We can determine the hydraulic gradient that brings a soil mass (essentially, coarse-grained soils) to static liquefaction. Static liquefaction, called quicksand condition, occurs when the seepage stress balances the vertical stress from the soil. The critical hydraulic gradient, i_{cr}, is

$$i_{cr} = \frac{G_s - 1}{1 + e} \qquad (4.26)$$

where G_s is specific gravity of the soil particles, and e is the void ratio. Since G_s is constant, the critical hydraulic gradient is solely a function of the void ratio of the soil. In designing structures that are subjected to steady-state seepage, it is absolutely essential to ensure that the critical hydraulic gradient cannot develop.

4.12.4 Porewater pressure distribution

The porewater pressure head at any point j within the flow domain (flownet) is

$$(h_p)_j = \Delta H - (N_d)_j \, \Delta h - h_z \tag{4.27}$$

and the porewater pressure is

$$u_j = (h_p)_j \gamma_w \tag{4.28}$$

A simple way to determine the porewater pressure head is as follows:

1. Measure the vertical distance from the upstream water level to the point of interest. This gives the total pressure head, H_t.
2. Subtract the total head loss up to that point H_t to get the pressure head. For example, let us say that the vertical distance from the downstream water level to point B (Figure 4.7) is 4 m. Then, the vertical distance from the upstream water level to point B is $H_t = 8\,\mathrm{m} + 4\,\mathrm{m} = 12\,\mathrm{m}$. The number of equipotential drops to point B is 16.5. You should note that N_d can be non-integer but N_f cannot. The pressure head is $12\,\mathrm{m} - 16.5\Delta h = 12\,\mathrm{m} - 16.5 \times 0.444 = 4.67\,\mathrm{m}$ and the porewater pressure is $9.81 \times 4.67 = 45.8\,\mathrm{kPa}$.

4.12.5 Uplift forces

Lateral and uplift forces due to groundwater flow can adversely affect the stability of structures such as dams and weirs. The uplift force per unit length (length is normal to the xz plane) is found by calculating the porewater pressure at discrete points along the base (in the x direction, Figure 4.8) and then finding the area under the porewater pressure distribution diagram, that is,

$$P_w = \sum_{j=1}^{n} u_j \Delta x_j \tag{4.29}$$

where P_w is the uplift force per unit length, u_j is the average porewater pressure over an interval Δx_j, and n is the number of intervals. It is convenient to use Simpson's rule to calculate P_w:

$$P_w = \frac{\Delta x}{3}\left(u_1 + u_n + 2\sum_{\substack{i=3 \\ \text{odd}}}^{n} u_i + 4\sum_{\substack{i=2 \\ \text{even}}}^{n} u_i \right) \tag{4.30}$$

Key points

1. A flownet is a graphical representation of a flow field that satisfies Laplace's equation and comprises a family of flow lines and equipotential lines.
2. From the flownet, we can calculate the flow rate, the distribution of heads, porewater pressures, and the maximum hydraulic gradient.
3. The critical hydraulic gradient should not be exceeded in design practice.

4.13 SUMMARY

Flow of water through soils is governed by Darcy's law, which states that the average veloc-ity is proportional to the hydraulic gradient. The proportionality constant is the hydraulic conductivity. The hydraulic conductivity depends on soil type, particle size, pore fluid prop-erties, void ratio, pore size, homogeneity, layering and fissuring, and entrapped gases. In coarse-grained soils the hydraulic conductivity is determined using a constant-head test, while for fine-grained soils a falling-head test is used. In the field, a pumping test is used to determine the hydraulic conductivity. The governing equation for flow of water through soils is Laplace's equation. A graphical technique, called flownet sketching, was used to solve Laplace's equation. A flownet consists of a network of flow and equipotential lines. From the flownet, we can calculate the flow rate, the distribution of heads, porewater pressures, and the maximum hydraulic gradient.

4.13.1 *Practical examples*

EXAMPLE 4.6 *Flownet for a Reservoir*

A preliminary design of a reservoir is shown in Figure E4.6a.

- (a) Draw the flownet.
- (b) Calculate the flow rate.
- (c) How much water will flow in a day from the upstream to the downstream side.
- (d) Calculate the porewater pressure at A.

Strategy Follow the procedures for drawing flownets and calculate the required parameters. Only one-half the flownet is necessary as the flow is symmetrical, but we will plot the flownet for the whole flow field.

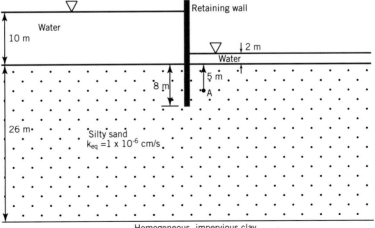

Figure E4.6a

Solution 4.6

Step 1: Draw the flow field to scale and select the datum.

You do not need to draw the structure and the water above the soil surface.

See *adhg* and *befc* (sheet pile) in Figure E4.6b.

Select the downstream end, *cd*, as the datum.

Step 2: Identify the impermeable and permeable boundaries.

With reference to Figure E4.6b, *ab* and *cd* are permeable boundaries. The heads along *ab* are equal (10 m). The heads along *cd* are equal (2 m). These are therefore equipotential lines; *befc* and *gh* are impermeable boundaries and are therefore flow lines.

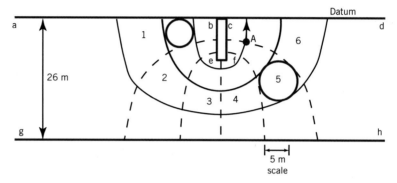

Figure E4.6b

Step 3: Sketch the flownet.

Draw about three more flow lines (remember *befc* and *gh* are flow lines) and then draw a suitable number of equipotential lines. Recall that flow lines are perpendicular to equipotential lines, and the area between two consecutive flow lines and two consecutive equipotential lines is approximately a square. Use a circle template to assist you in estimating the square. Adjust/add/subtract flow lines and equipotential lines until you are satisfied that the flownet consists essentially of curvilinear squares. See sketch of flownet in Figure E4.6b. This flownet can be improved by drawing more flow lines and equipotential lines to obtain larger numbers of curvilinear squares. However, the results for this case will not change significantly.

Step 4: Calculate the flow.

$$\Delta H = 10 - 2 = 8 \text{ m}$$

$$N_d = 6, \quad N_f = 4, \quad \Delta h = \frac{\Delta H}{N_d} = \frac{8}{6} = 1.33 \text{ m}$$

$$q = k_{eq} \Delta h N_f = 1 \times 10^{-8} \times 1.33 \times 4 \approx 5.32 \times 10^{-8} \text{ m}^3/\text{s}$$

Note: The flow rate is calculated for 1 m length of wall, and that is why the unit is m³/s. The unit for the hydraulic conductivity is converted from cm/s to m/s by dividing the ordinal value by 100 (1 m = 100 cm).

Step 5: Calculate the flow per day.

$$Q = qt = 5.32 \times 10^{-8} \times (24 \times 60 \times 60) \approx 4.6 \times 10^{-3} \text{ m}^3$$

Step 6: Determine the porewater pressure at A.

H_A = vertical distance from A to the upstream water level $= 10 + 5 = 15\,\mathrm{m}$

$(N_d)_A$ = number of equipotential drops to A $= 5$

$$(h_p)_A = H_A - (N_d)_A\,\Delta h = 15 - 5 \times 1.33 = 8.35\ \mathrm{m}$$

$$u_A = (h_p)_A\,\gamma_w = 8.35 \times 9.81 = 81.9\ \mathrm{kPa}$$

EXAMPLE 4.7 *Determining Flow into an Excavation*

A bridge pier is to be constructed in a riverbed by constructing a cofferdam, as shown in Figure E4.7a. A cofferdam is a temporary enclosure consisting of long, slender elements of steel, concrete, or timber members to support the sides of the enclosure. After construction of the cofferdam, the water within it will be pumped out.

(a) Draw the flownet.

(b) Calculate the minimum flow rate of a pump required to keep the water level at the base of the excavation.

(c) Calculate the porewater pressures at A and B.

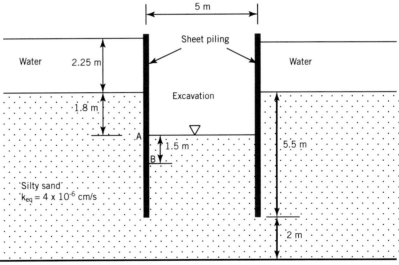

Figure E4.7a

Strategy Follow the procedures described for drawing flownets and calculate the required parameters. Only one-half the flownet is necessary as the flow is symmetrical.

Solution 4.7

Step 1: Draw the cofferdam to scale and sketch the flownet.

See Figure E4.7b. The base of the excavation is taken as datum.

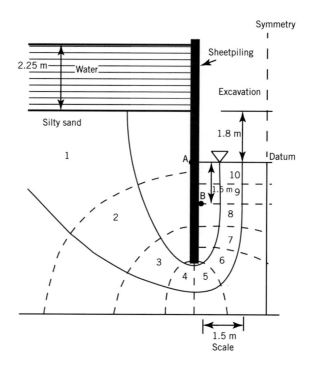

Figure E4.7b

Step 2: Calculate the flow.

$$\Delta H = 2.25 + 1.8 = 4.05 \text{ m}$$

$$N_d = 10, \quad N_f = 3, \quad \Delta h = \frac{\Delta H}{N_d} = \frac{4.05}{10} = 0.405 \text{ m}$$

$$q = 2(k\Delta h N_f) = 2(4 \times 10^{-8} \times 0.405 \times 3) = 9.7 \times 10^{-8} \text{ m}^3/\text{s}$$

Note: The flow rate is calculated for 1 m length of wall, and that is why the unit is m³/s. The unit for the hydraulic conductivity is converted from cm/s to m/s by dividing the ordinal value by 100 (1 m = 100 cm). The multiplier 2 is used because we have to consider both sides of the excavation.

The required minimum flow rate for the pump is $9.7 \times 10^{-8} \text{ cm}^3/\text{s}$

Step 3: Determine the porewater pressures at A and B

H_A = vertical distance from A to the upstream water level = 2.25 + 1.8 = 4.05 m

$(N_d)_A$ = number of equipotential drops to A ≈ 0.8

$$\left(h_p\right)_A = H_A - \left(N_d\right)_A \Delta h = 4.05 - 0.8 \times 0.405 = 3.73 \text{ m}$$

$$u_A = (h_p)_A \gamma_w = 3.73 \times 9.81 = 36.6 \text{ kPa}$$

H_B = vertical distance from B to the upstream water level = 1.5 + 1.8 + 2.25 = 5.55 m

$(N_d)_B$ = number of equipotential drops to B = 8

$$\left(h_p\right)_B = H_B - \left(N_d\right)_B \Delta h = 5.55 - 8 \times 0.405 = 2.31 \text{ m}$$

$$u_B = (h_p)_B \gamma_w = 2.31 \times 9.81 = 22.7 \text{ kPa}$$

EXERCISES

Concept understanding

4.1 What are the assumptions for Darcy's law to be valid? Discuss each of them.

4.2 It is customary to define a single void ratio for a soil. How does this definition affect the two-dimensional flow of water through the soil?

4.3 The dry unit weights of a sand and a clay are the same. Would you expect them to have the same hydraulic conductivity? Explain your answer.

4.4 Name two conditions that a flownet must satisfy (approximately) with respect to flow lines and equipotential lines.

4.5 What is quick sand? What causes it?

4.6 Does the critical hydraulic gradient in a soil depend on the hydraulic conductivity? Justify your answer.

4.7 What parameter is the gradient of the hydrostatic pressure? Justify your answer.

Problem solving

4.8 A porewater pressure transducer at a depth of 3 m registered a porewater pressure of 14.7 kPa. If a piezometer were to be installed at the same depth, how high would the water in the piezometer rise?

4.9 At an elevation of 2.5 m, the porewater pressure measured in a soil is 20 kPa. Determine the total head.

4.10 Two porewater pressure transducers, A and B, are located 3 m along a flow path. The pressures at A are B are 6 kPa and 5.5 kPa, respectively. (a) Determine the hydraulic gradient, and (b) calculate the velocity if the hydraulic conductivity is 15×10^{-5} cm/s.

4.11 The groundwater level in a soil layer 10 m thick is located at 3 m below the surface. (a) Plot the distribution of hydrostatic pressure with depth. (b) If the groundwater were to rise to the surface, plot on the same graph as (a), using a different line type, the distribution of hydrostatic pressure with depth. (c) Repeat (b), but the groundwater is now 2 m above the ground surface (flood condition). (d) Interpret and discuss these plots with respect to the effects of fluctuating groundwater levels. Neglect capillary action.

4.12 In a constant-head permeability test, a sample of soil 150 mm long and 50 mm diameter discharged 10 cm^3 of water in 10 minutes. The head difference in two piezometers, A and B, located at 5 mm and 125 mm, respectively, from the bottom of the sample is 20 mm. The average temperature was 20°C. (a) Determine the hydraulic conductivity of the soil. (b) What type of soil type was tested?

4.13 A constant-head test was conducted on a sample of soil 150 mm long and 1963 mm^2 in cross-sectional area. The quantity of water collected was 5 cm^3 in 20 seconds under a head difference of 220 mm. The average temperature was 24°C. (a) Calculate the hydraulic conductivity. (b) If the porosity of the sand is 55%, calculate the average velocity and the seepage velocity when the test was conducted.

4.14 A falling-head permeability test was carried out on a clay of diameter 100 mm and length 150 mm. In 1 hour the head in the standpipe of diameter 6 mm dropped from 700 mm to 540 mm. Calculate the hydraulic conductivity of this clay. Assume a temperature of 20°C.

4.15 A sieve analysis of a sand shows $D_{10} = 2$ mm. Estimate the hydraulic conductivity using Hazen's empirical equation with $C = 0.8$ and $C = 1.4$. Which value will you use and why? What are the limitations of your results?

4.16 A soil profile consists of three horizontal layers of sand, each 2 m thick. The hydraulic conductivities of the top, middle and bottom layers from constant head permeability tests are

8×10^{-1} cm/s, 4×10^{-3} cm/s and 5×10^{-5} cm/s. Assuming that the ratios of the lateral to the vertical hydraulic conductivities of the top, middle, and bottom layers are 2, 3, and 4 respectively, calculate the equivalent hydraulic conductivity of the sand.

Critical thinking and decision making

4.17 The flownet at a site of a reservoir is shown in Figure P4.17. (a) How many flow channels are present? (b) How many equipotential drops are present? (c) What is the total head loss? (d) Calculate the head loss between each equipotential line. (e) Calculate the average flow rate if $k_{eq} = 4 \times 10^{-6}$ cm/s (f) Calculate the porewater pressures at A and B located on opposite sides of the sheet pile retaining wall. (g) What would happen to the wall if the critical hydraulic gradient at the bottom of it (C) is exceeded?

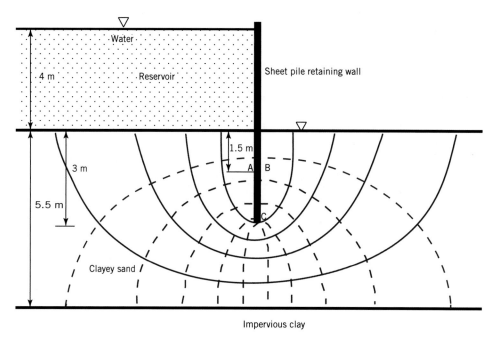

Figure P4.17

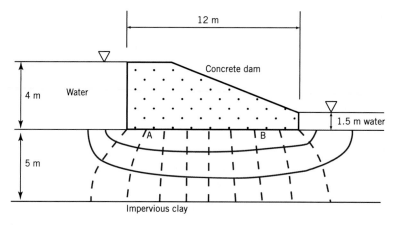

Figure P4.18

4.18 A sketch of a flownet under a dam is shown in Figure P4.18. (a) Determine the flow rate under the dam, if $k_{eq} = 2 \times 10^{-7}$ cm/s. (b) Determine the porewater pressures at A and B.

4.19 A trapezoidal excavation is required to construct a foundation, 3 m × 3 m, for an electric transformer station as shown in Figure P4.19. (a) Calculate the minimum flow rate of a pump to prevent a buildup of water in the excavation. (b) If the faces of the excavation are lined with an impermeable material, what must be the minimum weight of the transformer and its base to prevent uplift?

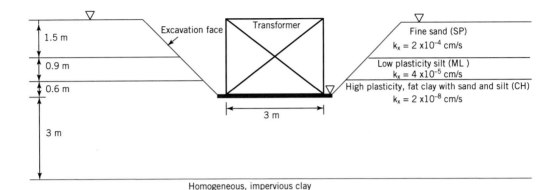

Figure P4.19

Chapter 5
Soil Compaction

5.1 INTRODUCTION

Soil compaction is the densification–reduction in void ratio of a soil through the expulsion of air. This is normally achieved by using mechanical compactors, rollers, and rammers with the addition of water. We will discuss the fundamentals of soil compaction, compaction tests, field compaction, and quality control in the field.

Learning outcomes

When you complete this chapter, you should be able to do the following:

- Understand the importance of soil compaction.
- Determine maximum dry unit weight and optimum water content.
- Specify soil compaction criteria for field applications.
- Identify suitable equipment for field compaction.
- Specify soil compaction quality control tests.

5.2 DEFINITION OF KEY TERMS

Compaction is the densification of soils by the expulsion of air.

Maximum dry unit weight ($\gamma_{d(max)}$) is the maximum unit weight that a soil can attain using a specified means of compaction.

Optimum water content (w_{opt}) is the water content required to allow a soil to attain its maximum dry unit weight following a specified means of compaction.

Soil Mechanics Fundamentals, First Edition. Muni Budhu.
© 2015 John Wiley & Sons, Ltd. Published 2015 by John Wiley & Sons, Ltd.
Companion website: www.wiley.com\go\budhu\soilmechanicsfundamentals

Degree of compaction (DC), also called relative compaction, is the ratio of the measured dry unit weight achieved to the desired dry unit weight.

5.3 BENEFITS OF SOIL COMPACTION

Compaction is an economical and popular technique for improving soils. The soil fabric is forced into a dense configuration by the expulsion of air using mechanical effort with or without the assistance of water. The benefits of compaction are:

1. Increased soil strength.
2. Increased load-bearing capacity.
3. Reduction in settlement (lower compressibility).
4. Reduction in the flow of water (water seepage).
5. Reduction in soil swelling (expansion) and collapse (soil contraction).
6. Increased soil stability.
7. Reduction in frost damage.

Improper compaction can lead to:

1. Structural distress from excessive total and differential settlements.
2. Cracking of pavements, floors, and basements.
3. Structural damage to buried structures, water and sewer pipes, and utility conduits.
4. Soil erosion.

5.4 THEORETICAL MAXIMUM DRY UNIT WEIGHT

From Equation (2.12), the dry unit weight of a soil is

$$\gamma_d = \left(\frac{G_s}{1+e}\right)\gamma_w = \frac{\gamma}{1+w} = \left(\frac{G_s}{1+wG_s/S}\right)\gamma_w \tag{5.1}$$

or

$$R_d = \frac{\gamma_d}{\gamma_w} = \frac{G_s}{1+wG_s/S} \tag{5.2}$$

Since G_s and γ_w are constants (the changes in the unit weight of water from changes in temperatures are small for geotechnical applications), the dry unit weight can increase only if the void ratio, e, is reduced. Also, since $e = wG_s/S$, the water content must be reduced.

A plot of Equation (5.1) showing the theoretical dry unit weight with $G_s = 2.7$ versus water content is shown in Figure 5.1. The theoretical dry unit weight decreases as the water content increases because the soil solids are heavier than water for the same volume occupied. The curve shown with $S = 100\%$ (Figure 5.1) is called the zero air voids curve.

5.5 PROCTOR COMPACTION TEST

A laboratory test, called the standard Proctor compaction test, was developed to deliver a standard amount of mechanical energy (compactive effort) to determine the maximum dry

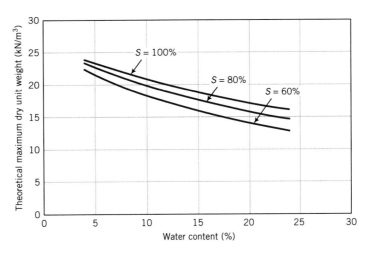

Figure 5.1 Theoretical maximum dry unit weight variation with water content.

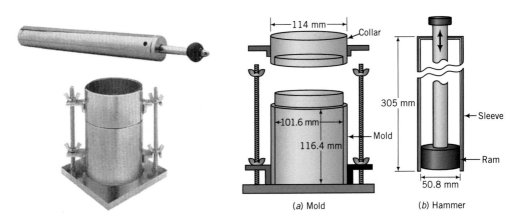

Figure 5.2 Compaction apparatus. (Photo courtesy of Geotest.)

unit weight of a soil. In the standard Proctor compaction test, a dry soil specimen is mixed with water and compacted in 3 layers in a cylindrical mold 101.6 mm internal diameter and 116.4 mm high (Figure 5.2). The volume of the standard Proctor mold is $9.433 \times 10^{-4} \, \mathrm{m}^3$. Each layer is subjected to 25 blows from a 2.5 kg hammer falling freely from a height of 305 mm. The energy imparted by the hammer is 594 kJ/m³.

For projects involving heavy loads, such as runways to support heavy aircraft loads, a modified Proctor compaction test was developed. In this test, the hammer has a 4.54 kg mass and falls freely from a height of 457 mm. The soil is compacted in 5 layers with 25 blows per layer in the standard Proctor mold. The compaction energy of the modified Proctor compaction test compaction test is 2695 kJ/m³, which is about 4.5 times the energy of the standard Proctor test.

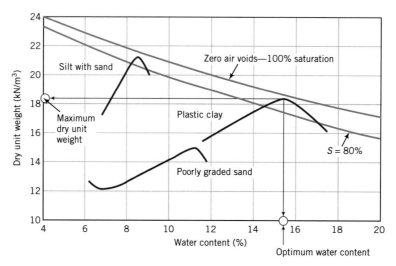

Figure 5.3 Dry unit weight–water content curves.

Four or more tests are conducted on the soil using different water contents. The last test is identified when additional water causes the bulk unit weight of the soil to decrease. The results are plotted as dry unit weight (ordinate) versus water content (abscissa). Typical dry unit weight–water content plots are shown in Figure 5.3.

Clays usually yield bell-shaped curves. Sands often show an initial decrease in dry unit weight, attributed to capillary tension that restrains the free movement of soil particles, followed by a hump. Some soils—those with liquid limit less than 30% and fine, poorly graded sands—may produce one or more humps before the maximum dry unit weight is achieved.

The water content at which the maximum dry unit weight, $\gamma_{d(max)}$, is achieved is called the optimum water content (w_{opt}). Typically the optimum water content for sand is less than 10%. The optimum water contents for clays are typically greater than 10%. At water contents below optimum (dry of optimum), air is expelled and water facilitates the rearrangement of soil grains into a denser configuration—the number of soil grains per unit volume of soil increases. At water contents just above optimum (wet of optimum), the compactive effort cannot expel more air and additional water displaces soil grains, thus decreasing the number of soil grains per unit volume of soil. Consequently, the dry unit weight decreases.

The modified Proctor test, using higher levels of compaction energy, achieves a higher maximum dry unit weight at a lower optimum water content than the standard test (Figure 5.4). The degree of saturation is also lower at higher levels of compaction than in the standard compaction test.

The dry unit weight is calculated from

$$\gamma_d = \frac{M/V}{1+w} = \frac{10.4M}{1+w} \quad \text{(unit: kN/m}^3\text{)} \tag{5.3}$$

where M is the mass of the wet soil (kg), V is the volume of the mold which is a constant $(9.433 \times 10^{-4}\,\text{m}^3)$, and w is the water content. The soil is invariably unsaturated at the maximum dry unit weight, that is, $S < 1$. We can determine the degree of saturation at any value of dry unit weight using

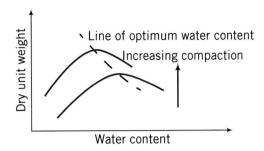

Figure 5.4   Effect of increasing compaction efforts on the dry unit weight–water content relationship.

$$S = \frac{wG_s}{(G_s\gamma_w/\gamma_d) - 1} \tag{5.4}$$

If G_s is unknown, you can substitute a value of 2.7 with little resulting error in most cases. The curve corresponding to $S = 1$ (100%) is the saturation line or the zero air voids line. For $G_s = 2.7$ and $\gamma_w = 9.81\,\text{kN/m}^3$, Equation (5.1) becomes, for the zero air voids line,

$$\gamma_d \approx \frac{26.5}{2.7w + 1}; \quad S = 1 \ (\text{unit: kN/m}^3) \tag{5.5}$$

Arbitrarily choose values of w and for each value find γ_d from Equation (5.5). Then plot the results of w versus γ_d.

What's next ... In the next section, you will learn how to interpret the Proctor test for practical applications.

5.6   INTERPRETATION OF PROCTOR TEST RESULTS

Knowledge of the optimum water content and the maximum dry unit weight of soils is very important for construction specifications of soil improvement by compaction. Specifications for earth structures (embankments, footings, etc.) usually call for a minimum of 95% of Proctor maximum dry unit weight. This level of compaction can be attained at two water contents: one before the attainment of the maximum dry unit weight, or dry of optimum, the other after attainment of the maximum dry unit weight, or wet of optimum (Figure 5.5). Normal practice is to compact the soil dry of optimum. Compact the soil wet of optimum for swelling (expansive) soils, soil liners for solid waste landfills, and projects where soil volume changes from changes in moisture conditions are intolerable.

Fine-grained soils compacted dry of optimum develop a flocculated structure independent of the method of compaction. The strength of soils compacted dry of optimum increases

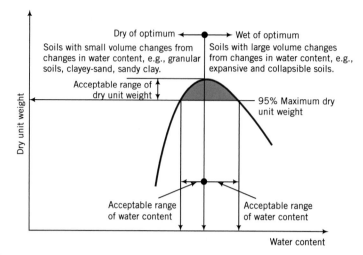

Figure 5.5 Illustration of compaction specification of soils in the field.

with increases in compactive energy. The maximum strength is obtained at the optimum water content. The structure of fine-grained soils compacted wet of optimum depends on the method of compaction. In general, a dispersed structure is created. There is very little change in the strength of soils from the optimum when soils are compacted wet of optimum.

When a heavily compacted soil mass (near to maximum dry unit weight) is sheared, it tends to expand (dilate) and gets looser. Usually this expansion is not uniform; some parts of the soil mass are looser than other parts. The flow rate of water in the soil will increase as water can easily (compared to the intact one) flow through the looser parts, possibly leading to catastrophic failure. Heavily compacted soils tend to show sudden decrease in strength when sheared. In engineering, if failure is to occur, we prefer that it occur gradually rather than suddenly so that mitigation measures can be implemented. In some earth structures (e.g., earth dams) you should try to achieve a level of compaction that would cause the soil to behave ductile (ability to deform without rupture). This may require compaction wet of optimum at levels less than 95% of the maximum dry unit weight (approximately 80% to 90% of maximum dry unit weight).

Some soil types such as poorly graded sand (SP) and poorly graded sand-silty sand (SP-SM) might not show any distinct maximum dry density and optimum water content from a standard Proctor compaction test. To compact these soils in the field, water contents within 50% to 75% saturation level is often used.

The water content of a soil to achieve say 95% of the standard Proctor compaction maximum dry unit weight compacted dry of optimum may result in compaction wet of optimum if the compaction energy is greater than the standard compaction test (see Figure 5.4). Because the hydraulic conductivity of a soil compacted dry of optimum is generally larger (after the soil is saturated) than the hydraulic conductivity of a soil compacted wet of optimum, higher compaction energy for a given compaction water content based on the standard Proctor compaction test would result in a reduced hydraulic conductivity. Therefore it is essential to control the amount of energy applied during field compaction to achieve the desired results.

Key points

1. Compaction is the densification of a soil by the expulsion of air and the rearrangement of soil particles.
2. The Proctor test is used to determine the maximum dry unit weight and the optimum water content, and it serves as the reference for field specifications of compaction.
3. Higher compactive effort than the standard Proctor test increases the maximum dry unit weight and reduces the optimum water content.
4. Compaction increases strength, lowers compressibility, and reduces the rate of flow water through soils.
5. Soils with low volume changes from changes in water content such as coarse-grained soils are usually compacted dry of optimum.
6. Soils with high volume changes from changes in water content such as expansive soils are usually compacted wet of optimum.

EXAMPLE 5.1   *Calculating Dry Unit Weight from Proctor Test Data*

In a standard Proctor test, the wet mass of the silty-clay in the mold was 1.82 kg. The corresponding water content was 8%. The volume of the standard Proctor test mold is $9.433 \times 10^{-4} \, \text{m}^3$. (a) Determine the bulk and dry unit weight. (b) Determine the degree of saturation, if $G_s = 2.7$.

Strategy From the wet weight and the volume of the Proctor mold (also the volume of the sample), you can calculate the bulk unit weight. Divide the bulk unit weight by 1 plus the water content to find the dry unit weight. Use Equation (5.4) to calculate the degree of saturation. The weight (Newton, N) is mass (kg) × acceleration due to gravity ($g = 9.81 \, \text{m/s}^2$).

Solution 5.1

(a)

Step 1: Find the bulk unit weight.

$$\gamma = \frac{M g}{V} = \frac{1.82 \times 9.81 \times 10^{-3}}{9.433 \times 10^{-4}} = 18.93 \, \text{kN/m}^3$$

Step 2: Find the dry unit weight.

$$\gamma_d = \frac{\gamma}{1+w} = \frac{18.93}{1+0.08} = 17.53 \, \text{kN/m}^3$$

Step 3: Check reasonableness of the answer.

By definition, dry unit weight is less than the bulk unit weight (see Equation 2.12)

The bulk and dry unit weights are within the ranges given for silts and clays in Table 2.1. The results are reasonable.

(b)

Step 3: Find the degree of saturation.

$$\text{Eq.(5.4): } S = \frac{w G_s}{(G_s \gamma_w / \gamma_d) - 1} = \frac{0.08 \times 2.7}{[(2.7 \times 9.81)/17.73] - 1} = 0.423 = 42.3\%$$

EXAMPLE 5.2   *Interpreting Compaction Data (1)*

The results of a standard Proctor test on a clay soil are shown in the table below.

Water content (%)	6.2	8.1	9.8	11.5	12.3	13.2
Bulk unit weight (kN/m³)	16.9	18.7	19.8	20.5	20.4	20.1

(a) Determine the maximum dry unit weight and optimum water content.

(b) Plot the zero air voids line.

(c) What is the dry unit weight and water content at 95% standard compaction, dry of optimum?

(d) Determine the degree of saturation at the maximum dry density, assuming that $G_s = 2.7$.

Strategy Compute γ_d and then plot the results of γ_d versus w (%). Extract the required information.

Solution 5.2

Step 1: Use a table or a spreadsheet program to tabulate γ_d.

Water content, w(%)	Bulk unit weight, γ (kN/m³)	Dry unit weight, γ_d $\gamma_d = \gamma/(1 + w)$ (kN/m³)	Zero air voids	
			Water content, w (%)	Dry unit weight $\gamma_d = G_s\gamma_w/(wG_s + 1)$ (kN/m³)
6.2	16.9	15.9	6	22.8
8.1	18.7	17.3	8	21.8
9.8	19.8	18.0	10	20.9
11.5	20.5	18.4	12	20
12.3	20.4	18.2	14	19.2
13.2	20.1	17.8		

Step 2: Plot water content versus dry unit weight graph and zero air voids line.

A plot of water content versus dry unit weight is shown in Figure E5.2a. You can also plot w versus R_d as shown in Figure E5.2b.

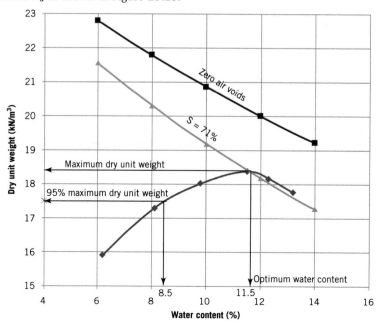

Figure E5.2a

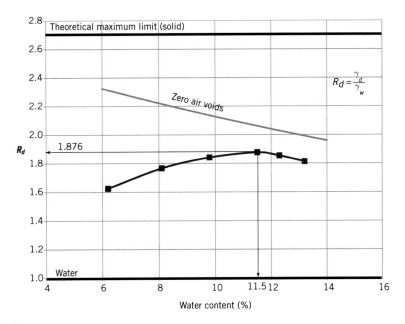

Figure E5.2b

Step 3: Extract the maximum dry unit weight and optimum water content.

$$\gamma_{d(max)} = 18.4 \text{ kN/m}^3, \quad w_{opt} = 11.5\%$$

Step 4: Check reasonableness of the maximum dry unit weight.

From Table 2.1 dry unit weights of clays range between 14 kN/m^3 and 21 kN/m^3. Also, all values of R_d lie within its limiting range $2.7 \geq R_d > 1$

$$\gamma_{d(max)} = 18.4 \text{ kN/m}^3 \text{ is reasonable.}$$

Step 5: Calculate and plot the 95% maximum dry unit weight.

At 95% compaction, $\gamma_d = 18.4 \times 0.95 = 17.5 \text{ kN/m}^3$ and $w = 8.5\%$ (from the graph, Figure E5.2a).

Step 6: Calculate the degree of saturation at maximum dry unit weight.

$$S = \frac{wG_s}{(G_s \gamma_w / \gamma_d) - 1} = \frac{0.115 \times 2.7}{(2.7 \times 9.81/18.4) - 1} = 0.706 \approx 71\%$$

EXAMPLE 5.3 *Interpreting Compaction Data (2)*

The detailed results of a standard Proctor test on a soil classified as CL-ML and group name silty-clay with sand are shown in the table below. Determine the maximum dry unit weight and optimum water content.

Diameter of mold = 101.6 mm

Height of mold = 116.4 mm

Volume of mold = $9.433 \times 10^{-4} \text{ m}^3$

Mass of mold, $M = 2.02$ kg

Unit weight data		Water content data	
Mass of wet soil and mold (kg)	Mass of can and wet soil (grams)	Mass of can and dry soil (grams)	Mass of can (grams)
M_{wm}	M_w	M_d	M_c
3.50	114.92	111.48	46.50
3.65	163.12	155.08	46.43
3.72	190.43	178.64	46.20
3.74	193.13	178.24	46.50
3.58	188.77	171.58	46.10

Strategy This example is similar to Example 5.2 except that you have to calculate the water content as you would do in an actual test.

Solution 5.3

Step 1: Set up a spreadsheet or a table to do the calculations.

Water content calculations				Dry unit weight calculations		
Mass of can and wet soil (grams)	Mass of can and dry soil (grams)	Mass of can (grams)	Water Content	Mass of wet soil and mold (kg)	Mass of wet soil (kg)	Dry unit weight
M_w	M_d	M_c	w (%)	M_{wm}	$M_w = M_{wm} - M$	γ_d (kN/m³)
114.92	111.48	46.50	5.3	3.50	1.48	14.6
163.12	155.08	46.43	7.4	3.65	1.63	15.8
190.43	178.64	46.20	8.9	3.72	1.70	16.2
193.13	178.24	46.50	11.3	3.74	1.72	16.1
188.77	171.58	46.10	13.7	3.58	1.56	14.3

$$\text{Dry unit weight} = \frac{\text{Weight of wet soil}}{\text{Volume of mold} \times (1 + \text{water content})} = 10.4 \times \frac{\text{Mass (kg) of wet soil}}{(1 + \text{water content})}$$

Step 2: Plot the dry unit weight versus water content curve.
See Figure E5.3.

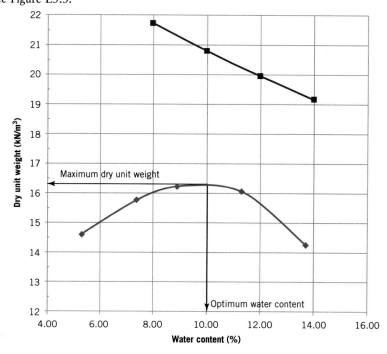

Figure E5.3

Step 3: Extract the results.

Maximum dry unit weight $= 16.4$ kN/m³; optimum water content $= 10\%$

Step 4: Check reasonableness of results.

The soil is likely a clay because clays normally show a bell-shaped Proctor curve. Because of the presence of silt and sand, the dry unit weight will tend to be on the low side of the range of values for clay (Table 2.1). The dry unit weight in step 3 is reasonable.

What's next ... In the next section, general guidelines to help you specify field compaction equipment are presented.

5.7 FIELD COMPACTION

A variety of mechanical equipment is used to compact soils in the field. Compaction is accomplished by static and vibratory vertical forces. Static vertical forces are applied by deadweights that impart pressure and/or kneading action to the soil mass. Sheepsfoot rollers (Figure 5.6a), grid rollers, rubber-tired rollers, drum rollers (Figure 5.6b), loaders, and scrapers are examples of equipment that apply static vertical forces. Vibratory vertical forces are applied by engine-driven systems with rotating eccentric weights or spring/piston mechanisms that impart a rapid sequence of blows to the soil surface. The soil is compacted by pressure and rearranging of the soil structure by either impact or vibration. Common types of vibrating equipment are vibrating plate compactors, vibrating rollers, and vibrating sheepsfoot rollers. Vibrating sheepsfoot and impact rammers are impact compactors.

The soil mass is compacted in layers called lifts. The lift thickness rarely exceeds 300 mm. Coarse-grained soils are compacted in lifts between 250 mm and 300 mm, while fine-grained soils are compacted in lifts about 150 mm. The stresses imparted by compactors, especially static compactors, decrease with lift depth. Consequently, the top part of the lift is subjected to greater stresses than the bottom and attains a higher degree of compaction. Lower lift thickness is then preferable for uniform compaction. A comparison of various types of (heavy) field compactors and the type of soils they are suitable for is shown in Table 5.1.

Table 5.1 Comparison of field compactors for various soil types.

		Compaction type				
		Static		Dynamic		
		Pressure with kneading	Kneading with pressure	Vibration	Impact	
Material	Lift thickness (mm)	Static sheepsfoot grid roller; scraper	Scraper; rubber-tired roller; loader; grid roller	Vibrating plate compactor; vibrating roller; vibrating sheepsfoot roller	Vibrating sheepsfoot rammer	Compactability
Gravel	300±	Not applicable	Very good	Good	Poor	Very easy
Sand	250±	Not applicable	Good	Excellent	Poor	Easy
Silt	150±	Good	Excellent	Poor	Good	Difficult
Clay	150±	Very good	Good	No	Excellent	Very difficult

(a)

(b)

Figure 5.6 Two types of machinery for field compaction. (a) Sheepsfoot roller and (b) drum type roller. (Photos courtesy of Volvo.)

For smaller, lighter equipment such as small vibratory plates and wacker hammers, smaller lift thickness less than 150 mm should be used. Generally, it is preferable to specify the amount of compaction desired based on the relevant Proctor test and let the contractor select the appropriate equipment. You will have to ensure that the contractor has the necessary experience.

What's next ... When you specify the amount of compaction desired for a project, you need to ensure that the specifications are met. In next section, three popular apparatuses for compaction quality control tests are discussed.

5.8 COMPACTION QUALITY CONTROL

A geotechnical engineer needs to check that field compaction meets specifications. A measure of the degree of compaction (DC), also called relative compaction, is the ratio of the measured dry unit weight achieved to the desired dry unit weight.

$$DC = \frac{\text{Measured dry unit weight}}{\text{Desired dry unit weight}} \tag{5.6}$$

The degree of compaction is not related to relative density. Various types of equipment are available to check the amount of compaction achieved in the field. Three popular pieces of equipment are (1) the sand cone, (2) the balloon, and (3) nuclear density meters.

5.8.1 Sand cone

A sand cone apparatus is shown in Figure 5.7. It consists of a glass or plastic jar with a funnel attached to the neck of the jar.

The procedure for a sand cone test is as follows:

1. Fill the jar with a standard sand—a sand with known density—and determine the mass of the sand cone apparatus with the jar filled with sand (M_1).
2. Determine the mass of sand to fill the cone (M_2).
3. Excavate a small hole in the soil and determine the mass of the excavated soil (M_3).
4. Determine the water content of the excavated soil (w).
5. Fill the hole with the standard sand by inverting the sand cone apparatus over the hole and opening the valve.
6. Determine the mass of the sand cone apparatus with the remaining sand in the jar (M_4).

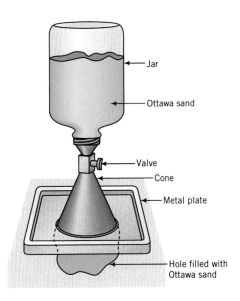

Figure 5.7 A sand cone apparatus.

7. Calculate the unit weight of the soil as follows:

$$\text{Mass of sand to fill hole} = M_s = M_1 - (M_2 + M_4)$$

$$\text{Volume of hole} = V = \frac{M_s g}{(\gamma_d)_{\text{Ottawa sand}}}$$

$$\text{Mass of dry soil} = M_d = \frac{M_3}{1 + w}$$

$$\text{Dry unit weight} = \gamma_d = \frac{M_d g}{V}$$

EXAMPLE 5.4 *Interpreting Sand Cone Test Results*

A sand cone test was conducted during the compaction of a roadway embankment. The data are as follows:

Calibration to find dry unit weight of the standard sand	
Mass of Proctor mold	4178 grams
Mass of Proctor mold and sand	5609 grams
Volume of mold	0.00095 m³
Calibration of sand cone	
Mass of sand cone apparatus and jar filled with sand	5466 grams
Mass of sand cone apparatus with remaining sand in jar	3755 grams
Sand cone test results	
Mass of sand cone apparatus and jar filled with sand	7387 grams
Mass of excavated soil	2206 grams
Mass of sand cone apparatus with remaining sand in jar	3919 grams
Mass content of excavated soil	9.2%

(a) Determine the dry unit weight.

(b) The standard Proctor maximum dry unit weight of the roadway embankment soil is 20 kN/m^3 at an optimum water content of 10.8%, dry of optimum. The specification requires a minimum dry unit weight of 95% of Proctor maximum dry unit weight. Is the specification met? If not, how can it be achieved?

Strategy Set up a spreadsheet to carry out the calculations following the method described for the sand cone test. Compare the measured field dry unit weight with the specification requirement to check satisfaction.

Solution 5.4

Step 1: Set up a spreadsheet or a table and carry out calculations following the method described for the sand cone test.

Calibration to find dry unit weight of standard sand	
Mass of Proctor mold, M_1	4178 grams
Mass of Proctor mold + sand, M_2	5609 grams
Volume of mold, V_1	0.00095 m³
Dry unit weight of sand in cone, $\gamma_{dc} = (M_2 - M_1)/V_1$	14.8 kN/m³
Calibration of sand cone	
Mass of sand cone apparatus + jar filled with sand, M_a	5466 grams

Mass of sand cone apparatus with remaining sand in jar, M_b	3755 grams
Mass of sand to fill cone, M_2	1711 grams
Sand cone test results	
Mass of sand cone apparatus + jar filled with sand, M_1	7387 grams
Mass of excavated soil, M_3	2206 grams
Mass of sand cone apparatus with remaining sand in jar, M_4	3919 grams
Mass of sand to fill hole, $M_s = M_1 - (M_2 + M_4)$	1757 grams
Volume of hole, $V = M_s g / \gamma_{dc}$	0.001166 m³
Water content of excavated soil, w	9.2%
Mass of dry soil, $M_d = M_3/(1 + w)$	2020 grams
Dry unit weight $= M_d g / V$	17.0 kN/m³

Step 2: Compare specification with sand cone results.

Minimum dry unit weight required $= 0.95 \times 20 = 19\,\text{kN/m}^3$.

The sand cone test result gives a dry unit weight of $17\,\text{kN/m}^3$. The degree of compaction is $DC = 17/19 \approx 89\%$ and the water content is near the optimum water content. Therefore, the specification is not met.

Step 3: Decide on how best to meet the specification.

The water content in the field is 9.2%, while the Proctor test gave an optimum water content of 10.8%. The compacted state of the soil is dry of optimum. Thus, water should be added to the embankment soil and re-compacted. Care should be taken to ensure that the water content does not exceed 10.8%. The dry unit weight and water content should be rechecked and the embankment re-compacted as needed.

5.8.2 Balloon test

The balloon test apparatus (Figure 5.8) consists of a graduated cylinder with a centrally placed balloon. The cylinder is filled with water. The procedure for the balloon test is as follows:

1. Fill the cylinder with water and record its volume, V_1.
2. Excavate a small hole in the soil and determine the mass of the excavated soil (M).
3. Determine the water content of the excavated soil (w).
4. Use the pump to invert the balloon to fill the hole.
5. Record the volume of water remaining in the cylinder, V_2.

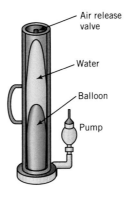

Figure 5.8   Balloon test device.

6. Calculate the unit weight of the soil as follows:

$$\gamma = \frac{Mg}{V_1 - V_2}, \quad \gamma_d = \frac{\gamma}{1+w}$$

The balloon test is not often used.

5.8.3 Nuclear density meter

The nuclear density apparatus (Figure 5.9) is a versatile device to rapidly obtain the unit weight and water content of the soil nondestructively. Soil particles cause radiation to scatter to a detector tube, and the amount of scatter is counted. The scatter count rate is inversely proportional to the unit weight of the soil. If water is present in the soil, the hydrogen in water scatters the neutrons, and the amount of scatter is proportional to the water content. The radiation source is either radium or radioactive isotopes of cesium and americium. The nuclear density apparatus is first calibrated using the manufacturer's reference blocks. This calibration serves as a reference to determine the unit weight and water content of a soil at a particular site.

There are two types of measurements:

1. Backscatter, in which the number of backscattered gamma rays detected by the counter is related to the soil's unit weight. The depth of measurement is 50 mm to 75 mm.
2. Direct transmission, in which the number of rays detected by the counter is related to the soil's unit weight. The depth of measurement is 50 mm to 200 mm.

Figure 5.9 Nuclear density meter. (Photo courtesy of Seaman Nuclear Corp.)

5.8.4 Comparisons among the three popular compaction quality control tests

A comparison among the three compaction quality control tests is shown in Table 5.2.

Table 5.2 Comparisons among the three popular compaction quality control tests.

Material	Sand cone	Balloon	Nuclear density meter
Advantages	▪ Low cost ▪ Accurate ▪ Large sample	▪ Low to moderate cost ▪ Fewer computational steps compared to sand cone ▪ Large sample	▪ Quick ▪ Direct measurement of unit weight and water content
Disadvantages	▪ Slow; many steps required ▪ Standard sand in hole has to be retrieved ▪ Unit weight has to be Computed ▪ Difficult to control density of sand in hole ▪ Possible void space under Plate ▪ Hole can reduce in size through soil movement ▪ Hole can cave in (granular materials)	▪ Slow ▪ Extra care needed to prevent damage to balloon, especially in gravelly materials ▪ Unit weight has to be computed ▪ Difficult to obtain accurate hole size ▪ Possible void space under plate ▪ Hole can reduce in size through soil movement ▪ Hole can cave in (granular materials)	▪ High cost ▪ Radiation certification required for operation ▪ Water content error can be significant ▪ Surface preparation needed ▪ Radiation backscatter can be hazardous

Key points

1. A variety of field equipment is used to obtain the desired compaction.
2. The sand cone apparatus, the balloon apparatus, and the nuclear density meter are three types of equipment used for compaction quality control in the field.
3. It is generally best to allow the contractor to select and use the appropriate equipment to achieve the desired compaction.

5.9 SUMMARY

Compaction—the densification of a soil by expulsion of air and forcing the soil particles closer together—is a popular method for improving soils. The laboratory test to investigate the maximum dry unit weight and the optimum water content is the Proctor test. This standard test is used in most applications. For heavy loads, the modified Proctor test is used. Various types of equipment are available to achieve specified compaction. You need to select the appropriate equipment based on the soil type and the availability of the desired equipment.

5.9.1 Practical example

EXAMPLE 5.5 *Interpreting Standard Proctor Test Results and Specifying Field Compaction Equipment*

The results of a standard Proctor test for a clay, classified as CL (lean clay with sand and traces of gravel) is to be used as a core for an earth dam is shown in Figure E5.5a.

(a) Specify the compaction criteria for the field. It is desired that shrinkage cracks be kept to the minimum.

(b) Recommend field compaction equipment that would achieve the desired compaction.

(c) Specify an appropriate quality control test.

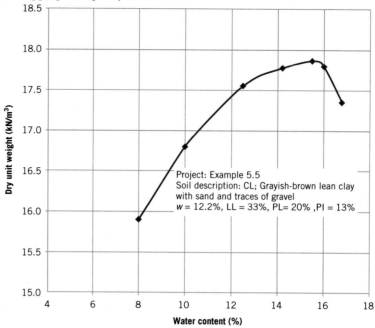

Figure E5.5a

Strategy Because the soil is low-plasticity clay with sand and with traces of gravel, a reasonable assumption is that it would not change volume significantly from changes with water content compared with a clay with group symbol, CH.

Solution 5.5

Step 1: Determine the maximum dry unit weight and optimum water content.

The maximum dry unit weight and optimum water content are $17.9 \, \text{kN/m}^3$ and 15.5%, respectively.

Step 2: Specify the dry unit weight and water content.

Specify the 90% maximum dry unit weight to be compacted dry of optimum (Figure E5.5b). The industry standard is 95% maximum dry unit weight. However, to reduce the potential of cracking, a lower degree of compaction can be used.

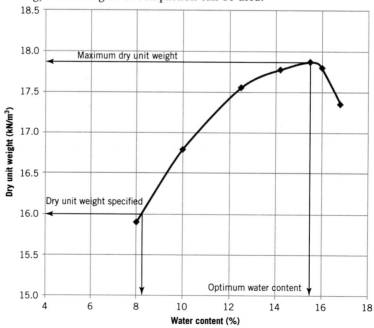

Figure E5.5b

$\gamma_d = 0.9 \times 17.9 \, \text{kN/m}^3 \approx 16 \, \text{kN/m}^3$ (minimum); $w = 8.1\%$ (tolerable limits 8.1% to 8.5% to keep the compaction close to 90%)

Step 3: Determine the field method of compaction.

The soil is a clay. From Table 5.1, either a sheepsfoot grid roller, rated as very good, or a vibrating sheepsfoot rammer, rated as excellent, can be used.

Step 4: Specify the quality control equipment.

Either the balloon test or the nuclear density meter is suitable.

EXERCISES

Assume $G_s = 2.7$ and the volume of the standard Proctor mold as $9.43 \times 10^{-4} \, \text{m}^3$, where necessary, for solving the following problems.

Concept understanding

5.1 Can you obtain the highest possible dry unit weight for a soil using either the standard or modified Proctor test? Justify your answer.

5.2 Would the addition of water to an unsaturated soil change its dry unit weight if the void ratio does not change? Justify your answer. Assume that the soil is nonexpansive or noncollapsible and that no external load is applied.

5.3 Is the soil usually saturated at maximum dry unit weight and optimum water content in either a standard or modified Proctor test? Justify your answer.

5.4 What would happen to the maximum dry unit weight and optimum water content determined from a standard Proctor test if a higher level of compactive energy is used to compact the soil? Justify your answer.

Problem solving

5.5 In a standard Proctor test, the wet mass of the clay in the mold was 1.75 kg. The corresponding water content was 8%. The volume of the standard Proctor test mold is $9.43 \times 10^{-4} \, \text{m}^3$. Determine (a) the bulk unit weight, (b) the dry unit weight, and (c) the degree of saturation if $G_s = 2.7$.

5.6 The maximum dry unit weight and optimum water content of a clay rich soil are 17.3 kN/m³ and 12%, respectively. Determine (a) the degree of saturation of the soil and (b) the dry unit weight for zero air voids at the optimum water content.

Critical thinking and decision making

5.7 The data from a standard Proctor test on a fat clay with traces of sand and silt with low expansion potential are as follows:

Diameter of mold = 101.6 mm

Height of mold = 116.4 mm

Mass of mold = 1.92 kg

Specific gravity, $G_s = 2.69$

Unit weight determination	Water content determination		
Mass of wet soil and mold (kg)	Mass of can and wet soil (grams)	Mass of can and dry soil (grams)	Mass of can (grams)
3.52	108.12	105.1	42.1
3.65	98.57	94.9	40.9
3.84	121.90	114.7	42.7
3.82	118.39	110.5	42.5
3.72	138.02	126.8	41.8

(a) Plot the dry unit weight–water content curve.

(b) Determine the maximum dry unit weight and optimum water content.

(c) If the desired compaction for a roadbed (subgrade) in the field is 95% of the standard Proctor maximum dry unit weight, what values of dry unit weight and water content would you specify? Explain why you select these values.

(d) What field equipment would you specify to compact the soil and why?

(e) How would you check that the specified dry unit weight and water content are achieved in the field?

5.8 A fine-grained soil has 60% clay with $LL = 220\%$, $PL = 45\%$, and a natural water content of 6%. A standard Proctor compaction test was carried out in the laboratory and the following data were recorded:

Diameter of mold = 101.6 mm

Height of mold = 116.4 mm

Mass of mold = 1.94 kg

Specific gravity, $G_s = 2.69$

Unit weight determination	Water content determination		
Mass of wet soil and mold (kg)	Mass of can and wet soil (grams)	Mass of can and dry soil (grams)	Mass of can (grams)
3.19	105.05	103.1	42.1
3.54	100.69	97.9	40.9
3.67	114.71	110.7	42.7
3.64	134.26	128.5	42.5
3.26	109.34	104.8	41.8

(a) Plot the dry unit weight–water content curve.

(b) Determine the maximum dry unit weight and optimum water content.

(c) If the desired compaction for a dam core in the field is a minimum of 90% of the standard Proctor maximum dry unit weight, what values of dry unit weight and range of water content would you specify? The expansion potential of the soil is high. Explain the values you select.

(d) What field equipment would you specify to compact the soil and why?

(e) How would you check that the specified dry unit weight and water content are achieved in the field?

5.9 Standard Proctor test results on a sandy clay (35% sand, 55% clay, and 10% silt), taken from a borrow pit, are given in the following table:

Water content (%)	3.8	5.1	7.8	9.2	12
Dry unit weight (kN/m³)	16.7	17.7	19.0	19.1	18.1

The sandy clay in the borrow pit has a porosity of 65% and a water content of 5.2%. A highway embankment is to be constructed using this soil.

(a) Specify the compaction (dry unit weight and water content) to be achieved in the field. Justify your specification.

(b) How many cubic meters (m³) of borrow pit soil are needed for 1 m³ of compacted highway fill?

(c) How much water per cubic meter is required to meet the specification?

(d) How many truckloads of soil will be required for a 100,000 m³ highway embankment? Each truck has a load capacity of 30 m³ and regulations require a maximum load capacity of 90%.

(e) Determine the cost for 100,000 m³ of compacted soil based on the following:

Purchase and load borrow pit material at site, haul 4 km round-trip, and spread with 200 HP dozer = $15/m³; extra mileage charge for each mile = $0.05/m³; round-trip distance = 20 km; compaction = $1.02/m³.

5.10 A sand cone test was conducted for quality control during the compaction of a sandy clay. The data are as follows:

Calibration to find dry unit weight of the standard sand	
Mass of Proctor mold	4178 grams
Mass of Proctor mold and sand	5609 grams
Volume of mold	0.00095 m³
Calibration of sand cone	
Mass of sand cone apparatus and jar filled with sand	5466 grams
Mass of sand cone apparatus with remaining sand in jar	3755 grams
Sand cone test results	
Mass of sand cone apparatus and jar filled with sand	7387 grams
Mass of excavated soil	2206 grams
Mass of sand cone apparatus with remaining sand in jar	3919 grams
Water content of excavated soil	5.9%

(a) Determine the dry unit weight.

(b) The standard Proctor maximum dry unit weight of the sandy clay is 18.5 kN/m³ at an optimum water content of 7.2%. The specification requires 95% Proctor dry unit weight at acceptable water contents ranging from 5% to 8.5%. Is the specification met? Justify your answer.

5.11 A soil at a mining site is classified as GW-GM.

(a) Would this soil be suitable for the base course of a road?

(b) What type of field compaction equipment would you recommend?

(c) How would you check that the desired compaction is achieved in the field?

(d) Would you specify compaction dry or wet of optimum? Why?

Chapter 6
Stresses from Surface Loads and the Principle of Effective Stress

6.1 INTRODUCTION

In this chapter, we consider the vertical stresses induced on soils from some common types of surface loads, and the principle of effective stresses, which is the most important principle in soil mechanics.

Learning outcomes

When you complete this chapter, you should be able to do the following:

- Understand how surface loads are distributed within soil as an elastic material.
- Understand the concept of effective stress.
- Be able to calculate total stress increase from surface loads and the effective stresses within soils.
- Understand and be able to determine the effects of seepage stresses on effective stresses within soils.

6.2 DEFINITION OF KEY TERMS

Stress, or intensity of loading, is the load per unit area. The fundamental definition of stress is the ratio of the force ΔP acting on a plane ΔS to the area of the plane ΔS when ΔS tends to zero; Δ denotes a small quantity.

Effective stress (σ') is the stress carried by the soil particles.

Total stress (σ) is the stress carried by the soil particles and the liquids and gases in the voids.

Soil Mechanics Fundamentals, First Edition. Muni Budhu.
© 2015 John Wiley & Sons, Ltd. Published 2015 by John Wiley & Sons, Ltd.
Companion website: www.wiley.com\go\budhu\soilmechanicsfundamentals

Stress (strain) state at a point is a set of stress (strain) vectors corresponding to all planes passing through that point. Mohr's circle is used to graphically represent stress (strain) state for two-dimensional bodies.

Porewater pressure (*u*) is the pressure of the water held in the soil pores.

Isotropic means the material properties are the same in all directions, and also the loadings are the same in all directions.

Elastic materials are ideal materials that return to their original configuration on unloading and obey Hooke's law.

6.3 VERTICAL STRESS INCREASE IN SOILS FROM SURFACE LOADS

Computer program utility

Access www.wiley.com\go\budhu\soilmechanicsfundamentals, and click on Chapter 6 and then STRESS.zip to download and run a computer application to obtain the stress increases and displacements due to surface loads. You can use this program to explore stress changes due to different types of loads, and prepare and print Newmark charts for vertical stresses beneath arbitrarily shaped loads. This computer program will also be helpful in solving problems in later chapters.

The distribution of stresses within a soil from applied surface loads or stresses is determined by assuming that the soil is a semi-infinite, homogeneous, linear, isotropic, elastic material. A semi-infinite mass is bounded on one side and extends infinitely in all other directions; this is also called an "elastic half-space." For soils, the horizontal surface is the bounding side. Because of the assumption of a linear elastic soil mass, we can use the principle of superposition. That is, the stress increases at a given point in a soil mass from different surface loads can be added together.

Surface loads are divided into two general classes, finite and infinite. However, these are qualitative classes, and they are subject to interpretation. Examples of finite loads are point loads, circular loads, and rectangular loads. Examples of infinite loads are fills and surcharges. The relative rigidity of the foundation (a system that transfers the load to the soil) to the soil mass influences the stress distribution within the soil. The elastic solutions presented are for flexible loads and do not account for the relative rigidity of the soil foundation system. If the foundation is rigid, the stress increases are generally lower (15% to 30% less for clays and 20% to 30% less for sands) than those calculated from the elastic solutions presented in this section. Traditionally, the stress increases from the elastic solutions are not adjusted because soil behavior is nonlinear and it is better to err on the conservative side. The increases in soil stresses from surface loads are total stresses. These increases in stresses are resisted initially by both the porewater and the soil particles.

6.3.1 Regular shaped surface loads on a semi-infinite half-space

Boussinesq (1885) presented a solution for the distribution of stresses for a point load applied on the soil surface. An example of a point load is the vertical load transferred to the soil from an electric power line pole. Boussinesq's solution was subsequently integrated to give solutions for distributed loads. A summary of the increase in vertical stresses on a homogeneous linearly elastic soil from some common types of surface loads is presented in Table 6.1. Charts based on these equations are shown in Figure 6.1, Figure 6.2, Figure 6.3, Figure 6.4, Figure 6.5, Figure 6.6, and Figure 6.7. The equations are only valid for a single semi-infinite soil layer. Many soil profiles consist of layered soils of finite thickness for which Boussinesq's solution can result in underestimation of the stress increase. A comprehensive set of equations for a variety of loading situations and finite soil thickness is available in Poulos and Davis (1974) and, to a limited extent, in the author's *Soil Mechanics and Foundations* (3rd ed., Wiley, 2011).

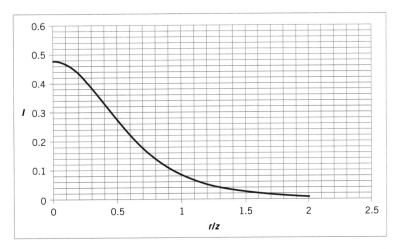

Figure 6.1 Stress influence chart for the increase in vertical total stress from a point load.

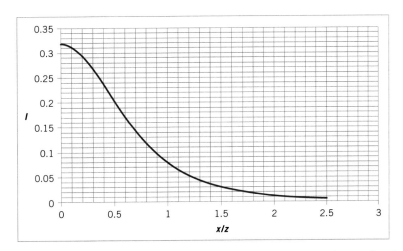

Figure 6.2 Stress influence chart for the increase in vertical total stress from a line load.

Table 6.1 Summary of increase in vertical stress from some common surface loads.

Surface load	Illustration	Example	Equation for vertical stress increase	Chart
Point load		Transmission pole	$\Delta\sigma_z = \dfrac{Q}{z^2} I$ $I = \dfrac{3}{2\pi}\dfrac{1}{\left[1+(r/z)^2\right]^{5/2}}$	Figure 6.1
Line load		Foundation of a fence	$\Delta\sigma_z = \dfrac{2Qz^3}{\pi(x^2+z^2)^2} = \dfrac{2Q}{z} I$ $I = \dfrac{1}{\pi\left[(x/z)^2+1\right]^2}$	Figure 6.2
Strip load	 L >> B	Base of a retaining wall	$\Delta\sigma_z = \dfrac{q_s}{\pi}[\alpha + \sin\alpha\cos(\alpha+2\beta)] = q_s I_s$ $I_s = \dfrac{1}{\pi}[\alpha + \sin\alpha\cos(\alpha+2\beta)]$	Figure 6.3

Circular load

Tank foundation

Under center: $\Delta\sigma_z = q_s I_c$

$$I_c = \left[1 - \left(\frac{1}{1 + (r_o/z)^2}\right)^{3/2}\right]$$

Figure 6.4

Rectangular load

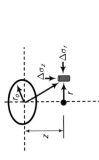

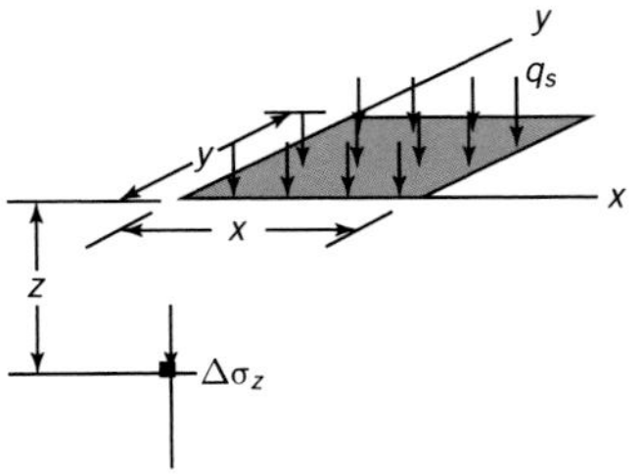

Column foundation

Under corner: $\Delta\sigma_z = q_s I_z$

$$I_z = \frac{1}{4\pi}\left[\frac{2mn\sqrt{m^2+n^2+1}}{m^2+n^2+m^2n^2+1}\left(\frac{m^2+n^2+2}{m^2+n^2+1}\right)\right.$$
$$\left. + \tan^{-1}\left(\frac{2mn\sqrt{m^2+n^2+1}}{m^2+n^2-m^2n^2+1}\right)\right];$$

$$(mn)^2 \geq (m^2+n^2+1)$$

$m = B/z$ and $n = L/z$

Approximate method: under center of loaded area

$$\Delta\sigma_z = \frac{q_s B L}{(B+z)(L+z)}$$

The approximate method is reasonably accurate (compared with Boussinesq's elastic solution) when $z > B$.

Figure 6.5 (square)
Figure 6.6 (rectangle)

Approximate method

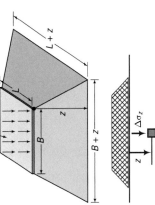

$q_s = \dfrac{Q}{B \times L}$; Q is total load

Embankment

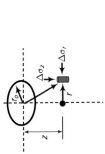

Embankment (infinite load)

Combination of a rectangular and two triangular strip loads

Figure 6.7

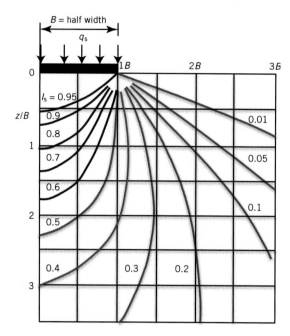

Figure 6.3 Stress influence chart for the increase in vertical total stress from a strip load imposing a uniform surface stress. The isobars (equal pressures) are called pressure bulbs. (Source: Jurgenson, 1934. Reprinted by permission of Boston Society of Civil Engineers/ASCE.)

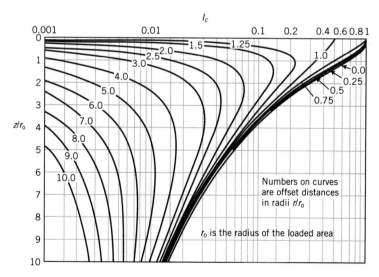

Figure 6.4 Stress influence chart for the increase in vertical total stress from a uniformly loaded circular area. Note that the abscissa is logarithmic scale. (Source: Foster and Ahlvin, 1954. Reprinted by permission of the American Association of State Highway and Transportation Officials.)

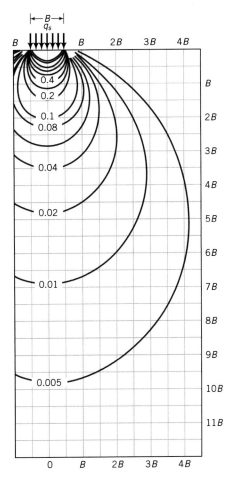

Figure 6.5 Stress influence chart for the increase in vertical total stress from a uniformly stressed square loaded area. The isobars (equal pressures) are called pressure bulbs. (Source: NAV-FAC-DM 7.1.)

6.3.2 *How to use the charts*

The charts used two normalized parameters. One is the normalized depth factor, which is the depth of interest divided by the width B, or length L, or radius r, of the surface load. This factor is usually the ordinate. The other is the normalized lateral distance factor, which is the lateral distance x or r from either a corner (rectangular load) or center (circular load) divided by the width or length or radius of the surface load. This is plotted as the abscissa. Because for each depth factor there are many lateral distance factors, the charts have a set of curves (contours), one for each lateral distance factor. For example, if you wish to find the vertical stress increase at a depth z, say, 5 m under the center of a circular uniformly distributed surface stress of intensity 100 kPa and radius r_o, 10 m along the center (r or $x = 0$), then the depth factor is $z/r_o = 5/10 = 0.5$ and the lateral distance factor is $r/r_o = 0/10 = 0$. You now

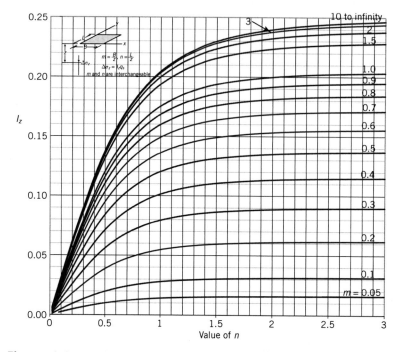

$n > 3$	
m	I_z
0.05	0.016
0.1	0.032
0.2	0.062
0.3	0.090
0.4	0.115
0.5	0.137
0.6	0.156
7	0.172
0.8	0.185
0.9	0.196
1	0.205
1.5	0.230
2	0.240
3	0.247
10	0.250

Figure 6.6   Stress influence chart for the increase in vertical total stress under the corner of a uniformly stressed rectangular loaded area. (Modified from NAV-FAC-DM 7.1.)

look up the stress intensity factor I from the chart using the curve corresponding to the lateral distance factor of 0 and a depth factor of 0.5. In Figure 6.4, $I = I_c \approx 0.91$. The vertical stress increase is then $\Delta\sigma_z = q_s I_c = 100 \times 0.91 = 91\,\text{kPa}$.

You need to be extra careful with rectangular surface loads. The equation and chart are for vertical stress increase *at a corner of the rectangle*. To use the equation or chart for the vertical stress increase at any point other than at the corner, you have to subdivide the original rectangular loaded area into smaller or larger rectangular areas such that the point at which the vertical stress increase is desired is directly under the corner of any of the subdivided rectangular area. Example 6.3 provides the general methodology that can be adopted.

Regardless of the type of surface loads, the vertical stress increase decreases with depth and distance away from the center of the loaded area. In fact, for depth factors greater than about 2, the vertical stress increase is generally less than 10% of the applied surface stress. The isobars (a system of lines of equal pressures) shown in, for example, Figure 6.5 are similar to onion bulbs and indicate how the stress intensity decreases. The boundaries of the pressure bulb signify the extent of the soil mass that provides the bearing pressure for a foundation system such as a footing (a concrete slab) for a column

6.3.3 *Infinite loads*

Uniform loads of large lateral extent such as fills and surcharges are assumed to be transferred to the soil as a uniformly distributed vertical stress throughout the depth. For example,

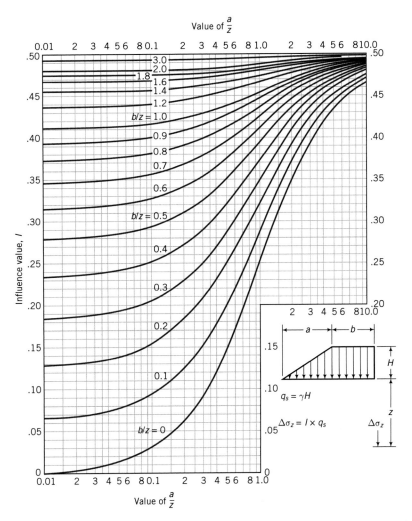

Figure 6.7 Stress influence chart for the increase in vertical total stress due to an embankment. (Source: NAV-FAC-DM 7.1.)

if a fill of unit weight $20\,kN/m^3$ and height $5\,m$ is placed on the surface of a soil, then the increase in vertical stress at any depth below the surface is $5 \times 100 = 500\,kPa$.

6.3.4 *Vertical stress below arbitrarily shaped areas*

Newmark (1942) developed a chart to determine the increase in vertical stress due to a uniformly loaded area of any shape. The chart consists of concentric circles divided by radial lines (Figure 6.8). The area of each segment represents an equal proportion of the applied surface stress at a depth z below the surface. If there are 10 concentric circles and 20 radial

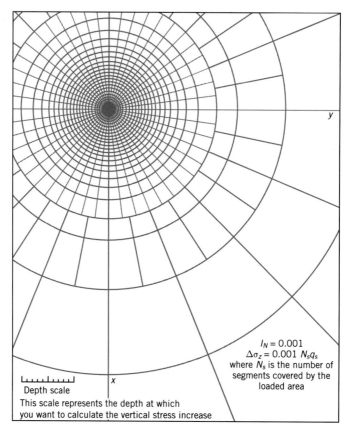

Figure 6.8 Newmark's chart for increase in vertical total stress for irregular loaded areas. (Source: NAV-FAC-DM 7.1.)

lines, the stress on each circle is $q_s/10$ and on each segment is $q_s/(10 \times 20)$. The chart is normalized to the depth; that is, all dimensions are scaled by a factor initially determined for the depth. Every chart should show a scale and an influence factor I_N. The influence factor for Figure 6.8 is 0.001.

The procedure for using Newmark's chart is as follows:

1. Set the scale, shown on the chart, equal to the depth at which the increase in vertical total stress is required. We will call this the depth scale.
2. Identify the point below the loaded area where the increase in vertical total stress is required. Let us say this point is A.
3. Plot the loaded area, scaling its plan dimension using the depth scale with point A at the center of the chart.
4. Count the number of segments (N_s) covered by the scaled loaded area. If certain segments are not fully covered, you can estimate what fraction is covered.
5. Calculate the increase in vertical total stress as $\Delta\sigma_z = q_s I_N N_s$.

EXAMPLE 6.1 *Vertical Stress Increase Due to an Electric Power Transmission Pole*

A Douglas fir electric power transmission pole is 12 m above ground level and embedded 3 m into the ground. The butt diameter is 450 mm and the tip diameter (the top of the pole) is 300 mm. The weight of the pole, cross arms, and wires is 35 kN. Assume that the pole transmits the load as a point load with the soil surface at the embedded base. Determine the stress increase at a depth 1 m below the embedded base of the pole along the center and at 1 m from the center.

Strategy This is a straightforward application of Boussinesq's equation. We will use the vertical stress increase equation rather than the chart.

Solution 6.1

Step 1: Calculate vertical stress increase under the center of the pole.

At center of pole, $r = 0$, $r/z = 0$.

$$I = \frac{3}{2\pi} \frac{1}{\left[1 + (r/z)^2\right]^{5/2}} = \frac{3}{2\pi} \frac{1}{\left[1 + (0)^2\right]^{5/2}} = \frac{3}{2\pi} = 0.48$$

$$\Delta\sigma_z = \frac{Q}{z^2} I = \frac{35}{1^2} \times 0.48 = 16.8 \text{ kPa}$$

Step 2: Determine the vertical stress increase at the radial distance of 1 m.

$$r = 1 \text{ m}, z = 1 \text{ m}, \quad \frac{r}{z} = \frac{1}{1} = 1, \quad I = \frac{3}{2\pi} \frac{1}{\left[1 + (1)^2\right]^{5/2}} = 0.085$$

$$\Delta\sigma_z = \frac{35}{1^2} \times 0.085 = 3 \text{ kPa}$$

Step 3: Determine the reasonableness of the answers.

The vertical stress increase decreases with depth and with radius away from the center of the loaded area.

The applied vertical stress at the base (butt diameter = 450 mm) is

$$\frac{\text{Vertical force}}{\text{Butt area}} = \frac{35}{\pi \dfrac{0.45^2}{4}} = 220 \text{ kPa}$$

This stress is assumed to be transmitted as a point load. The calculated vertical stress increases are all lower than the applied vertical stress. The answers are reasonable.

EXAMPLE 6.2 *Vertical Stress Increase Due to a Ring Load*

The total vertical load on a ring foundation shown in Figure E6.2a is 4000 kN. (a) Plot the vertical stress increase with depth up to 8 m under the center of the ring (point O, Figure E6.2a). (b) Determine the maximum vertical stress increase and its location.

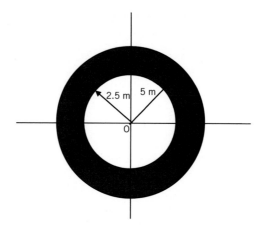

Figure E6.2a

Strategy To use the equation for a uniform circular area to simulate the ring foundation, you need to create two artificial circular foundations, one with an outer radius of 5 m and the other with an outer radius of 2.5 m. Both foundations must be fully loaded with the applied uniform, vertical surface stress. By subtracting the vertical stress increase of the smaller foundation from the larger foundation, you would obtain the vertical stress increase from the ring foundation. You are applying here the principle of superposition.

Solution 6.2

Step 1: Identify the loading type.

It is a uniformly loaded ring foundation.

Step 2: Calculate the imposed surface stress.

$$r_2 = 5 \text{ m}, \quad r_1 = 2.5 \text{ m}$$

$$\text{Area} = \pi(r_2^2 - r_1^2) = \pi(5^2 - 2.5^2) = 58.9 \text{ m}^2$$

$$q_s = \frac{Q}{A} = \frac{4000}{58.9} = 67.9 \text{ kPa}$$

Step 3: Create two solid circular foundations of radii 5 m and 2.5 m.

See Figure E6.2b. Let "large" denotes the foundation of radius 5 m and "small" denotes the foundation of radius 2.5 m.

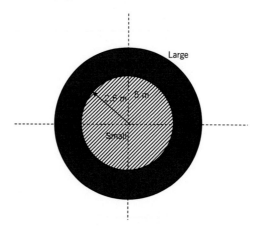

Figure E6.2b

Step 4: Create a spreadsheet to do the calculations.

Ring load	
Load	4000 kN
Outer radius	5 m
Inner radius	2.5 m
Area	58.9 m²
q_s	67.9 kPa

z (m)	Large		Small		I_{diff}	$\Delta\sigma_z$ (kPa)
	r/z	$(I_c)_{large}$	r_o/z	$(I_c)_{small}$	$(I_c)_{large} - (I_c)_{small}$	$q_s \times I_{diff}$
1	5.00	0.992	2.50	0.949	0.044	3.0
2	2.50	0.949	1.25	0.756	0.193	13.1
3	1.67	0.864	0.83	0.547	0.317	21.5
4	1.25	0.756	0.63	0.390	0.366	24.9
5	1.00	0.646	0.50	0.284	0.362	24.6
6	0.83	0.547	0.42	0.213	0.333	22.6
7	0.71	0.461	0.36	0.165	0.296	20.1
8	0.63	0.390	0.31	0.130	0.260	17.6

Step 5: Plot the increase in vertical stress with depth.

See Figure E6.2c.

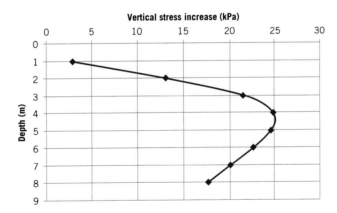

Figure E6.2c

Step 6: Check reasonableness of results.

From Figure 6.4, the increase in vertical stress from a circular loaded area placed on the ground surface will be distributed parabolically with depth. The shape of the increase of vertical stress with depth (Figure E6.2c) is similar to that shown in Figure 6.4. The increase in vertical stress also has to be less than the applied vertical stress of 67.9 kPa. The results are then reasonable.

Step 7: Determine the maximum vertical stress increase and depth of occurrence.

From Figure E6.2c, the maximum vertical stress increase is about 25 kPa and the depth of occurrence is about 4 m from the surface.

EXAMPLE 6.3 *Vertical Stress Increase Due to a Rectangular Load*

A rectangular concrete slab, 3 m × 4.5 m, rests on the surface of a soil mass. The load on the slab is 2025 kN. Determine the vertical stress increase at a depth of 3 m **(a)** under the center of the slab, point A (Figure E6.3a); **(b)** under point B (Figure E6.3a); **(c)** at a distance of 1.5 m from a corner, point C (Figure E6.3a); and **(d)** compare the results from (a) with an estimate using the approximate method.

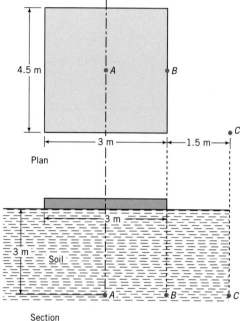

Figure E6.3a

Strategy The slab is rectangular and the equations for a uniformly loaded rectangular area are for the corner of the area. You should divide the area so that the point of interest is a corner of a rectangle(s). You may need to extend the loaded area if the point of interest is outside it (loaded area). The extension is fictitious, so you have to subtract the fictitious increase in vertical stress for the extended area.

Solution 6.3

Step 1: Identify the loading type.

It is a uniformly loaded rectangle.

Step 2: Divide the rectangle so that the center is a corner.

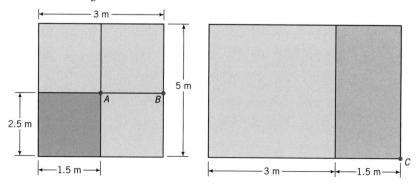

Figure E6.3b, c

In this problem, all four rectangles, after the subdivision, are equal (in Figure E6.3b, point C is excluded for simplicity), so you only need to find the vertical stress increase for one rectangle of size $B = 1.5$ m, $L = 2.25$ m, and multiply the results by 4.

$$m = \frac{B}{z} = \frac{1.5}{3} = 0.5; \quad n = \frac{L}{z} = \frac{2.25}{3} = 0.75$$

From the chart in Figure 6.6, $I_z = 0.105$.

Step 3: Find the vertical stress increase at the center of the slab (point A, Figure E6.3b).

$$q_s = \frac{Q}{A} = \frac{2025}{3 \times 4.5} = 150 \text{ kPa}$$

$$\Delta\sigma_z = 4q_s I_z = 4 \times 150 \times 0.105 = 63 \text{ kPa}$$

Step 4: Find the vertical stress increase for point B.

Point B is at the corner of two rectangles, each of width 3 m and length 2.25 m. You need to find the vertical stress increase for one rectangle and multiply the result by 2.

$$m = \frac{3}{3} = 1; \quad n = \frac{2.25}{3} = 0.75$$

From the chart in Figure 6.6, $I_z = 0.158$.

$$\Delta\sigma_z = 2q_s I_z = 2 \times 150 \times 0.158 = 47.4 \text{ kPa}$$

You should note that the vertical stress increase at B is lower than at A, as expected.

Step 5: Find the stress increase for point C.

Stress point C is outside the rectangular slab. You have to extend the rectangle to C (Figure E6.3c) and find the vertical stress increase for the large rectangle of width $B = 4.5$ m, length $L = 4.5$ m, and then subtract the vertical stress increase for the smaller rectangle of width $B = 1.5$ m and length $L = 4.5$ m.

Large rectangle: $m = \dfrac{4.5}{3} = 1.5, \quad n = \dfrac{4.5}{3} = 1.5$; from chart in Figure 6.6, $I_z = 0.22$

Small rectangle: $m = \dfrac{1.5}{3} = 0.5, \quad n = \dfrac{4.5}{3} = 1.5$; from chart in Figure 6.6, $I_z = 0.13$

$$\Delta\sigma_z = q_s \Delta I_z = 150 \times (0.22 - 0.13) = 13.5 \text{ kPa}$$

Step 6: Find the vertical stress increase at point A using approximate methods.

$$\Delta\sigma_z = \frac{Q}{(B+z)(L+z)} = \frac{2025}{(3+3)(4.5+3)} = 45 \text{ kPa}$$

This increase is about 30% less than that obtained by Boussinesq's elastic solution. The approximate method is only used to estimate the vertical stress increase under the center of rectangular (or square) surface loads.

Step 7: Check reasonableness of results.

The vertical stress increase at any depth has to be less than the applied vertical stress of 150 kPa. The vertical stress increase at C has to be lower than the vertical stress increase at B, which is the case. The results are then reasonable.

EXAMPLE 6.4 *Vertical Stress Increase Due to an Irregular Loaded Area*

The plan of a foundation of uniform thickness for a small monument is shown in Figure E6.4a. Determine the vertical stress increase at a depth of 4 m below the centroid. The foundation applies a vertical stress of 200 kPa on the soil surface.

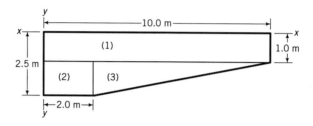

Figure E6.4a

Strategy You need to locate the centroid of the foundation, which you can find using the given dimensions. The shape of the foundation does not fit neatly into one of the standard shapes (e.g., rectangles or circles) discussed. The convenient method to use for this (odd) shape foundation is Newmark's chart.

Solution 6.4

Step 1: Find the centroid.

Divide the loaded area into a number of regular shapes. In this example, we have three. Take the sum of moments of the areas about y–y (Figure E6.4a) and divide by the sum of the areas to get $\bar{x}$. Take moments about x–x (Figure E6.4a) to get $\bar{y}$.

$$\bar{x} = \frac{(1.0 \times 10.0 \times 5.0) + (1.5 \times 2.0 \times 1.0) + \left[\frac{1}{2} \times 8.0 \times 1.5 \times \left(2 + \frac{1}{3} \times 8.0\right)\right]}{(1.0 \times 10.0) + (1.5 \times 2.0) + \frac{1}{2} \times 8.0 \times 1.5} = \frac{81}{19} = 4.26 \text{ m}$$

$$\bar{y} = \frac{(1 \times 10 \times 0.5) + (1.5 \times 2 \times 1.75) + \left[\frac{1}{2} \times 8.0 \times 1.5 \times \left(1.0 + \frac{1.5}{3}\right)\right]}{(1.0 \times 10.0) + (1.5 \times 2.0) + \frac{1}{2} \times 8.0 \times 1.5} = \frac{19.25}{19} \approx 1 \text{ m}$$

Step 2: Scale and plot the foundation on a Newmark chart.

The scale on the chart is set equal to the depth, which in this case is 4 m. If the depth required were, say, 8 m, then the scale on the chart would be set to 8 m. The centroid is located at the center of the chart and the foundation is scaled using the depth scale (Figure E6.4b).

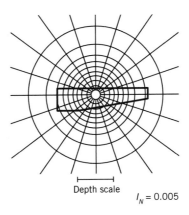

Depth scale $I_N = 0.005$

Figure E6.4b

Step 3: Count the number of segments covered by the foundation.

$$N_s = 61$$

Step 4: Calculate the vertical stress increase.

$$\Delta\sigma_z = q_s I_N N_s = 200 \times 0.005 \times 61 \approx 61 \text{ kPa}$$

Step 5: Check reasonableness of results.

The vertical stress increase at any depth has to be less than the applied vertical stress of 200 kPa. The results are then reasonable.

Key points

1. The increases in stresses below a surface load are found by assuming that the soil is an elastic, semi-infinite mass.
2. Various equations are available for the increases in stresses from surface loading.
3. The stress increase at any depth depends on the shape and distribution of the surface load.
4. A stress applied at the surface of a soil mass by a loaded area decreases with depth and lateral distance away from the center of the loaded area.
5. The vertical stress increases are generally less than 10% of the surface stress when the depth-to-width ratio is greater than 2.

What's next ... The stresses we have calculated are for soils as solid elastic materials. We have not accounted for the fluid and gas pressure within the soil pore spaces. In the next section, we will discuss the principle of effective stresses that accounts for the pressures within the soil pores. This principle is the most important principle in soil mechanics.

6.4 TOTAL AND EFFECTIVE STRESSES

6.4.1 The principle of effective stress

The deformations of soils are similar to the deformations of structural framework such as a truss. The truss deforms from changes in loads carried by each member. If the truss is loaded in air or submerged in water, the deformations under a given load will remain unchanged. Deformations of the truss are independent of hydrostatic pressure. The same is true for soils.

Let us consider an element of a *saturated soil* subjected to a normal stress σ applied on the horizontal boundary, as shown in Figure 6.9. The stress σ is called the *total stress*, and for equilibrium (Newton's third law) the stresses in the soil must be equal and opposite to σ The resistance or reaction to σ is provided by a combination of the stresses from the solids, called *effective stress* (σ'), and from water in the pores, called *porewater pressure* (u). We will denote effective stresses by a prime (′) following the symbol for normal stress, usually σ. The equilibrium equation is

$$\sigma = \sigma' + u \tag{6.1}$$

so that

$$\sigma' = \sigma - u \tag{6.2}$$

Equation (6.2) is called the *principle of effective stress* and was first recognized by Terzaghi (1883–1963) in the mid-1920s during his research into soil consolidation (Chapter 7). *The principle of effective stress is the most important principle in soil mechanics. Deformations of soils are a function of effective stresses, not total stresses. The principle of effective stresses applies only to normal stresses and not to shear stresses. Also, Equation (6.2) applies only to saturated soils.* The porewater cannot sustain shear stresses, and therefore, the soil solids must resist the shear forces. Thus, $\tau = \tau'$, where τ is the total shear stress and τ' is the effective shear stress. The effective stress is not the contact stress between the soil solids. Rather, it is the average stress on a plane through the soil mass.

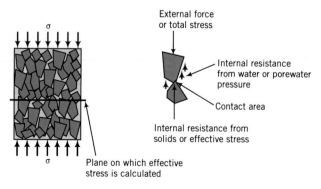

Figure 6.9 Effective stress.

Soils cannot sustain tension. Consequently, the effective stress cannot be less than zero. Porewater pressure can be positive or negative. The latter is sometimes called suction or suction pressure.

For unsaturated soils, the effective stress (Bishop et al., 1960) is

$$\sigma' = \sigma - u_a + \chi(u_a - u_w) \tag{6.3}$$

where u_a is the pore air pressure, $u_w = u$ is the porewater pressure, and χ is a factor depending on the degree of saturation. The expression $(u_a - u_w)$ is called the matrix suction. For dry soil, $\chi = 0$; for saturated soil, $\chi = 1$ and $u_a = 0$. Values of χ for a silt are shown in Figure 6.10.

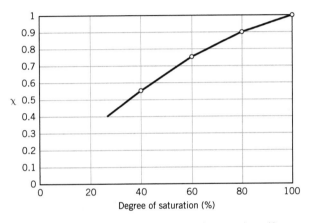

Figure 6.10 Values of χ for a silt at different degrees of saturation. (Source: Bishop et al., 1960.)

6.4.2 *Total and effective stresses due to geostatic stress fields*

The effective stress in a soil mass not subjected to external loads is found from the unit weight of the soil and the depth of groundwater. Consider a soil element at a depth z below the ground surface, with the groundwater level (GWL) at ground surface (Figure 6.11a). The total vertical stress is

$$\sigma = \gamma_{sat} z \tag{6.4}$$

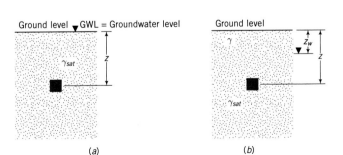

Figure 6.11 Soil element at a depth z with groundwater level (a) at ground level and (b) below ground level.

The porewater pressure is

$$u = \gamma_w z \tag{6.5}$$

and the effective stress is

$$\sigma' = \sigma - u = \gamma_{sat} z - \gamma_w z = (\gamma_{sat} - \gamma_w) z = \gamma' z \tag{6.6}$$

If the GWL is at a depth z_w below ground level (Figure 6.11b), then

$$\sigma = \gamma z_w + \gamma_{sat}(z - z_w) \quad \text{and} \quad u = \gamma_w(z - z_w)$$

The effective stress is

$$\sigma' = \sigma - u = \gamma z_w + \gamma_{sat}(z - z_w) - \gamma_w(z - z_w)$$
$$= \gamma z_w + (\gamma_{sat} - \gamma_w)(z - z_w) = \gamma z_w + \gamma'(z - z_w)$$

We have tacitly assumed that the air pressure (relative to atmospheric pressure or gauge air pressure) is zero within the unsaturated zone, which is the zone above the groundwater level.

6.4.3 Effects of capillarity

In silts and fine sands, the soil above the groundwater can be saturated by capillary action. You would have encountered capillary action in your physics course when you studied menisci. We can get an understanding of capillarity in soils by idealizing the continuous void spaces as capillary tubes. Consider a single idealized tube, as shown in Figure 6.12.

The height at which water will rise in the tube can be found from statics. Summing forces vertically (upward forces are negative), we get

$$\sum F_z = \text{weight of water} - \text{tension forces from capillary action}$$

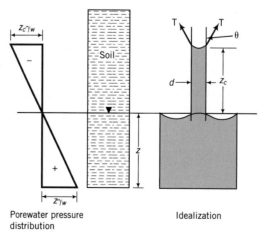

Figure 6.12 Capillary simulation in soils.

that is,

$$\frac{\pi d^2}{4} z_c \gamma_w - \pi dT \cos \theta = 0 \qquad (6.7)$$

Solving for z_c, we get

$$z_c = \frac{4T \cos \theta}{d\gamma_w} \qquad (6.8)$$

where T is the surface tension (force per unit length), θ is the contact angle, z_c is the height of capillary rise, and d is the diameter of the tube representing the diameter of the void space. The surface tension of water is $0.073\,\text{N/m}$ at $15.6°\text{C}$ and the contact angle of water with a clean glass surface is 0. Since T, θ, and γ_w are constants,

$$z_c \propto \frac{1}{d} \qquad (6.9)$$

For soils, d is assumed to be equivalent to $0.1\,D_{10}$ where D_{10} is the effective size. The interpretation of Equation (6.9) is that the smaller the soil pores, the higher the capillary zone. The capillary zone in fine sands will be larger than for medium or coarse sands.

The porewater pressure due to capillarity is negative (suction), as shown in Figure 6.12, and is a function of the size of the soil pores and the water content. At the groundwater level, the porewater pressure is zero and decreases (becomes negative) as you move up the capillary zone. The effective stress increases because the porewater pressure is negative. For example, for the capillary zone, z_c, the porewater pressure at the top is $-z_c\gamma_w$ and the effective stress is $\sigma' = \sigma - (-z_c\gamma_w) = \sigma + -z_c\gamma_w$.

The approach we have taken to interpret capillary action in soils is simple, but it is sufficient for most geotechnical applications.

6.4.4 Effects of seepage

In Chapter 4, we discussed the flow of water through soils. As water flows through soil, it exerts a frictional drag on the soil particles, resulting in head losses. The frictional drag is called seepage force in soil mechanics. It is often convenient to define seepage as the seepage force per unit volume (it has units similar to unit weight), which we will denote by j_s. If the head loss over a flow distance, L, is Δh, the seepage force is

$$j_s = \frac{\Delta h \gamma_w}{L} = i\gamma_w \qquad (6.10)$$

If seepage occurs downward (Figure 6.13a), then the seepage stresses are in the same direction as the gravitational effective stresses. From static equilibrium, the resultant vertical effective stress is

$$\sigma'_z = \gamma' z + iz\gamma_w = \gamma' z + j_s z \qquad (6.11)$$

If seepage occurs upward (Figure 6.13b), then the seepage stresses are in the opposite direction to the gravitational effective stresses. From static equilibrium, the resultant vertical effective stress is

$$\sigma'_z = \gamma' z - iz\gamma_w = \gamma' z - j_s z \qquad (6.12)$$

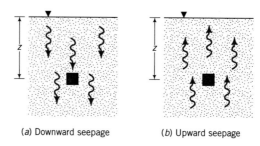

(a) Downward seepage   (b) Upward seepage

Figure 6.13 Seepage in soils.

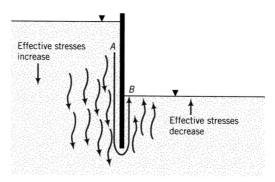

Figure 6.14 Effects of seepage on the effective stresses near a retaining wall.

Seepage forces play a very important role in destabilizing geotechnical structures. For example, a cantilever retaining wall, shown in Figure 6.14, depends on the depth of embedment for its stability. The retained soil (left side of wall) applies an outward lateral pressure to the wall, which is resisted by an inward lateral resistance from the soil on the right side of the wall. If a steady quantity of water is available on the left side of the wall, for example, from a broken water pipe, then water will flow from the left side to the right side of the wall. The path followed by a particle of water is depicted by AB in Figure 6.14, and as water flows from A to B, head loss occurs. The seepage stresses on the left side of the wall are in the direction of the gravitational stresses. The effective stress increases and, consequently, an additional outward lateral force is applied on the left side of the wall. On the right side of the wall, the seepage stresses are upward and the effective stress decreases. The lateral resistance provided by the embedment is reduced. Seepage stresses in this problem play a double role (increase the lateral disturbing force and reduce the lateral resistance) in reducing the stability of a geotechnical structure.

> ### Key points
>
> 1. The effective stress in a saturated represents the average stress carried by the soil solids and is the difference between the total stress and the porewater pressure.
> 2. The effective stress principle applies only to normal stresses and not to shear stresses.

3. Deformations of soils are due to effective stress not total stress.
4. Soils, especially silts and fine sands, can be affected by capillary action.
5. Capillary action results in negative porewater pressures (suction) and increases the effective stresses.
6. Downward seepage increases the resultant effective stress; upward seepage decreases the resultant effective stress.

EXAMPLE 6.5   *Calculating Vertical Effective Stress*

Calculate the effective stress for a soil element at depth of 5 m in a uniform deposit of soil, as shown in Figure E6.5. Assume that the pore air pressure is zero.

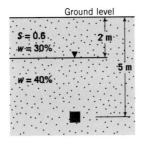

Figure E6.5

Strategy You need to get unit weights from the given data, and you should note that the soil above the groundwater level is not saturated.

Solution 6.5

Step 1: Calculate unit weights.

Above groundwater level

$$\gamma = \left(\frac{G_s + Se}{1+e}\right)\gamma_w = \frac{G_s(1+w)}{1+e}\gamma_w$$

$$Se = wG_s, \quad \therefore e = \frac{0.3 \times 2.7}{0.6} = 1.35$$

$$\gamma = \frac{2.7(1+0.3)}{1+1.35} \times 9.81 = 14.6 \text{ kN/m}^3$$

Below groundwater level

Soil is saturated, $S = 1$.

$$e = wG_s = 0.4 \times 2.7 = 1.08$$

$$\gamma_{sat} = \left(\frac{G_s + e}{1+e}\right)\gamma_w = \left(\frac{2.7+1.08}{1+1.08}\right) \times 9.81 = 17.8 \text{ kN/m}^3$$

Step 2: Calculate the effective stress.

$$\text{Total stress: } \sigma_z = 2\gamma + 3\gamma_{sat} = 2 \times 14.6 + 3 \times 17.8 = 82.6 \text{ kPa}$$

$$\text{Porewater pressure: } u = 3\gamma_w = 3 \times 9.81 = 29.4 \text{ kPa}$$

$$\text{Effective stress: } \sigma_z' = \sigma_z - u = 82.6 - 29.4 = 53.2 \text{ kPa}$$

Alternatively:

$$\sigma_z' = 2\gamma + 3(\gamma_{sat} - \gamma_w) = 2\gamma + 3\gamma' = 2 \times 14.6 + 3(17.8 - 9.81) = 53.2 \text{ kPa}$$

EXAMPLE 6.6 *Calculating and Plotting Vertical Effective Stress Distribution*

A borehole at a site reveals the soil profile shown in Figure E6.6a. The negative sign on the depth indicates below ground level. Plot the distribution of vertical total and effective stresses with depth up to 20 m. Assume pore air pressure is zero.

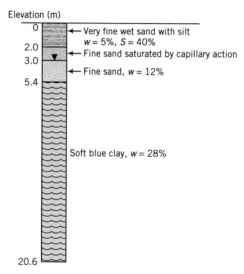

Figure E6.6a

Strategy From the data given, you will have to find the unit weight of each soil layer to calculate the stresses. You are given that the 1.0 m of fine sand above the groundwater level is saturated by capillary action. Therefore, the porewater pressure in this 1.0 m zone is negative.

Solution 6.6

Step 1: Calculate the unit weights.

0–2.0 m

$$S = 40\% = 0.4; w = 0.05$$

$$e = \frac{wG_s}{S} = \frac{0.05 \times 2.7}{0.4} = 0.34$$

$$\gamma = \frac{G_s(1+w)}{1+e}\gamma_w = \frac{2.7(1+0.05)}{1+0.34} \times 9.81 = 20.8 \text{ kN/m}^3$$

2.0–5.0 m

$$S = 1; w = 0.2$$

$$e = wG_s = 0.2 \times 2.7 = 0.54$$

$$\gamma_{sat} = \left(\frac{G_s + e}{1+e}\right)\gamma_w = \left(\frac{2.7 + 0.54}{1 + 0.54}\right) \times 9.81 = 20.6 \text{ kN/m}^3$$

5.0–20.0 m

$$S = 1; w = 0.28$$

$$e = wG_s = 0.28 \times 2.7 = 0.76$$

$$\gamma_{sat} = \left(\frac{2.7 + 0.76)}{1 + 0.76}\right) \times 9.81 = 19.3 \text{ kN/m}^3$$

Step 2: Calculate the stresses using a table or a spreadsheet program.

Depth (m)	Thickness (m)	σ_z (kPa)	u (kPa)	$\sigma'_z = \sigma - u$ (kPa)
0.0	0	0	0	0
2.0	2	20.8 × 2 = 41.6	−1 × 9.81 = −9.81	51.4
3.0	1	41.6 + 20.6 × 1 = 62.2	0	62.2
5.0	2	62.2 + 20.6 × 2 = 103.4	2 × 9.81 = 19.6	83.8
20.0	15	103.4 + 19.3 × 15 = 392.9	17 × 9.81 = 166.8	226.1

Step 3: Plot the stresses versus depth; see Figure E6.6b.

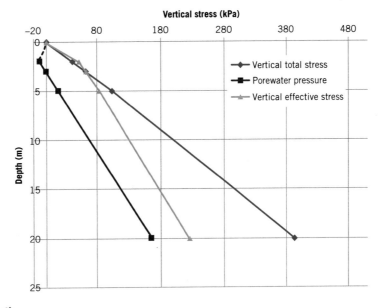

Figure E6.6b

The porewater pressure above the capillary zone is shown as a broken curved line because the pore air pressure is not known. By assuming the pore air to be zero, you would have a sudden change in negative porewater pressure from $-9.81\,\text{kPa}$ to zero at the top of the capillary zone. It is unlikely that such sudden change can occur in the soil. If the soil suction in the soil layer above the capillary zone is larger than the negative pore water pressure at the top of the capillary zone, the curvature of the broken line shown will be reversed. Of course, the vertical effective stress line above the capillary zone should also be drawn as a broken line with a curvature. For simplicity and practical purposes, the distribution of both the porewater pressure and the vertical effective stress above the capillary zone are shown as straight lines.

EXAMPLE 6.7 *Effects of Seepage on Vertical Effective Stress*

Water is seeping downward through a saturated soil layer, as shown in Figure E6.7. Two piezometers (A and B) located 2 m apart (vertically) showed a head loss of 0.2 m. Calculate the resultant vertical effective stress for a soil element at a depth of 6 m.

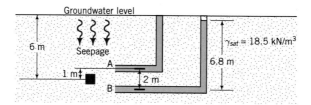

Figure E6.7

Strategy You have to calculate the seepage stress. But to obtain this you must know the hydraulic gradient, which you can find from the data given.

Solution 6.7

Step 1: Find the hydraulic gradient.

$$\Delta H = 0.2 \text{ m}; \quad L = 2 \text{ m}; \quad i = \frac{\Delta H}{L} = \frac{0.2}{2} = 0.1$$

Step 2: Determine the effective stress.

Assume that the hydraulic gradient is the average for the soil mass; then

$$\sigma'_z = (\gamma_{sat} - \gamma_w)z + i\gamma_w z = (18.5 - 9.81) \times 6 + 0.1 \times 9.81 \times 6$$
$$= 52.1 + 5.9 = 58 \text{ kPa}$$

Step 3: Check reasonableness of answer.

The seepage pressure is downward. Therefore, the vertical effective stress will be higher than that without seepage (52.1 kPa). The answer is reasonable.

EXAMPLE 6.8   *Effects of Groundwater Condition on Vertical Effective Stress*

(a) Plot the vertical total and effective stresses and porewater pressure with depth for the soil profile shown in Figure E6.8a for steady-state seepage condition. A porewater pressure transducer installed at the top of the sand layer gives a pressure of 58.8 kPa. Assume that $G_s = 2.7$ and neglect pore air pressure.

(b) If a borehole were to penetrate the sand layer, how far would the water rise above the groundwater level?

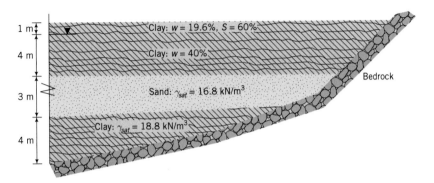

Figure E6.8a   (not to scale)

Strategy You have to calculate the unit weight of the top layer of clay. From the soil profile, the groundwater appears to be under artesian condition, so the vertical effective stress would change sharply at the interface of the top clay layer and the sand. It is best to separate the soil above the groundwater from the soil below the groundwater. So, divide the soil profile into artificial layers.

Solution 6.8

Step 1:   Divide the soil profile into artificial layers.
See Figure E6.8b.

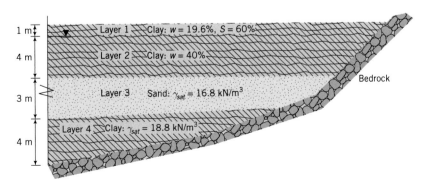

Figure E6.8b

Step 2: Find the unit weight of the top clay layers.

Above groundwater level: $\gamma = \dfrac{G_s + Se}{1+e}\gamma_w = \dfrac{G_s(1+w)}{1+(wG_s/S)}\gamma_w = \dfrac{2.7(1+0.196)}{1+[(0.196\times2.7)/0.6]}\times9.81$

$= 16.8\ \text{kN/m}^3$

Below groundwater level: $\gamma_{\text{sat}} = \dfrac{G_s+e}{1+e}\gamma_w = \dfrac{G_s(1+w)}{1+wG_s}\gamma_w = \dfrac{2.7(1+0.4)}{1+0.4\times2.7}\times9.81$

$= 17.8\ \text{kN/m}^3$

Step 3: Determine the effective stress.

See spreadsheet. Note: The porewater pressure at the top of the sand is $58.8\,\text{kPa}$.

Layer	Depth (m)	Thickness (m)	γ (kN/m³)	σ_z (kPa)	u (kPa)	σ'_z (kPa)
1, top	0	1	16.8	0	0	0
1, bottom	1			16.8	0.0	16.8
2, top	1	4	17.8	16.8	0.0	16.8
2, bottom	5			88.1	39.2	48.9
3, top	5	3	16.8	88.1	**58.8**	29.3
3, bottom	8			138.5	88.2	50.3
4, top	8	4	18.8	138.5	88.2	50.3
4, bottom	12			213.7	127.5	86.3

Step 4: Plot vertical stress and porewater pressure distributions with depth.

See Figure E6.8c.

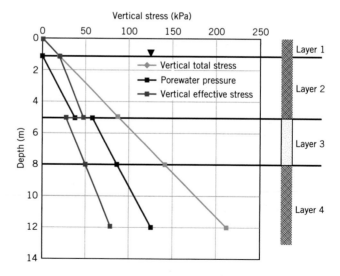

Figure E6.8c

Note:

1. The vertical effective stress changes abruptly at the top of the sand layer due to the artesian condition.

2. For each layer or change in condition (groundwater or unit weight), the vertical stress at the bottom of the preceding layer acts a surcharge, transmitting a uniform vertical stress of equal magnitude to all subsequent layers. For example, the vertical total stress at the bottom of layer 2 is 88.1 kPa. This stress is transferred to both layers 3 and 4. Thus, the vertical total stress at the bottom of layer 3 from its own weight is $3 \times 16.8 = 50.4$ kPa, and adding the vertical total stress from the layers above gives $88.1 + 50.4 = 138.5$ kPa.

Step 5: Calculate the height of water.

$$h = \frac{58.8}{9.81} = 6 \text{ m}$$

Height above existing groundwater level = $6 - 4 = 2$ m, or 1 m above ground level.

What's next ... We have considered only vertical stresses. But an element of soil in the ground is also subjected to lateral stresses. Next, we will introduce an equation that relates the vertical and lateral effective stresses.

6.5 LATERAL EARTH PRESSURE AT REST

The ratio of the horizontal principal effective stress to the vertical principal effective stress is called the lateral earth pressure coefficient at rest (K_o), that is,

$$K_o = \frac{\sigma_3'}{\sigma_1'} \tag{6.13}$$

The at-rest condition implies that no change in deformation. Remember, K_o applies only to effective principal stress not total principal stress. To find the lateral total stress, you must add the porewater pressure. Remember also that the porewater pressure is hydrostatic and that, at any given depth, the porewater pressure in all directions is equal.

For a soil that was never subjected to vertical effective stresses higher than its current vertical effective stress (normally consolidated soil), $K_o = K_o^{nc}$ is reasonably predicted by an equation suggested by Jaky (1944) as

$$K_o^{nc} \approx 1 - \sin \phi_{cs}' \tag{6.14}$$

where ϕ_{cs}' is a fundamental frictional soil constant (angle of internal friction) that will be discussed in Chapter 8.

The value of K_o^{nc} is constant. During unloading or reloading, the soil stresses must adjust to be in equilibrium with the applied stress. This means that stress changes take place not only vertically but also horizontally. For a given surface stress, the changes in horizontal total stresses and vertical total stresses are different, but the porewater pressure changes in every direction are the same. Therefore, the current effective stresses are different in different directions. A soil in which the current effective stress is lower than the past maximum stress is called an overconsolidated soil (to be discussed further in Chapter 7). The K_o values for overconsolidated soils are not constants. We will denote K_o for overconsolidated soils as K_o^{oc}.

Various equations have been suggested linking K_o^{oc} to K_o^{nc}. One equation that is popular and found to match test data reasonably well is an equation proposed by Meyerhof (1976) as

$$K_o^{oc} = K_o^{nc}(OCR)^{1/2} = (1 - \sin\phi_{cs}')(OCR)^{1/2} \tag{6.15}$$

where OCR is the overconsolidation ratio (see Chapter 7 for more information), defined as the ratio of the past vertical effective stress to the current vertical effective stress.

EXAMPLE 6.9 *Calculating Horizontal Effective and Total Stresses*

Calculate the horizontal effective stress and the horizontal total stress for the soil element at 5 m in Example 6.5 if $K_o = 0.5$.

Strategy The stresses on the horizontal and vertical planes on the soil element are principal stresses (no shear stress occurs on these planes). You need to apply K_o to the effective principal stress and then add the porewater pressure to get the lateral total principal stress.

Solution 6.9

Step 1: Calculate the horizontal effective stress.

$$K_o = \frac{\sigma_3'}{\sigma_1'} = \frac{\sigma_x'}{\sigma_z'}; \quad \sigma_x' = K_o\sigma_z' = 0.5 \times 53.2 = 26.6 \text{ kPa}$$

Step 2: Calculate the horizontal total stress.

$$\sigma_x = \sigma_x' + u = 26.6 + 29.4 = 56 \text{ kPa}$$

6.6 FIELD MONITORING OF SOIL STRESSES

Stresses within soil in practice are measured using earth pressure sensors. One type of earth pressure sensor is an earth pressure cell (Figure 6.15). This cell consists of an oil filled cavity between two thin circular metal plates welded at their periphery. When this cell is buried

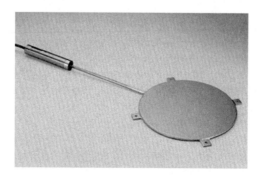

Figure 6.15 An earth pressure cell. (Courtesy of Geokon, Inc.)

within a soil, the earth pressure squeezes the plates together introducing a pressure in the fluid, which is then measured by a pressure transducer.

6.7 SUMMARY

The distribution and amount of vertical total stress transmitted to a soil by surface loads is determined using Boussinesq's elastic solution. The vertical total stress increase from surface loads are distributed such that their magnitudes decrease with depth and distance away from their points of application. The most important principle in soil mechanics is the principle of effective stress. Soil deformation is due to effective, not total, stresses. Seepage stresses can increase or decrease the effective stresses depending on the seepage direction.

6.7.1 Practical example

EXAMPLE 6.10 *Vertical Stress Increase on a Box Culvert*

A developer proposes to construct a warehouse 10 m wide and 40 m long directly above an existing box culvert, 2 m × 2 m, as shown in Figure E6.10a. The foundation for the warehouse must be located at least 1 m below the ground surface as required by local building code (regulations). (a) Determine the vertical stress increase on the top of the box culvert at the middle of the warehouse and at the edge. (b) What are some of the possible negative effects of placing the warehouse over the box culvert? The warehouse will impose a surface stress of 100 kPa.

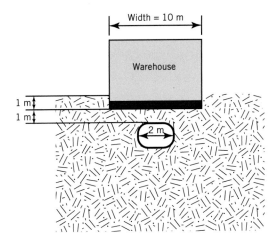

Figure E6.10a

Strategy The solution of this problem requires the calculation of the vertical stress increase from a rectangular load. Although the culvert is located 2 m below the ground surface, the construction of the building will require the removal of 1 m of soil above it. So the depth for calculation of the vertical total stress increase is 1 m, not 2 m.

Solution 6.10

Step 1: Make a sketch of the problem.

Figure E6.10b shows a plan (footprint) of the warehouse.

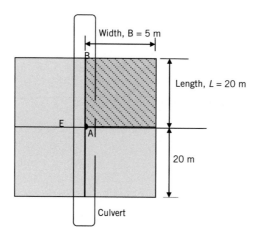

Figure E6.10b

Step 2: Determine vertical stress increase over the top of the culvert.

Divide the warehouse footprint into 4 parts, each of width $B = 5$ m and length 20 m to find the vertical total stress increase at the center. Since we are also required to find the vertical stress increase at the edge, the warehouse footprint is divided into two equal rectangles with $B = 5$ m and $L = 40$ m, and B is the common corner.

The depth is $z = 1$ m

Center, A: $\dfrac{B}{z} = \dfrac{5}{1} = 5$; $\dfrac{L}{z} = \dfrac{20}{1} = 20$

From Figure 6.6, $I_z = 0.25$ (Note: This is the maximum stress influence value.)

Edge, B: $\dfrac{B}{z} = \dfrac{5}{1} = 5$; $\dfrac{L}{z} = \dfrac{40}{1} = 40$

From Figure 6.6, $I_z = 0.25$ (Note: This is the maximum stress influence value.)

At A: $\Delta\sigma_z = 4q_sI_z = 4 \times 100 \times 0.25 = 100$ kPa (Note: This is equal to the surface stress.)

At B: $\Delta\sigma_z = 2q_sI_z = 2 \times 100 \times 0.25 = 50$ kPa

Step 3: Determine possible negative effects.
The vertical stress increase is not uniform over the length where the box culvert crosses the proposed building. This could lead to overstressing of the soil, nonuniform settlement (more settlement at the center than at the edge because the stress increase is higher at the center) and increased bending moment and shear on the culvert.

EXERCISES

Stresses in soil from surface loads

Concept understanding

6.1 A soil at a site consists of 2 m of sand above 10 m of clay. (a) If a circular surface load, 2 m diameter, is applied to the sand, would Boussinesq's solution for the vertical stress increase from surface loads apply to find the vertical stress increase in the clay? Justify your answer. (b) Would your answer be the same if the diameter of the load is 1 m? Explain your answer.

6.2 The soil at a site consists of 20 m of clay above bedrock. The unit weight of the top 10 m of the clay is lower than the bottom 10 m. Would Boussinesq's solution apply to find the vertical stress increase at any depth from a surface load? Justify your answer.

6.3 Does the increase in vertical stress at a certain soil depth from an applied surface load a total stress increase or an effective stress increase? Explain your answer.

Problem solving

6.4 Calculate the increase in vertical total stress at a depth of 2 m under the center of a water tank, 5 m diameter, and 10 m high, filled with water. The self-weight of the tank and its foundation is 300 kN and the unit weight of water is 9.81 kN/m³. Assume the base of the tank foundation is at the soil surface.

6.5 A strip foundation (a long foundation in which the length is much longer than the width) of width 1 m is used to transmit a load of 40 kN/m from a block wall to the soil. Determine the increase in total vertical stress at a depth of 1 m under the center and at the edge of the foundation.

6.6 The inner radius of a ring foundation resting on the surface of a soil is 3 m and the external radius is 5 m. The total vertical load including the weight of the ring foundation is 4000 kN. (a) Plot the vertical stress increase with depth up to 10 m under the center of the ring. (b) Determine the maximum vertical stress increase and its location.

6.7 A column transfers a load of 150 kN to a square concrete foundation 1.5 m × 1.5 m. (a) Plot the increase of vertical total stress with depth up to a depth of 5 m under the center of the foundation. (b) At what depth is the increase in vertical total stress less than 10% of the surface stress?

6.8 A column transfers a load of 200 kN to a rectangular concrete foundation 1.5 m × 2 m. (a) Plot the increase of vertical total stress with depth up to a depth of 5 m under the center of the foundation. (b) At what depth is the increase in vertical total stress less than 10% of the surface stress?

6.9 A rectangular foundation 1.5 m × 2 m (Figure P6.9) transmits a stress of 100 kPa on the surface of a soil deposit. (a) Plot the distribution of increases of vertical total stress with depth under points *A*, *B*, and *C* up to a depth of 5 m. (a) At what depth is the increase in vertical stress below *A* less than 10% of the surface stress?

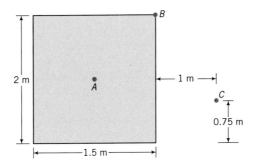

Figure P6.9

6.10 A foundation $7.5\,m \times 5\,m$ is proposed near an existing one, $3\,m \times 5\,m$ as shown in Figure P6.10. (a) Determine the vertical total stress increase at A, B, and C at a depth of $5\,m$ below the ground surface of the existing foundation before construction of the proposed foundation. (b) Determine the total increase in vertical stress at A, B, and C at a depth of $5\,m$ below the ground surface of the existing foundation after the construction of the proposed foundation.

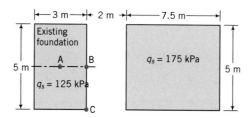

Figure P6.10

6.11 Determine the increase in vertical total stress at a depth of $4\,m$ below the centroid of the foundation shown in Figure P6.11.

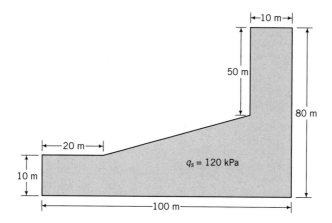

Figure P6.11

Critical thinking and decision making

6.12 A farmer requires two steel silos to store wheat. Each silo is $8\,m$ in external diameter and $10\,m$ high. The foundation for each silo is a circular concrete slab thickened at the edge. The total load of each silo filled with wheat is $9000\,kN$. The soil consists of a $30\,m$ thick deposit of medium clay above a deep deposit of very stiff clay. The farmer desires that the silos be a distance of $1.6\,m$ apart and hires you to recommend whether this distance is satisfactory. (a) Plot the distribution of vertical stress increase at the edges and at the center of one of the silos up to a depth of $20\,m$. Assume the medium clay layer is semi-infinite and the concrete slab is flexible. Use a spreadsheet to tabulate and plot your results. (b) Assuming that any porewater pressure developed from the loading is dissipated so that the increase in vertical total stress is equal to the increase in vertical effective stress and that the increase in settlement is proportional to the increase in vertical effective stress, explain the effects of the stress distribution on the settlement of the silos.

Effective stresses in soils

Concept understanding

6.13 Would the principle of effective stress apply equally to a saturated sand and a saturated clay? Justify your answer.

6.14 How would a change in void ratio affect the effective stress of a saturated soil at a given depth? Justify your answer.

6.15 The groundwater level at a site is at the ground surface. (a) What would be the change in vertical effective stress at a depth of 1 m below the surface if the site becomes flooded with water up to a height 2 m above the ground surface? (b) If the water level were to drop to 1 m below the ground surface and the soil at the site remained saturated, what would be the change in vertical effective stress?

6.16 Can the lateral stress in a soil at rest be greater than the vertical stress? Justify your answer.

Problem solving

6.17 The saturated unit weight of a 10 m thick clay layer is 20 kN/m^3. Calculate the vertical effective stress at a depth of 5 m if the groundwater level is at the surface.

6.18 The saturated unit weight of a 10 m thick clay layer is 20 kN/m^3. Calculate the vertical effective stress at a depth of 5 m if the groundwater level (a) is at 2 m below the surface and the clay is saturated over the full thickness, (b) rises to the ground surface, and (c) rises 1 m above the ground surface (flood condition). From your results, would you expect the vertical effective stresses at any depth in a river to change from a rise in river water level? Justify your answer.

6.19 Plot the distribution of total stress, effective stress, and porewater pressure with depth for the soil profile shown in Figure P6.19. Neglect capillary action and pore air pressure.

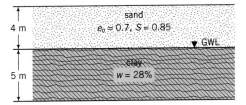

Figure P6.19

Critical thinking and decision making

6.20 A box culvert, 1.5 m × 1.5 m, is to be constructed in a bed of sand ($e = 0.5$) for drainage purposes. The roof of the culvert will be located 2 m below ground surface. Currently, the groundwater level is at ground surface. But, after installation of the culvert, the groundwater level is expected to drop to 1.5 m below ground surface. The sand above the groundwater level is saturated. (a) Calculate the change in vertical effective stress on the roof of the culvert after installation. (b) Calculate the change in lateral effective stress assume the at-rest lateral earth pressure coefficient is 0.5. (c) Plot the changes in effective stresses (vertical and lateral as appropriate) along the top and side of the culvert. (d) What effects would these changes have on the structural integrity of the culvert? Explain your answer,

6.21 A soil profile and the results of water content and degree of saturation (the latter for the stiff, gray clay) for five samples are shown in Figure P6.21. Assume that the results represent the average at the depth at which the soil sample was taken. (a) Plot the distribution of vertical total and effective stress, and porewater pressure with depth. The negatives for the depths in Figure P6.21 indicate measurements below datum (ground surface in this case). Neglect pore air pressure. (Hint: Divide the soil profile into artificial layers such that each test result represents the average over the layer, e.g., sample 1 result would represent a layer from 0 to 3 m.)

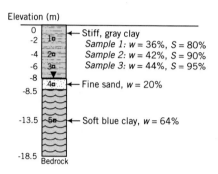

Figure P6.21

6.22 (a) Plot the distribution of the at-rest lateral total stress and lateral effective stress for the soil profile shown in Figure P6.21 if K_o for the sand is 0.5 and K_o for both clays is 0.5. (b) Explain why the lateral stresses for the soil above the groundwater may not be correct?

6.23 (a) Plot the vertical effective stress with depth along section A–B shown in Figure P6.23. (b) If the river level were to rise by 1 m, would the vertical effective stress distribution change? Explain your answer. (c) If the river level were to drop by 1 m, how much would the vertical effective stress at B change?

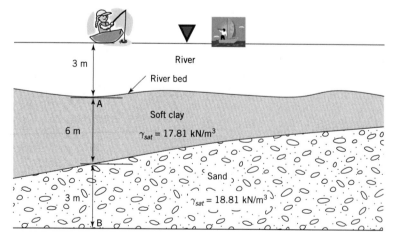

Poorly graded gravel with sand

Figure P6.23

6.24 (a) Plot the at-rest lateral total stress and lateral effective stress with depth that the soil will impose on the back face of the retaining wall shown in Figure P6.24. The soil at the front face of the wall is not shown in Figure P6.24. The angle of internal friction of the soil is $\phi'_{cs} = 30°$ and its saturated unit weight is 19.9 kN/m³. If the groundwater were to drop below the bottom the wall, plot the at-rest lateral total stress and the lateral effective stress. (c) Would the at-rest lateral total stress and lateral effective stress increase or decrease along the wall? Explain your answer.

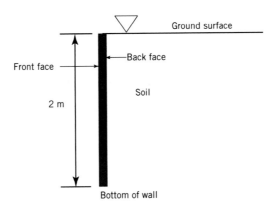

Figure P6.24

Chapter 7
Soil Settlement

7.1 INTRODUCTION

It is practically impossible to prevent soil settlement. For any geosystem, we have to make an estimate of the amount and rate of settlement during construction and over the lifetime of the system. Geosystems may settle uniformly or nonuniformly. The latter condition is called differential settlement and is often the crucial design consideration. In this chapter, we will consider one-dimensional settlement of soils.

Learning outcomes

When you complete this chapter, you should be able to do the following:

- Understand the types of settlement that can occur in soils.
- Know the differences in settlement between coarse-grained and fine-grained soils.
- Have a basic understanding of soil consolidation under vertical loads.
- Be able to calculate the amount and time rate of settlement for soils.

7.2 DEFINITIONS OF KEY TERMS

Elastic settlement is the settlement of a geosystem that can be recoverable upon unloading

Consolidation is the time-dependent settlement of soils resulting from the expulsion of water from the soil pores.

Primary consolidation is the change in volume of a fine-grained soil caused by the expulsion of water from the voids and the transfer of stress from the excess porewater pressure to the soil particles.

Soil Mechanics Fundamentals, First Edition. Muni Budhu.
© 2015 John Wiley & Sons, Ltd. Published 2015 by John Wiley & Sons, Ltd.
Companion website: www.wiley.com\go\budhu\soilmechanicsfundamentals

Secondary compression is the change in volume of a fine-grained soil caused by the adjustment of the soil fabric (internal structure) after primary consolidation has been completed.

Excess porewater pressure (Δu) is the porewater pressure in excess of the current equilibrium porewater pressure. For example, if the porewater pressure in a soil is u_0 and a load is applied to the soil so that the existing porewater pressure increases to u_1, then the excess porewater pressure is $\Delta u = u_1 - u_0$.

Drainage path (H_{dr}) is the longest vertical path that a water particle will take to reach the drainage surface.

Past maximum vertical effective stress (σ'_{zc}) is the maximum vertical effective stress that a soil was subjected to in the past.

Normally consolidated soil is one that has never experienced vertical effective stresses greater than its current vertical effective stress ($\sigma'_{zo} = \sigma'_{zc}$).

Overconsolidated soil is one that has experienced vertical effective stresses greater than its existing vertical effective stress ($\sigma'_{zo} = \sigma'_{zc}$).

Overconsolidation ratio (OCR) is the ratio by which the current vertical effective stress in the soil was exceeded in the past ($OCR = \sigma'_{zc}/\sigma'_{zo}$).

Compression index (C_c) is the slope of the normal consolidation line in a plot of the logarithm of vertical effective stress versus void ratio.

Unloading/reloading index or *recompression index (C_r)* is the average slope of the unloading/reloading curves in a plot of the logarithm of vertical effective stress versus void ratio.

Modulus of volume compressibility (m_v) is the slope of the curve between two stress points in a plot of vertical effective stress versus vertical strain.

Elastic or Young's modulus (E) is the slope of the stress–strain response of an elastic soil.

7.3 BASIC CONCEPT

The time rate of settlement of coarse-grained and fine-grained soils is different. Free draining, coarse sand and gravel with fines <5% generally have good drainage qualities (high hydraulic conductivity), so any excess porewater pressure developed by loading can easily dissipate relatively (relative to fine-grained soils) quickly. Thus, most of the settlement of these soils occurs during construction (short-term loading condition). Fine-grained soils have poor drainage qualities (low hydraulic conductivity). Excess porewater pressures developed during loading can take decades to dissipate. Thus, the settlement of fine-grained soils is time dependent and occurs over the life of the geosystems (long-term loading condition). The time dependent settlement or densification of soils, essentially fine-grained soils, by the expulsion of water from the voids is called consolidation. Recall that compaction is the densification of soils by the expulsion of air. Coarse sand with fines >10%, fine sand and medium sand are not free-draining. Settlement in these soils can occur well beyond the construction period.

Settlement is divided into rigid body or uniform settlement (Figure 7.1), tilt or distortion (Figure 7.1b), and nonuniform settlement (Figure 7.1c). Most damage from uniform settlement is limited to surrounding drainage systems, attached buildings, and utilities. Nonuniform settlement, also called differential settlement, may lead to serious structural problems such as cracking and structural distress in members. Distortion is caused by differential settlement. Distortion is quantified by the ratio δ/l where δ is the maximum differential settlement and l is the length over which the maximum differential settlement occurs. Thus, distortion is an angular measurement (radians) and is often referred to as angular distortion. Limiting values vary from $l/150$ to $l/5000$ depending on the type of structure. While

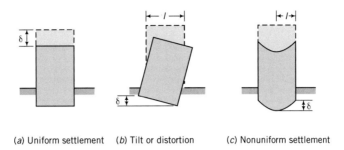

(a) Uniform settlement (b) Tilt or distortion (c) Nonuniform settlement

Figure 7.1 Types of settlement.

distortion may not cause serious structural distress, it can cause the building to be unserviceable.

Because of the variability of soils and the complexity of their behavior, it is difficult to estimate settlement unless simplifying assumptions are made. One of these assumptions is that the soil is an elastic material. An elastic soil is an idealization. Elastic materials return to their initial geometry upon unloading. A linear elastic soil is one that has a linear stress–strain relationship (Figure 7.2). For one-dimensional loading, say in the vertical (z) direction, the slope of line in Figure 7.2 gives Young's modulus E. Young's modulus then relates the normal stress (σ_z) to the normal strain (ε_z). A nonlinear elastic soil is one in which the normal stress–normal strain response is not linear (Figure 7.2). Two Young's moduli are interpreted from the response shown in Figure 7.2. One, the initial Young's modulus, E, is the initial slope of the nonlinear normal stress–normal strain response. The other is the secant Young's modulus, E_{sec}, which is the slope of a line linking the origin of the normal stress–normal strain response to a desired normal stress or normal strain level. Usually, the maximum normal stress or a normal strain of 1% is used. Traditionally, E_{sec} is used in practice.

The shear modulus, G, links shear stress (τ) to shear strain (γ) (Figure 7.3). Although, we will not be using G in this chapter, G is related to E, and this relationship allows you to estimate E if you know G. Recall from your Mechanics of Materials course that shear distorts a material. Similar to the interpretation of Young's modulus, the shear modulus for a linear elastic soil is G and the secant shear modulus is G_{sec}. The shear modulus is related to Young's modulus as

$$G = \frac{E}{2(1+\nu)} \tag{7.1}$$

The assumption of elastic behavior allows us to calculate settlement from just knowing the Young's modulus and Poisson's ratio, ν. When the letter E or ν is followed by a prime, it denotes effective stress condition. Table 7.1 and Table 7.2 give typical values of Young's moduli, shear moduli, and Poisson's ratio (based on effective stress) for soils. These values are intended for problem solving in this textbook.

Presently, the estimation of soil settlement, especially for coarse-grained soils, is based mainly on empirical or semi-empirical relationships, and requires significant field experience.

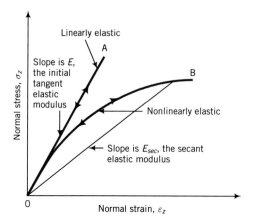

Figure 7.2 Normal stress–normal strain curves of linear and nonlinear elastic materials.

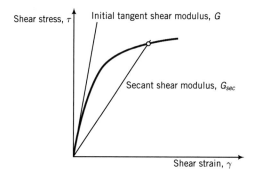

Figure 7.3 Shear stress–shear strain curve of a nonlinear elastic material.

Table 7.1 Typical values of E'_{sec} and G_{sec}.

Soil type	Description	E'_{sec} (MPa)	G_{sec} (MPa)
Clay	Soft	1–15	0.5–5
	Medium	15–50	5–15
	Stiff	50–100	15–40
Sand	Loose	10–20	2–10
	Medium	20–50	10–15
	Dense	50–100	15–40
Gravel	Loose	20–75	2–20
	Medium	75–100	20–40
	Dense	100–200	40–75
Estimated from standard penetration test (SPT)*		E'_{sec} (MPa)	G_{sec} (MPa)
Gravels and gravels with sands		$1.2N_{60}$*	$0.45N_{60}$
Coarse sands, sands with gravel (<10%)		$0.95N_{60}$	$0.35N_{60}$
Fine and medium sands, clean and with fines (<10%)		$0.7N_{60}$	$0.25N_{60}$
Silts and sandy silt		$0.4N_{60}$	$0.15N_{60}$

Continued

Estimated from cone penetrometer test (CPT)*	E'_{sec} (kPa)	G_{sec} (kPa)
Fine and medium sands, clean and with fines (<10%)	$3q_c^*$	$1.2q_c$
Clayey silt and silty sand	$5q_c$	$2q_c$
Clays	$7q_c$	$2.5q_c$

Note: *N_{60} is the SPT N values corrected for 60% energy (see Chapter 3); q_c is the cone tip resistance in kPa.

Table 7.2 Typical values of Poisson's ratio.

Soil type	Description	ν'
Clay	Soft	0.35–0.4
	Medium	0.3–0.35
	Stiff	0.2–0.3
Sand	Loose	0.15–0.25
	Medium	0.25–0.3
	Dense	0.25–0.35

Note: For all soils at constant volume, $\nu = 0.5$ (total stress condition).

7.4 SETTLEMENT OF FREE-DRAINING COARSE-GRAINED SOILS

The settlement of free draining coarse-grained soils (e.g., medium sand with fines less than 5%, clean, coarse sand) is generally calculated assuming that these soils behave as elastic materials. For a rectangular flexible loaded area with a uniform surface stress, q_s, the settlement of a homogeneous coarse-grained soil (constant E value) can be calculated (Giroud, 1968) from

$$\rho_e = \frac{q_s B (1-(v')^2)}{E'} I_s \tag{7.2}$$

where ρ_e is the elastic settlement, E' is Young's modulus based on effective stresses, v' is Poisson's ratio based on effective stresses, and I_s is a settlement influence factor that is a function of the L/B ratio (L is length and B is width of the foundation. I_s can be approximated as

$$\text{Center of the rectangle:} \quad I_s \approx 0.62\ln\left(\frac{L}{B}\right)+1.12 \tag{7.3}$$

$$\text{Corner of the rectangle:} \quad I_s \approx 0.31\ln\left(\frac{L}{B}\right)+0.56 \tag{7.4}$$

The vertical elastic settlement due to a circular flexible loaded area with a uniform surface stress, q_s, is

$$\rho_e = \frac{q_s D (1-(v')^2)}{E'} I_{ci} \tag{7.5}$$

$$\text{Center of the circular area:} \quad I_{ci} = 1 \tag{7.6}$$

$$\text{Edge of circular area:} \quad I_{ci} = \frac{2}{\pi} \tag{7.7}$$

where D is the diameter of the loaded area. In practice, Equations (7.2) and (7.5) have to be modified to account for nonelastic behavior, variations of Young's modulus with depth, foundation stiffness, and embedment. The most important parameter for elastic settlement is Young's modulus. However, this parameter is difficult to obtain with good accuracy for most soils especially coarse-grained soils. A plethora of empirical expressions have been proposed to estimate Young's modulus from field test data especially the SPT and the cone penetrometer tests (see Chapter 3 for description of these tests and Table 7.1 for some of the suggested relationships). These expressions give large variation in the estimation of the elastic modulus. For practical applications, experience is needed to judge a reasonable value E from empirical expressions. Also, the application of Equations (7.2) and (7.5) to estimate the settlement of foundations (structures used to transfer structural loads to the soil) requires modifications for such factors as the relative stiffness of the foundation (e.g., concrete) to the soil, and for the variation of the soil's Young's modulus with depth. In practice, the secant Young's modulus is used in Equations (7.2) and (7.5).

EXAMPLE 7.1   *Estimating Elastic Settlement of a Rectangular Loaded Area on a Sand*

A column is pinned to a concrete slab 1 m × 1 m that serves as the foundation. The foundation rests on the surface of a sand with $E'_{sec} = 40$ MPa and $\nu' = 0.35$. The vertical load from the column and the self-weight of the foundation is 75 kN. Estimate the settlement under the center of the foundation.

Strategy The solution is a direct application of Equation (7.2).

Solution 7.1

Step 1:   Determine the elastic settlement.

$$\text{Center of the rectangle: } I_s \approx 0.62 \ln\left(\frac{L}{B}\right) + 1.12 \approx 0.62 \ln\left(\frac{2}{2}\right) + 1.12 \approx 1.12$$

$$\rho_e = \frac{q_s B (1 - (\nu')^2)}{E'_{sec}} I_s = \frac{[75/(1 \times 1)] \times 1(1 - 0.35^2)}{40{,}000} \times 1.12 = 1.8 \times 10^{-3} \text{ m} = 1.8 \text{ mm}$$

7.5   SETTLEMENT OF NON–FREE-DRAINING SOILS

There are two common modes of settlement of non–free-draining soils (fine-grained soils, fine sand, and medium sand with fines greater than 10%) of importance to geoengineers. One is the natural drainage of water from the soil due to a hydraulic gradient. This is called self-weight consolidation. An example is the natural deposition process of sediments and the passage of time with drainage of the water from the voids. The other is the application of a load that immediately increases the porewater pressure above its current value. This is called excess porewater pressure. When this excess porewater pressure drains from the soil, the soil fabric is forced into a denser configuration. Geoengineers are particularly concerned with the vertical movement (settlement) because the footprints of human-made structures generally apply loads over a very small portion of the Earth's surface. For example, a column transmitting loads from a building to the soil may do so through a concrete slab that is only say 1 m × 1 m in plan. This is extremely small compared with the lateral dimensions of the

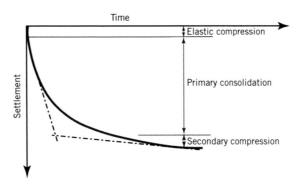

Figure 7.4   Illustration of settlement of fine-grained soils.

Earth's surface. Any displacement in the lateral direction will result in very small lateral strains that geoengineers can neglect. Of course, there are exceptions such as excavations and embankments.

The settlement of non–free-draining soils consists of three parts (Figure 7.4).

1.   Elastic compression (short term; occurs during construction).
2.   Primary consolidation (long term; occurs during the design life of the structure).
3.   Secondary compression or creep (long term; occurs during the design life of the structure but more pronounced at the late stages of primary consolidation and beyond).

The elastic compression can be calculated using Equation (7.2) or Equation (7.5). The parameters to estimate settlement from primary consolidation and secondary compression are found from a one-dimensional consolation test.

7.6 THE ONE-DIMENSIONAL CONSOLIDATION TEST

This test is generally performed on fine-grained soils. As a summary, a disk of soil is enclosed in a stiff metal ring and placed between two porous stones in a cylindrical container filled with water, as shown in Figure 7.5a. A metal load platen mounted on top of the upper porous stone transmits the applied vertical stress (vertical total stress) to the soil sample. Both the metal platen and the upper porous stone can move vertically inside the ring as the soil settles under the applied vertical stress. The ring containing the soil sample can be fixed to the container by a collar (fixed ring cell, Figure 7.5b) or is unrestrained (floating ring cell, Figure 7.5c).

Incremental loads, including unloading sequences, are applied to the platen, and the settlement of the soil at various fixed times under each load increment is measured by a displacement gauge. Each load increment is allowed to remain on the soil until the change in settlement is negligible, and the excess porewater pressure developed under the current load increment has dissipated. For many soils, this usually occurs within 24 hours, but longer monitoring times may be required for exceptional soil types, for example, montmorillonite. Each load increment is doubled. The ratio of the load increment to the previous load is called the load increment ratio (LIR); conventionally, $LIR = 1$. To determine soil rebound (uplift), the soil sample is unloaded using a load decrement ratio—load decrement divided by current load—of 2.

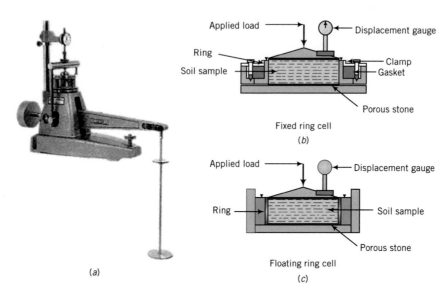

Figure 7.5 (a) A typical consolidation apparatus, (b) a fixed ring cell, and (c) a floating ring cell. (Photo courtesy of Geotest.)

At the end of the consolidation test, the apparatus is dismantled, and the water content of the sample is determined. It is best to unload the soil sample to a small pressure before dismantling the apparatus because, if you remove the final consolidation load completely, a negative excess porewater pressure that equals the final consolidation pressure will develop. This negative excess porewater pressure can cause water to flow into the soil and increase the soil's water content. Consequently, the final void ratio calculated from the final water content will be erroneous.

The data obtained from the one-dimensional consolidation test are as follows:

1. Initial height of the soil, H_o, which is fixed by the height of the ring.
2. Current height of the soil at various time intervals under each loading (time–settlement data).
3. Water content at the beginning and at the end of the test, and the dry weight of the soil at the end of the test.

Let us consider what happens when we load the soil in the one-dimensional consolidation test. A simulation of the sample and boundary condition is shown in Figure 7.6a. The porous stones are used to facilitate drainage of the porewater from the top and bottom faces of the soil. The top half of the soil will drain through the top porous stone and the bottom half of the soil will drain through the bottom porous stone. A platen on the top porous stone transmits applied loads to the soil. Expelled water is transported by plastic tubes to a burette. A valve is used to control the flow of the expelled water into the burette. Three porewater pressure transducers are mounted in the side wall of the cylinder to measure the excess porewater pressure near the porous stone at the top (A), at a distance of one-quarter the height (B), and at mid-height of the soil (C). Excess porewater pressure is the additional porewater pressure induced in a soil mass by loads. A displacement gauge with its stem on the platen keeps track of the vertical settlement of the soil.

We will assume that the porewater and the soil particles are incompressible, and the initial porewater pressure is zero. The volume of excess porewater that drains from the soil is then

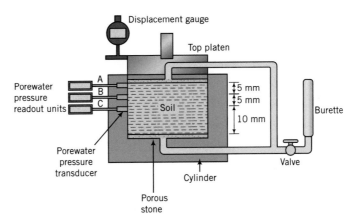

Figure 7.6a Experimental setup for illustrating basic concepts on consolidation.

a measure of the volume change of the soil resulting from the applied loads. Since the sidewall of the container is rigid, no radial displacement can occur. The lateral and circumferential strains are then equal to zero and the volumetric strain is equal to the vertical strain, $\varepsilon_z = \Delta H / H_o$, where ΔH is the change in height or thickness and H_o is the initial height or thickness of the soil.

7.6.1 Drainage path

The distance of the longest vertical path taken by a particle to exit the soil is called the length of the drainage path. Because we allowed the soil to drain on the top and bottom faces (double drainage), the length of the drainage path, H_{dr}, is

$$H_{dr} = \frac{H_{av}}{2} = \frac{H_o + H_f}{4} \tag{7.8}$$

where H_{av} is the average height and H_o and H_f are the initial and final heights, respectively, under the current loading. If drainage is permitted from only one face of the soil, then $H_{dr} = H_{av}$. A short drainage path will cause the soil to complete its settlement in a shorter time than a long drainage path. You will see later that, for single drainage, our soil sample will take four times longer to reach a particular settlement than for double drainage.

7.6.2 Instantaneous load

Let us now apply a load P to the soil through the load platen and keep the valve closed. Since no excess porewater can drain from the soil, the change in volume of the soil is zero ($\Delta V = 0$) and no load or stress is transferred to the soil particles ($\Delta\sigma_z' = 0$). The porewater carries the total load. The initial excess porewater pressure in the soil (Δu_o) is then equal to the change in applied vertical stress, $\Delta\sigma_z = P/A$, where A is the cross-sectional area of the soil. For our thin soil layer, we will assume that the initial excess porewater pressure is distributed uniformly with depth so that at every point in the soil layer, the initial excess porewater pressure is equal to the applied vertical stress. For example, if $\Delta\sigma_z = 100\,\text{kPa}$, then $\Delta u_o = 100\,\text{kPa}$, as shown in Figure 7.6b.

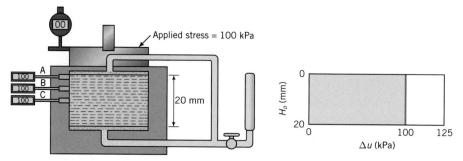

Figure 7.6b Instantaneous or initial excess porewater pressure when a vertical load is applied.

7.6.3 Consolidation under a constant load: primary consolidation

Let us now open the valve and allow the initial excess porewater to drain. The total volume of soil at time t_1 decreases by the amount of excess porewater that drains from it, as indicated by the change in volume of water in the burette (Figure 7.6c). At the top and bottom of the soil sample, the excess porewater pressure is zero because these are the drainage boundaries. The decrease of initial excess porewater pressure at the middle of the soil (position C) is the slowest because a water particle must travel from the middle of the soil to either the top or bottom boundary to exit the system. The distribution of excess porewater pressure at any time t_1 is called an isochrone (a Greek word meaning equal time).

You might have noticed that the settlement of the soil (ΔH) with time t (Figure 7.6c) is not linear. A significant amount of settlement occurs shortly after the valve is opened. The rate of settlement, $\Delta H/t$, is also much faster soon after the valve is opened compared with later times. Before the valve is opened, an initial hydraulic head, $\Delta u_o/\gamma_w$, is created by the applied vertical stress. When the valve is opened, the initial excess porewater is forced out of the soil by this initial hydraulic head. With time, the initial hydraulic head decreases and, consequently, smaller amounts of excess porewater are forced out. An analogy can be drawn with a pipe containing pressurized water that is ruptured. A large volume of water gushes out as soon as the pipe is ruptured, but soon after, the flow becomes substantially reduced. We will call the initial settlement response soon after the valve is opened the early time response, or primary consolidation. Primary consolidation is the change in volume of the soil caused by the expulsion of water from the voids and the transfer of load from the excess porewater pressure to the soil particles. In general, the soil will undergo an initial elastic compression and then primary consolidation occurs.

7.6.4 Effective stress changes

Since the applied vertical stress (vertical total stress) remains constant, then according to the principle of effective stress ($\Delta\sigma_z' = \Delta\sigma_z - \Delta u$), any reduction of the initial excess porewater pressure must be balanced by a corresponding increase in vertical effective stress. Increases in vertical effective stresses lead to soil settlement caused by changes to the soil fabric. As time increases, the initial excess porewater continues to dissipate and the soil continues to settle (Figure 7.6c). At time $t = t_1$ in Figure 7.6c the excess porewater pressure remaining in the middle of the soil sample is 60 kPa. The change in effective stress at the middle of the soil sample is $100 - 60 = 40$ kPa.

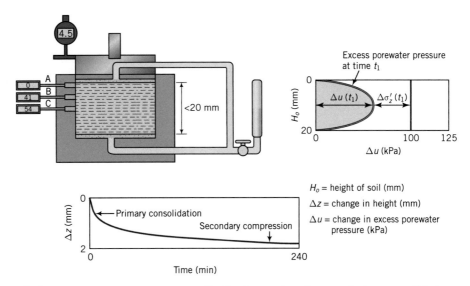

Figure 7.6c Excess porewater pressure distribution and settlement during consolidation.

After some time, usually within 24 hours for many small soil samples tested in the laboratory, the initial excess porewater pressure in the middle of the soil reduces to approximately zero, and the rate of decrease of the volume of the soil becomes very small. When the initial excess porewater pressure becomes zero, then, from the principle of effective stress, all of the applied vertical stress is transferred to the soil; that is, the vertical effective stress is equal to the vertical total stress ($\Delta\sigma_z' = \Delta\sigma_z$).

When the excess porewater pressure from a loading has dissipated, another increment of loading is applied and the process repeated until the applied load is approximately 1.5 to 2 times the in situ effective stress (overburden pressure) or higher. Figure 7.7 shows the typical response of soils during loading in a one-dimensional consolidation test. The line AB is called the primary consolidation line or normal consolidation line, NCL. Figure 7.7b is same data from Figure 7.7a plotted with the abscissa as logarithmic scale because the range of vertical stresses is generally large.

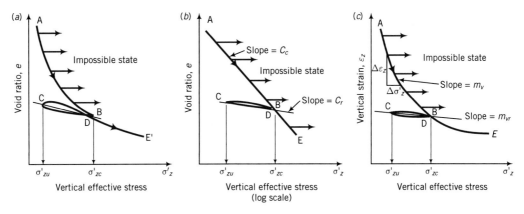

Figure 7.7 Three plots of settlement data from soil consolidation.

When the soil is unloaded incrementally, it expands (increase in void ratio) as shown by BC in Figure 7.7. Usually, the soil is unloaded to its original overburden pressure and then reloaded past the prior maximum vertical effective stress during loading (CDE in Figure 7.7). A hysteresis unloading/reloading loop is typical observed. This is likely due to the nonlinearities of soil behavior. In practice, a straight line CD approximates the unloading/reloading loop. The line CD is called the unloading/reloading line, URL.

When the past maximum vertical effective stress during loading is exceeded, the vertical effective stress-void ratio response follows the NCL. The loading-unloading-reloading response will repeat itself at any further past maximum vertical effective stress.

7.6.5 Effects of loading history

During reloading the soil follows the normal consolidation line when the past maximum vertical effective stress is exceeded. The history of loading of a soil is locked in its fabric, and the soil maintains a memory of the past maximum effective stress. To understand how the soil will respond to loads, we have to unlock its memory. If a soil were to be consolidated to effective stresses below its past maximum vertical effective stress, the settlement would be small because the soil fabric was permanently changed by a higher stress in the past. However, if the soil were to be consolidated beyond its past maximum vertical effective stress, the settlement would be large because the soil fabric would now undergo further change from a current loading that is higher than its past maximum effective stress. The ratio of the past maximum vertical effective stress, σ'_{zc}, to the in situ vertical effective stress (overburden pressure), σ'_{zo}, is called the overconsolidation ratio.

The overconsolidation ratio or degree of overconsolidation is defined as

$$OCR = \frac{\sigma'_{zc}}{\sigma'_{zo}} \tag{7.9}$$

A soil with an $OCR = 1$ is normally consolidated That is, the in situ vertical effective stress or overburden effective stress is about equal to the past maximum vertical effective stress. Normally consolidated soils follow paths similar to ABE (Figure 7.7). If the past maximum vertical effective stress is greater than the current vertical effective stress or overburden pressure, the soil is overconsolidated soil. An overconsolidated soil will follow a vertical effective stress versus void ratio path similar to CDE (Figure 7.7) during loading. The overconsolidation ratio of soils has been observed to decrease with depth, eventually reaching a value of 1 (normally consolidated state).

The practical significance of loading history is that if the loading imposed on the soil by a structure is such that the vertical effective stress in the soil does not exceed its past maximum vertical effective stress, the settlement of the structure would be small; otherwise, significant permanent settlement would occur. The past maximum vertical effective stress defines the approximate limit of elastic behavior. For stresses that are lower than the past maximum vertical effective stress, the soil will follow the URL, and we can reasonably assume that the soil will behave like an elastic material. For stresses greater than the past maximum vertical effective stress, the soil would behave like an elastoplastic material.

7.6.6 Effects of soil unit weight or soil density

The graphs shown in Figure 7.7 are traditionally used in practice to obtain soil parameters to estimate soil settlement from consolidation. Rather than void ratio, we can consider the unit weight ratio, $R_d = \gamma_{sat}/\gamma_w$ (see Chapter 2). There are four primary advantages of considering unit weight rather than void ratio. First, unit weight is a physical property of all matter (gases, liquids, solids, plasma). Therefore, we can treat fine-grained and coarse-grained soils within the same conceptual framework. Second, the unit weight of soils can be readily determined. Third, unit weight is related to both strength and settlement. Fourth, we can consider the volume change in the soil rather than a one-dimensional change.

The data for the graphs in Figure 7.7 can be used to show how R_d varies with loading as depicted in Figure 7.8. If the soil is dense or overconsolidated, the densification of the soil (increase in unit weight from decrease in voids), *ab* in Figure 7.8, will be small until the past maximum vertical effective stress is reached. After that, the densification from the applied stress and the drainage of excess porewater pressure will be comparatively large, *bc* in Figure 7.8. Throughout the process from *a* to *c*, new structural arrangements of the soil particles (soil fabrics) are created to resist the applied vertical stresses. As the void spaces decrease, there is less opportunity for the soil particles to move around and, consequently, lower chance for new soil fabrics to develop. Thus, as the soil approaches the upper unit weight ratio (equals the specific gravity of the solids), the soil fabric would reach a critical configuration (approximately at *c*) that only allows small changes under increased applied vertical stresses. At any point of unloading between *b* and *c*, the unit weight ratio will decrease as the soil mobilizes a different soil fabric from stress relaxation.

For a normally consolidated soil, R_d increases continuously from its initial value, $R_d = (\gamma_{sat})_o/\gamma_w$, where $(\gamma_{sat})_o$ is the initial (in situ) unit weight, until the soil fabric reaches a critical configuration. From *a* or *o* to *c*, the soil strength increases. If the soil is sheared (to be considered in Chapter 8) at any vertical effective (consolidation) stress level between *a*

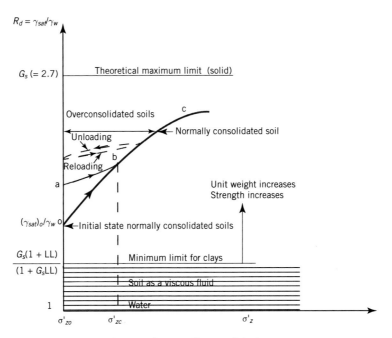

Figure 7.8 Unit weight ratio changes from soil consolidation.

or o and c, soil fabrics different from consolidation will be created and the soil strength mobilized will be different. For clays, the minimum value of R_d is

$$(R_d)_{min} = \frac{G_s(1+LL)}{(1+G_sLL)}$$

For example, if LL is 60% and $G_s = 2.7$, $(R_d)_{min} = 1.64$. Below $(R_d)_{min}$, the soil will behave as a viscous fluid.

All overconsolidated soils will have unit weight ratio–vertical effective stress states that lie to the left of the normally consolidated line shown in Figure 7.8. The consolidation response of a particular soil will be bounded by the upper and minimum value of R_d.

7.6.7 Determination of void ratio at the end of a loading step

In the one-dimensional consolidation test, the water content (w) of the soil sample is determined. Using these data, initial height (H_o), and the specific gravity (G_s) of the soil sample, you can calculate the void ratio for each loading step as follows:

1. Calculate the final void ratio, $e_{fin} = wG_s$, where w is the water content determined at the end of the test.
2. Calculate the total consolidation settlement of the soil sample during the test, $(\Delta H)_{fin} = H_{fin} - H_i$, where H_{fin} is the final displacement gauge reading and H_i is the displacement gauge reading at the start of the test.
3. Back-calculate the initial void ratio, using Equation (7.10), as

$$e_o = \frac{e_{fin} + [(\Delta H)_{fin}/H_o]}{1 - [(\Delta H)_{fin}/H_o]} \tag{7.10}$$

4. Calculate e for each loading using

$$e_i = e_o - \frac{(\Delta H)_i}{H_o}(1+e_o) \tag{7.11}$$

where e_i is the void ratio and $(\Delta H)_i$ is the change in height for the ith loading.

7.6.8 Determination of compression and recompression indexes

The slope of the normal consolidation line, AB, in Figure 7.7b gives the compression index, C_c (dimensionless), and the slope of the line CD gives the recompression index, C_r (dimensionless). Thus

$$C_c = -\frac{e_2 - e_1}{\log[(\sigma_z')_2/(\sigma_z')_1]} \quad \text{(no units)} \tag{7.12}$$

and

$$C_r = -\frac{e_4 - e_3}{\log[(\sigma_z')_4/(\sigma_z')_3]} \quad \text{(no units)} \tag{7.13}$$

where the subscripts 1 and 2 denote two arbitrarily selected points on the NCL, line AB (Figure 7.7b) and 3 and 4 denote two arbitrarily selected points on the URL, line CD (Figure 7.7b).

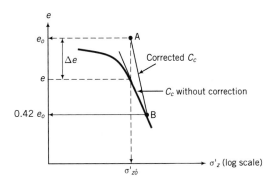

Figure 7.9 Schmertmann's method to correct C_c for soil disturbances.

Disturbances from sampling and sample preparation tend to decrease C_c. Schmertmann (1953) suggested a correction to the laboratory curve to obtain a more representative in situ value of C_c. His method is as follows. Locate a point A at coordinate (σ'_{zo}, e_o) and a point B at ordinate $0.42e_o$ on the laboratory e versus σ'_z (log scale) curve, as shown in Figure 7.9. The slope of the line AB is the corrected value for C_c. Typical ranges of values for C_c and C_r are given in Table 7.3 in Section 7.12.

7.6.9 *Determination of the modulus of volume change*

The modulus of volume compressibility, m_v, is found from plotting the vertical strain, $\varepsilon_z = \Delta H/H_o$ versus the vertical effective stress (Figure 7.7c) and determining the slope, as shown in this figure. You do not need to calculate void ratio to determine m_v. You need the final change in height at the end of each loading (ΔH), and then you calculate the vertical strain, $\varepsilon_z = \Delta H/H_o$ where H_o is the initial height. In the one-dimensional consolidation test, $\varepsilon_z = \Delta H/H_o = \Delta e/(1 + e_o)$. The modulus of volume compressibility is not constant but depends on the range of vertical effective stress that is used in the calculation. A representative value for m_v can be obtained by finding the slope between the current vertical effective stress and the final vertical effective stress $(\sigma'_{zo} + \Delta\sigma_z)$ at the center of the soil layer in the field or 100 kPa, whichever is less.

The modulus of volume recompressibility, m_{vr}, is expressed as (Figure 7.7c)

$$m_{vr} = -\frac{(\varepsilon_z)_2 - (\varepsilon_z)_1}{(\sigma'_z)_2 - (\sigma'_z)_1} \left(\frac{\text{m}^2}{\text{kN}}\right) \tag{7.14}$$

where the subscripts 1 and 2 denote two arbitrarily selected points on the URL. From Hooke's law, the constrained Young's modulus is

$$E'_c = \frac{\Delta\sigma'_z}{\Delta\varepsilon_z} = \frac{E'(1-v')}{(1+v')(1-2v')} \tag{7.15}$$

where the subscript c denotes constrained because we are constraining the soil to settle only in one direction (one-dimensional consolidation), E' is Young's modulus (unit: kPa or MPa) based on effective stresses, and v' is Poisson's ratio based on effective stresses. We can rewrite Equation (7.15) as

$$E'_c = \frac{1}{m_{vr}} \tag{7.16}$$

7.6.10 Determination of the coefficient of consolidation

The rate of consolidation for a homogeneous fine-grained soil depends on its hydraulic conductivity (permeability), the thickness, and the length of the drainage path. As the hydraulic conductivity decreases, the soil will take longer to drain the initial excess porewater, and settlement will proceed at a slower rate. A measure of the rate of consolidation is the coefficient of consolidation, C_v (unit: cm²/min or m²/yr). From the theory of one-dimensional consolidation (Terzaghi, 1925; see Appendix A), the coefficient of consolidation is expressed as

$$T_v = \frac{C_v t}{H_{dr}^2} \tag{7.17}$$

where T_v is known as the time factor; it is a dimensionless term. The time factor is related to the average degree of consolidation, U, which is the average excess porewater pressure dissipated (change in effective stress, $\Delta\sigma_z'$) divided by the initial excess porewater pressure (Δu_o) or the settlement that occurred divided by the expected settlement when all the excess porewater pressure dissipated.

$$U = \left(1 - \frac{\Delta u}{\Delta u_o}\right) = \frac{\Delta\sigma_z'}{\Delta u_o} \tag{7.18}$$

where Δu is the excess porewater pressure at time t.

The theoretical relationships between T_v and U for a uniform excess porewater pressure distribution and a triangular excess porewater pressure distribution are shown in Figure 7.10. A convenient set of equations for double drainage and uniform excess porewater pressure distribution, found by the curve fitting Figure 7.10, is

$$T_v = \frac{\pi}{4}\left(\frac{U}{100}\right)^2 \quad \text{for } U < 60\% \tag{7.19}$$

and

$$T_v = 1.781 - 0.933\log(100 - U) \quad \text{for } U \geq 60\% \tag{7.20}$$

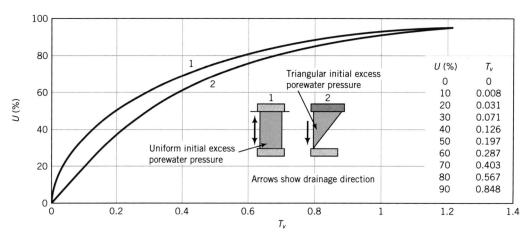

Figure 7.10 Relationship between time factor and average degree of consolidation for a uniform distribution and a triangular distribution of initial excess porewater pressure.

The time factor corresponding to every 10% of average degree of consolidation for double drainage conditions is shown in the inset table in Figure 7.10. The time factors corresponding to 50% and 90% consolidation are often used in interpreting consolidation test results. You should remember that $T_v = 0.848$ for 90% consolidation, and $T_v = 0.197$ for 50% consolidation.

There are two popular methods that can be used to calculate C_v. Taylor (1942) proposed one method, called the root time method. Casagrande and Fadum (1940) proposed the other method, called the log time method. The root time method utilizes the early time response, which theoretically should appear as a straight line in a plot of square root of time versus displacement gauge reading.

7.6.10.1 Root time method (square root time method)

Let us arbitrarily choose a point C on the displacement versus square root of time factor gauge reading, as shown in Figure 7.11. We will assume that this point corresponds to 90% consolidation ($U = 90\%$) for which $T_v = 0.848$ (Figure 7.11). If point C were to lie on a straight line, the theoretical relationship between U and T_v would be $U = 0.98\sqrt{T_v}$; that is, if you substitute $T_v = 0.848$, you get $U = 90\%$.

At early times, the theoretical relationship between U and T_v is given by Equation (7.19); that is,

$$U = \sqrt{\frac{4}{\pi}T_v} = 1.13\sqrt{T_v}, \quad U < 0.6$$

The straight line OA in Figure 7.11 represents the laboratory early time response. You should note that O is below the initial displacement gauge reading because there is an initial

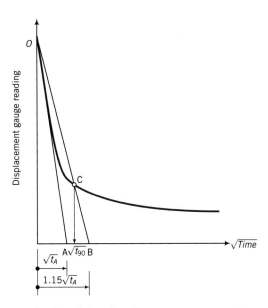

Figure 7.11 Root time method to determine C_v.

compression of the soil before consolidation begins. This compression can be due to the breaking of particle bonds in lightly cemented soils. The ratio of the gradient of OA and the gradient of the theoretical early time response, line OCB, is

$$\frac{1.13\sqrt{T_v}}{0.98\sqrt{T_v}} = 1.15$$

We can use this ratio to establish the time when 90% consolidation is achieved in the one-dimensional consolidation test.

The procedure, with reference to Figure 7.11, is as follows:

1. Plot the displacement gauge readings versus square root of times.
2. Draw the best straight line through the initial part of the curve intersecting the ordinate (displacement reading) at O and the abscissa ($\sqrt{\text{time}}$) at A.
3. Note the time at point A; let us say it is $\sqrt{t_A}$.
4. Locate a point B, $1.15\sqrt{t_A}$, on the abscissa.
5. Join OB.
6. The intersection of the line OB with the curve, point C, gives the displacement gauge reading and the time for 90% consolidation (t_{90}). You should note that the value read off the abscissa is $\sqrt{t_{90}}$. Now when $U = 90\%$, $T_v = 0.848$ (Figure 7.10) and from Equation (7.17) we obtain

$$C_v = \frac{0.848 H_{dr}^2}{t_{90}} \tag{7.21}$$

where H_{dr} is the length of the drainage path.

7.6.10.2 Log time method

In the log time method, the displacement gauge readings are plotted against the times (log scale). The logarithm of times is arbitrary and is only used for convenience. A typical curve obtained is shown in Figure 7.12. The theoretical early time settlement response in a plot of logarithm of times versus displacement gauge readings is a parabola. The experimental early time curve is not normally a parabola, and a correction is often required.

The procedure, with reference to Figure 7.12, is as follows:

1. Project the straight portions of the primary consolidation and secondary compression to intersect at A. The ordinate of A, d_{100}, is the displacement gauge reading for 100% primary consolidation.
2. Correct the initial portion of the curve to make it a parabola. Select a time t_1, point B, near the head of the initial portion of the curve ($U < 60\%$) and then another time t_2, point C, such that $t_2 = 4t_1$.
3. Calculate the difference in displacement reading, $\Delta d = d_2 - d_1$ between t_2 and t_1. Plot a point D at a vertical distance Δd from B. The ordinate of point D is the corrected initial displacement gauge reading, d_0, at the beginning of primary consolidation.
4. Calculate the ordinate for 50% consolidation as $d_{50} = (d_{100} + d_0)/2$. Draw a horizontal line through this point to intersect the curve at E. The abscissa of point E is the time for 50% consolidation, t_{50}.
5. You will recall (Figure 7.10) that the time factor for 50% consolidation is 0.197, and from Equation (7.17) we obtain

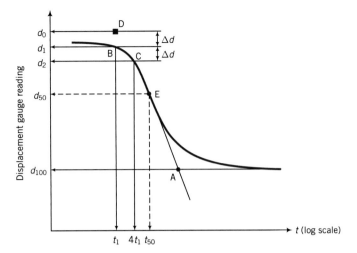

Figure 7.12 Log time method to determine C_v.

$$C_v = \frac{0.197 H_{dr}^2}{t_{50}} \tag{7.22}$$

The log time method makes use of the early (primary consolidation) and later time responses (secondary compression), while the root time method only utilizes the early time response, which is expected to be a straight line. In theory, the root time method should give good results except when nonlinearities arising from secondary compression cause substantial deviations from the expected straight line. These deviations are most pronounced in fine-grained soils with organic materials. Field observations indicate that, in many instances, the predictions of the rate of settlement using C_v from lab test may be as much as 4 times lower than the field. This is because of the complexities of the drainage conditions in the field and the nonlinearities of soil behavior that are not replicated by the simple, linear soil behavior and one dimensional drainage that are used to theoretically represent soil response to loading. Typical values for C_v for some clays are given below in Table 7.5 in Section 7.12.

7.6.11 Determination of the past maximum vertical effective stress

Many methods have been proposed to determine the past maximum vertical effective stress, σ'_{zc}. Three methods will be presented in this textbook. The actual values of σ'_{zc} for real soils are practically impossible to determine using laboratory methods and the procedures of the three methods. Degradation of the soil from its intact condition caused by sampling, transportation, handling, and sample preparation can lead to significant error in estimating σ'_{zc}. The three methods allow for comparison. The difference in results for σ'_{zc} should not exceed about 20%. The Casagrande (1936) method is established in practice but the procedures are subjective.

7.6.11.1 Casagrande's method

From the calculation of e for each loading step, we can plot a graph of the void ratio versus the vertical effective stress (log scale), as shown in Figure 7.13. The log scale is only used

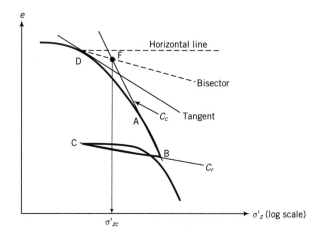

Figure 7.13 Determination of past maximum vertical effective stress using Casagrande's method.

for convenience because the range of vertical effective stresses is generally large. We will call Figure 7.13 the e versus σ'_z (log scale) curve.

The Casagrande procedure, with reference to Figure 7.13, is as follows:

1. Identify the point of maximum curvature, point D, on the initial part of the e versus σ'_z (log scale) curve.
2. Draw a horizontal line through D.
3. Draw a tangent to the e versus σ'_z (log scale) curve at D.
4. Bisect the angle formed by the tangent and the horizontal line at D.
5. Extend backward the straight portion of the e versus σ'_z (log scale) curve (the normal consolidation line), BA, to intersect the bisector line at F.
6. The abscissa of F is the past maximum vertical effective stress, σ'_{zc}.

7.6.11.2 Brazilian method

The identification of the point of maximum curvature in the Casagrande method is subjective. A method, known as the Brazilian method (Pacheco Silva, 1970), gives similar results to the Casagrande method but removes the subjectivity of the point of maximum curvature. The procedure with reference to Figure 7.14 is as follows:

1. Draw a horizontal line, AB, starting at the initial void ratio, e_o,
2. Extend backward the straight portion of the e versus σ'_z (log scale) curve (the normal consolidation line), DC, to intersect the line AB at X.
3. Draw a vertical line to intersect the e versus σ'_z (log scale) curve at E.
4. Draw a horizontal line from E to intersect the line CX at F.
5. The abscissa of F is the past maximum vertical effective stress, σ'_{zc}.

7.6.11.3 Strain energy method

Another method (Becker et al., 1987), called the strain energy or work method, uses the cumulated work per unit volume at the end of each load increment. The work done over an increment of loading is

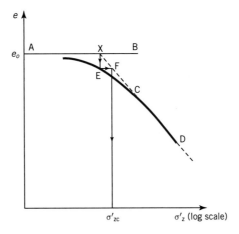

Figure 7.14 Determination of the past maximum vertical effective stress using the Brazilian method.

$$\Delta W = \left| \frac{(\sigma'_z)_{i+1} + (\sigma'_z)_i}{2} \right| \times \left[(\varepsilon_z)_{i+1} - (\varepsilon_z)_i \right] \quad \text{(unit: kN m per unit volume)} \tag{7.23}$$

where $(\sigma'_z)_{i+1}$ and $(\varepsilon_z)_{i+1}$ are the vertical effective stresses and vertical strain at the end of the $i + 1$ increment and $(\sigma'_z)_i$ and $(\varepsilon_z)_i$ are the vertical effective stresses and vertical strain at the end of the i increment. The procedure for the strain energy method is as follows:

1. Calculate the incremental work for each loading step using Equation (7.23).
2. Calculate the cumulative work by summing the incremental work in step 1.
3. Plot the cumulative work (ordinate, arithmetic scale) versus the vertical effective stress (abscissa, arithmetic scale). You would normally get two distinct averaged straight lines (see Figure 7.15).
4. Project the upper averaged straight line to intersect the projection of the lower averaged straight line.
5. The vertical effective stress at the intersection of the two lines in step 4 is the past maximum vertical effective stress, σ'_{zc}.

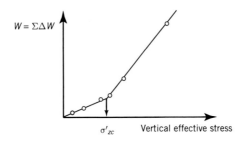

Figure 7.15 Determination of the past maximum vertical effective stress using the strain energy method.

The last two methods have the advantage over Casagrande's method in that subjectivity regarding the maximum curvature is removed. The strain energy method has the additional advantage in that you need not calculate the void ratio to determine the past maximum vertical effective stress.

7.6.12 Determination of the secondary compression index

Theoretically, primary consolidation ends when $\Delta u_o = 0$. The later time settlement response is called secondary compression, or creep. Secondary compression is the change in volume of a fine-grained soil caused by the adjustment of the soil fabric (internal structure) after primary consolidation has been completed. The term consolidation is reserved for the process in which settlement of a soil occurs from changes in effective stresses resulting from decreases in excess porewater pressure. The rate of settlement from secondary compression is very slow compared with that from primary consolidation.

We have separated primary consolidation and secondary compression. In reality, the distinction is not clear because secondary compression occurs as part of the primary consolidation phase, especially in soft clays. The mechanics of consolidation is still not fully understood, and to make estimates of settlement, it is convenient to separate primary consolidation and secondary compression. A measure of secondary compression is the secondary compression index, C_α.

Primary consolidation is assumed to end at the intersection of the projection of the two straight parts of the curve (Figure 7.16). The secondary compression index is

$$C_\alpha = -\frac{(e_t - e_p)}{\log(t/t_p)} = \frac{|\Delta e|}{\log(t/t_p)}, \quad t > t_p \tag{7.24}$$

where (t_p, e_p) is the coordinate at the intersection of the tangents to the primary consolidation and secondary compression parts of the logarithm of time versus void ratio curve, and (t, e_t) is the coordinate of any point on the secondary compression curve, as shown in Figure 7.16. The value of C_α usually varies with the magnitude of the applied loads and other factors such as the LIR. Typical values for C_α are given in Table 7.3 in Section 7.12.

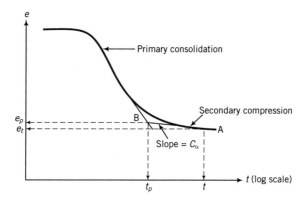

Figure 7.16 Secondary compression.

Key points

1. When a load is applied to a saturated soil, all of the applied stress is supported initially by the porewater (initial excess porewater pressure); that is, at $t = 0$, $\Delta u_o = \Delta \sigma_z$. The change in effective stress is zero ($\Delta \sigma_z' = 0$).

2. If drainage of porewater is permitted, the initial excess porewater pressure decreases and soil settlement (ΔH) increases with time; that is, $\Delta u(t) < \Delta u_o$ and $\Delta H > 0$. Since the change in total stress is zero, the change in effective stress is equal to the change in excess porewater pressure [$\Delta u_o - \Delta u(t)$].

3. When $t \rightarrow \infty$, the change in volume and the change in excess porewater pressure of the soil approach zero; that is, $\Delta V \rightarrow 0$ and $\Delta u_o \rightarrow 0$. The change in vertical effective stress is $\Delta \sigma_z' = \Delta \sigma_z$.

4. Soil settlement is not linearly related to time except very early in the consolidation process.

5. The change in volume of the soil is equal to the volume of initial excess porewater expelled.

6. The rate of settlement depends on the hydraulic conductivity (permeability) of the soil.

7. Path AB (Figure 7.6), called the normal consolidation line (NCL), describes the response of a normally consolidated soil—a soil that has never experienced a vertical effective stress greater than its current vertical effective stress. The NCL is approximately a straight line in a plot of log σ_z' versus e and is defined by a slope, C_c, called the compression index.

8. A normally consolidated soil would behave like an elastoplastic material. That is, part of the settlement under the load is recoverable, while the other part is permanent.

9. An overconsolidated soil has experienced vertical effective stresses greater than its current vertical effective stress.

10. An overconsolidated soil will follow paths such as CDE (Figure 7.6). For stresses below the past maximum vertical effective stress, an overconsolidated soil would behave approximately like an elastic material, and settlement would be small. However, for stresses greater than the past maximum vertical effective stress, an overconsolidated soil will behave like an elastoplastic material, similar to a normally consolidated soil.

EXAMPLE 7.2 *Change in Vertical Effective Stress at a Given Degree of Consolidation*

A soft clay layer 0.5 m thick is sandwiched between layers of sand. The initial vertical total stress at the center of the clay layer is 50 kPa and the porewater pressure is 20 kPa. The increase in vertical stress at the center of the clay layer from a building foundation is 25 kPa. The vertical stresses and pressures at the center of the clay layer are assumed to be the average vertical stresses and pressures of the clay layer. What are the vertical effective stress and excess porewater pressure at the center of the clay layer when 60% consolidation occurs?

Strategy You are given the increment in applied stress and the degree of consolidation. You can calculate the excess porewater pressure remaining in the soil and then the change in vertical effective stress.

Solution 7.2

Step 1:   Calculate the initial excess porewater pressure.

$$\Delta u_o = \Delta \sigma_z = 25 \text{ kPa}$$

Step 2:   Calculate the current excess porewater pressure at 60% consolidation.

$$\Delta u = \Delta u_o(1 - U) = 25(1 - 0.6) = 10 \text{ kPa}$$

Step 3:   Check reasonableness of answer.

The excess porewater pressure after consolidation is initiated (10 kPa) must be lower than the initial excess porewater pressure (25 kPa)

The answer is then reasonable.

Step 4:   Calculate the increase in vertical effective stress at 60% consolidation.

$$\Delta \sigma'_z = \Delta \sigma_z - \Delta u = 25 - 10 = 15 \text{ kPa}$$

Step 5:   Check reasonableness of answer.

Since 60% consolidation has occurred, the current excess porewater pressure (10 kPa) must be less than the excess porewater dissipated (15 kPa).

The answer is then reasonable.

Step 6:   Calculate the current vertical effective stress.

$$\sigma'_{zo} = \sigma_{zo} - u_o = 50 - 20 = 30 \text{ kPa}$$

$$\sigma'_z = \sigma'_{zo} + \Delta \sigma'_z = 30 + 15 = 45 \text{ kPa}$$

$$u = u_o + \Delta u = 20 + 10 = 30 \text{ kPa}$$

Step 7:   Check reasonableness of answer.

The current total stress is $50 + 25 = 75$ kPa.

$$\sigma_z = \sigma'_z + u = 45 + 30 = 75 \text{ kPa}$$

The answer is then reasonable.

EXAMPLE 7.3   *Calculating C_c, C_r, and OCR from Test Data*

The results of a one-dimensional consolidation test on a clay taken from a depth of 4 m is shown in Figure E7.3a. The initial overburden (effective) pressure was 40 kPa. The water content after the consolidation test was completed was 18.9%. The specific gravity of the solids was 2.65. The initial sample thickness was 20 mm. and the final thickness was 15.08 mm.

(a) Determine the initial voids ratio.
(b) Determine the compression index, C_c, of the soil.
(c) Determine the recompression index, C_r, of the soil.
(d) Determine the past maximum vertical effective stress using the Brazilian method.
(e) Determine the overconsolidation ratio if the initial vertical effective stress is 40 kPa.

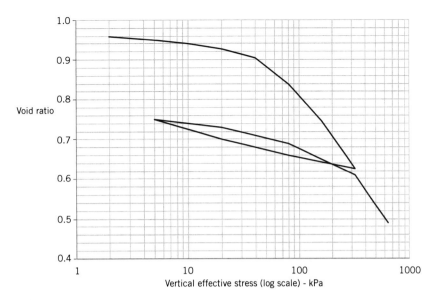

Figure E7.3a

Strategy Use the appropriate equation and procedure to determine the required parameters.

Solution 7.3

Step 1: Determine e_o.

$$(\Delta H)_{fin} = H_o - H_{fin} = 20 - 15.08 = 4.92 \text{ mm}$$

$$e_{fin} = wG_s = 0.189 \times 2.65 = 0.5$$

$$e_o = \frac{e_{fin} + [(\Delta H)_{fin}/H_o]}{1 - [(\Delta H)_{fin}/H_o]} = \frac{0.5 + (4.92/20)}{1 - (4.92/20)} = 0.989$$

Step 2: Determine C_c.

C_c is the slope AB shown in Figure E7.3b.

Void ratio at 100 kPa on AB $= e_{100} = 0.83$

Void ratio at 1000 kPa on AB $= e_{1000} = 0.41$

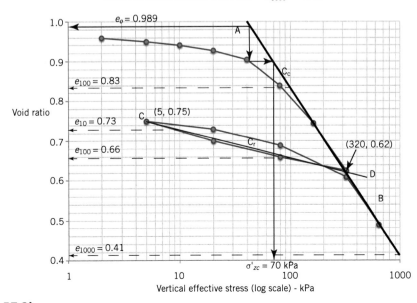

Figure E7.3b

By selecting the void ratios at vertical effective stresses of 1000 and 100, the value of C_c is simply the difference in the void ratio, since log (1000/100) = 1.

$$C_c = 0.83 - 0.41 = 0.42$$

Step 3: Determine C_r.

C_r is the slope of CD in Figure E7.3b.

$$\text{Void ratio at 10 kPa on CD} = e_{10} = 0.73$$

$$\text{Void ratio at 100 kPa on CD} = e_{100} = 0.66$$

$$C_r = 0.73 - 0.66 = 0.07$$

Step 4: Check reasonableness of answers.

From Table 7.3, the values of C_c and C_r are within typical ranges.

The answers are reasonable.

Step 4: Determine the past maximum vertical effective stress.

Following the procedure in Section 7.6 on the determination of the past maximum vertical effective stress, the past maximum vertical effective stress is $\sigma'_{zc} = 70$ kPa.

Note: A small change in e_o or a slight change in the inclination of the line AB (slope C_c) can result in a change of σ'_{zc} of ± 10 kPa.

Step 5: Determine the overconsolidation ratio.

Past maximum vertical effective stress: $\sigma'_{zc} = 70$ kPa

Current vertical effective stress (overburden pressure): $\sigma'_z = 40$ kPa

$$OCR = \frac{\sigma'_{zc}}{\sigma'_z} = \frac{70}{40} = 1.75 \text{ (approximately 2)}$$

EXAMPLE 7.4 *Determination of Past Maximum Vertical Effective Stress Using the Casagrande Method, the Brazilian Method, and the Strain Energy Method*

One-dimensional consolidation test on a soft clay gave the results shown in table below.

Vertical effective stress (kPa)	Vertical strain	Void ratio
0	0	2.200
13	0.0056	2.182
26	0.0110	2.165
52	0.0148	2.153
104	0.0235	2.125
208	0.0798	1.945
416	0.1760	1.637

Compare the past maximum vertical effective stress using the Casagrande method, the Brazilian method, and the stain energy method.

Strategy From the data, you can find the cumulated work for the strain energy method. Follow the procedure in Section 7.6 on the determination of the past maximum vertical effective stress.

Solution 7.4

Step 1: Determine the past maximum vertical effective stress using the Casagrande method.

You need to plot the void ratio versus the vertical effective stress (log scale) and then follow the procedures for the Casagrande method. Figure E7.4a gives a past maximum effective stress of about 160 kPa.

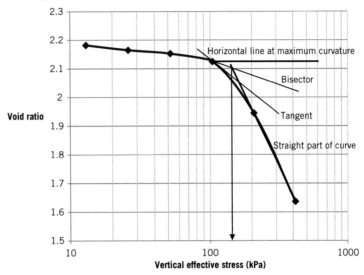

Figure E7.4a

Step 2: Determine the past maximum vertical effective stress using the Brazilian method.

You need to plot the void ratio versus the vertical effective stress (log scale) and then follow the procedures for the Brazilian method. Figure E7.4b gives a past maximum effective stress of about 160 kPa.

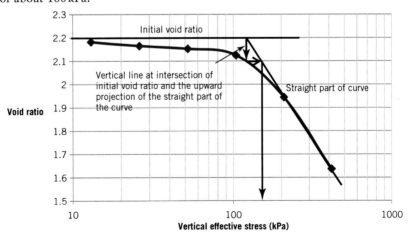

Figure E7.4b

Step 3: Determine the past maximum vertical effective stress using the strain energy method.

You need to plot cumulate work versus the vertical effective stress and then follow the procedures for the strain energy method. Calculate the cumulated strain energy as shown in the table below.

Vertical effective stress	Vertical strain	Averaged vertical effective stress	Vertical strain difference	$\Delta W =$ column (3) × column (4)	$W = \sum \Delta W$
kPa		kPa			
(1)	(2)	(3)	(4)	kN m/m³	kN m/m³
0	0	0	0	0	0
13	0.0056	6.5	0.0056	0.0364	0.0364
26	0.0110	19.5	0.0054	0.1053	0.1417
52	0.0148	39	0.0038	0.1482	0.2899
104	0.0235	78	0.0087	0.6786	0.9685
208	0.0798	156	0.0563	8.783	9.7512
416	0.1760	312	0.0962	30.014	39.766

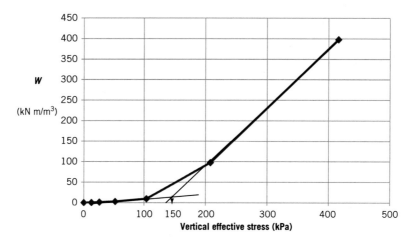

Figure E7.4c

Figure E7.4c gives a past maximum effective stress of about 150 kPa.

These results are close to each other. For practical purpose, the past maximum vertical effective stress ranges from 150 and 160 kPa. In practice, the differences among these procedures for the estimation of the past maximum vertical effective stress can be large (>20%).

EXAMPLE 7.5 *Determination of Elastic Parameters, m_{vr} and E'_c*

For the one-dimensional consolidation test result shown in Figure E7.3a, determine m_{vr} and E'_c.

Strategy From the data, you can find the vertical strain. You know the increase in vertical effective stress, so the appropriate equations to use to calculate m_{vr} and E'_c are Equations (7.14) and (7.16).

Solution 7.5

Step 1: Calculate the vertical strain.

With reference to Figure E7.3b,

$$\text{Void ratio at 5 kPa on CD} = 0.75$$

$$\text{Void ratio at 320 kPa on CD} = 0.62$$

$$\Delta e = 0.75 - 0.62 = 0.13$$

$$\Delta \varepsilon_{zr} = \frac{\Delta e}{1 + e_o} = -\frac{0.13}{1 + 0.62} = 0.08$$

It is best to restart the initial state at each reversal of loading because the soil fabric is different for the new direction of loading.

Step 2: Calculate the modulus of volume recompressibility.

$$m_{vr} = \frac{\Delta \varepsilon_{zr}}{\Delta \sigma'_z} = \frac{0.08}{320 - 5} = 2.5 \times 10^{-4} \ \text{m}^2/\text{kN}$$

Step 3: Calculate the constrained elastic modulus.

$$E'_c = \frac{1}{m_{vr}} = \frac{1}{2.5 \times 10^{-4}} = 4 \times 10^3 \ \text{kPa}$$

EXAMPLE 7.6   *Determination of C_v Using Root Time Method*

The following readings were taken for an increment of vertical stress of 20 kPa in a one-dimensional consolidation test on a saturated clay sample 75 mm diameter and 20 mm thick. Drainage was permitted from the top and bottom boundaries.

Time (min)	0.25	1	2.25	4	9	16	25	36	64	144	24 hours
ΔH (cm)	0.0047	0.009	0.0133	0.0169	0.023	0.0268	0.0291	0.03	0.032	0.033	0.035

Determine the coefficient of consolidation using the root time method.

Strategy Plot the data in a graph of displacement reading versus $\sqrt{\text{time}}$ and follow the procedures for the root time method.

Solution 7.6

Step 1: Make a plot of settlement (decrease in thickness) versus $\sqrt{\text{time}}$, as shown in Figure E7.6.

Step 2: Follow the procedures outlined in Section 7.6.10 on the root time method to find t_{90}.

From Figure E7.6, $\sqrt{t_{90}} = 5 \ \text{min}^{1/2}$; $t_{90} = 25 \ \text{min}$

Step 3: Calculate C_v from Equation (7.21).

$$C_v = \frac{0.848 H_{dr}^2}{t_{90}}$$

where H_{dr} is the length of the drainage path. The height of the sample at the end of the consolidation test for the increment of loading is $2.0 - 0.035 = 1.965$ cm. From Equation (7.8),

$$H_{dr} = \frac{H_o + H_f}{4} = \frac{2.0 + 1.965}{4} = 0.991 \ \text{cm}$$

$$\therefore C_v = \frac{0.848 \times 0.991^2}{25} = 33.6 \times 10^{-3} \text{ cm}^2/\text{min}$$

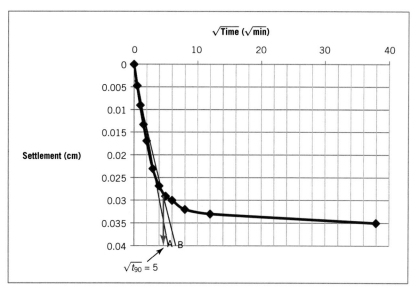

Figure E7.6

Step 4: Check reasonableness of answer.

$$C_v = 33.6 \times 10^{-3} \text{ cm}^2/\text{min} = \frac{33.6 \times 10^{-3}}{60} = 5.6 \times 10^{-4} \text{ cm}^2/\text{s} \text{ is reasonable}$$

based on the range of values in Table 7.5 in Section 7.12.

What's next ... We have described the consolidation test of a small sample of soil and the soil consolidation parameters that can be obtained. What is the relationship between this small test sample and the soil in the field? Can you readily calculate the settlement of the soil in the field based on the results of your consolidation test? The next section provides the relationship between the small test sample and the soil in the field.

7.7 RELATIONSHIP BETWEEN LABORATORY AND FIELD CONSOLIDATION

The time factor (T_v) provides a useful expression to estimate the settlement in the field from the results of a laboratory consolidation test. If two layers of the same clay have the same degree of consolidation, then their time factors and coefficients of consolidation are the same. Hence,

$$T_v = \frac{(C_v t)_{lab}}{(H_{dr}^2)_{lab}} = \frac{(C_v t)_{field}}{(H_{dr}^2)_{field}} \tag{7.25}$$

and, by simplification,

$$\frac{t_{field}}{t_{lab}} = \frac{\left(H_{dr}^2\right)_{field}}{\left(H_{dr}^2\right)_{lab}} \tag{7.26}$$

EXAMPLE 7.7   *Time-Settlement Calculations*

A one-dimensional consolidation test was performed on a sample, 50 mm in diameter and 20 mm high, taken from a clay layer 1 m thick. During the test, drainage was allowed at the upper and lower boundaries. It took the laboratory sample 75 minutes to reach 50% consolidation.

(a) If the clay layer in the field has the same drainage condition as the laboratory sample, calculate how long it will take the 1 m clay layer to achieve 50% and 90% consolidation.

(b) How much more time would it take the 1 m clay layer to achieve 50% consolidation if drainage existed on only one boundary?

Strategy You are given all the data to directly use Equation (7.26). For part (a), there is double drainage in the field and the lab, so the drainage path is one-half the soil thickness. For part (b), there is single drainage in the field, so the drainage path is equal to the soil thickness.

Solution 7.7

Step 1: Calculate the drainage path.

(a)

$$(H_{dr})_{lab} = \frac{0.02}{2} = 0.01 \text{ m}; \; (H_{dr})_{field} = \frac{1}{2} = 0.5 \text{ m}$$

Step 2: Calculate the field time using Equation (7.26).

Time for 50% consolidation in the field is

$$t_{field} = \frac{t_{lab}\left(H_{dr}^2\right)_{field}}{\left(H_{dr}^2\right)_{lab}} = \frac{75 \times 0.5^2}{0.01^2} = 18.75 \times 10^4 \text{ min} = 130 \text{ days}$$

Time for 90% consolidation in the field is

$$t_{90} = \frac{(T_v)_{90}\, t_{50}}{(T_v)_{50}} = \frac{0.848 \times 130}{0.197} = 560 \text{ days}$$

Step 3: Calculate the drainage path.

(b)

$$(H_{dr})_{lab} = \frac{0.02}{2} = 0.01 \text{ m}; \; (H_{dr})_{field} = 1 \text{ m}$$

Step 4: Calculate field time using Equation (7.26).

$$t_{field} = \frac{t_{lab}\left(H_{dr}^2\right)_{field}}{\left(H_{dr}^2\right)_{lab}} = \frac{75 \times 1^2}{1^2} = 75 \times 10^4 \text{ min} = 520 \text{ days}$$

You should take note that if drainage exists on only one boundary rather than both boundaries of the clay layer, the time taken for a given percent consolidation in the field is four times longer.

> *What's next ...* Next, we will consider how to use the basic concepts to calculate one-dimensional settlement.

7.8 CALCULATION OF PRIMARY CONSOLIDATION SETTLEMENT

7.8.1 *Effects of unloading/reloading of a soil sample taken from the field*

We want to take a soil sample from the field at a depth z (Figure 7.17a). We will assume that the groundwater level is at the surface. The current vertical effective stress or overburden pressure is

$$\sigma'_{zo} = (\gamma_{sat} - \gamma_w)z = \gamma'z$$

and the current void ratio can be found from γ_{sat} using Equation 2.11. On a plot of σ'_z (log scale) versus e, the current vertical effective stress can be represented as A, as depicted in Figure 7.17b.

To obtain a sample, we would have to make a borehole and remove the soil above it. The act of removing the soil and extracting the sample reduces the total stress to zero; that is, we have fully unloaded the soil. From the principle of effective stress, $\sigma'_z = -\Delta u$. Since σ' cannot be negative—that is, soil cannot sustain tension—the porewater pressure must be negative. As the porewater pressure dissipates with time, volume changes (swelling) occur. Using the basic concepts of consolidation described in Section 7.6, the sample will follow the unloading path AB (Figure 7.17b). The point B does not correspond to zero effective stress because we cannot represent zero on a logarithmic scale. If we were to reload our soil sample, the reloading path followed would depend on the *OCR*. If *OCR* = 1 (normally

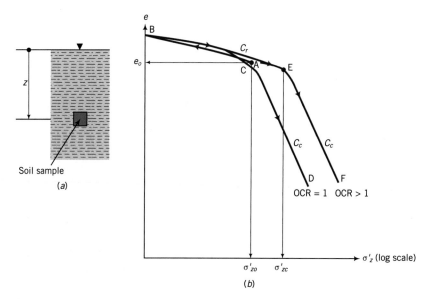

Figure 7.17 (a) Soil sample at a depth z below ground surface. (b) Expected one-dimensional consolidation response.

consolidated soil), the path followed during reloading would be BCD (Figure 7.17b). The average slope of ABC is C_r. Once σ'_{zo} is exceeded, the soil will follow the normal consolidation line, CD, of slope C_c. If the soil were overconsolidated, $OCR > 1$, the reloading path followed would be BEF because we have to reload the soil beyond σ'_{zc} before it behaves like a normally consolidated soil. The average slope of ABE is C_r and the slope of EF is C_c. The point E marks the past maximum vertical effective stress.

7.8.2 Primary consolidation settlement of normally consolidated fine-grained soils

Let us consider a site consisting of a normally consolidated soil on which we wish to construct a building. We will assume that the increase in vertical total stress due to the building at depth z, where we took our soil sample, is $\Delta\sigma_z$. Further, we will assume that all the excess porewater pressure, due to the increase in vertical total stress, dissipated. So the increase in vertical effective stress is equal to the increase in vertical total stress. The final vertical stress is

$$\sigma'_{fin} = \sigma'_{zo} + \Delta\sigma_z \tag{7.27}$$

The increase in vertical stress will cause the soil to settle following the NCL, and the primary consolidation settlement is

$$\rho_{pc} = H_o \frac{\Delta e}{1+e_o} = \frac{H_o}{1+e_o} C_c \log\frac{\sigma'_{fin}}{\sigma'_{zo}}, \quad OCR = 1 \tag{7.28}$$

where $\Delta e = C_c \log(\sigma'_{fin}/\sigma'_{zo})$.

7.8.3 Primary consolidation settlement of overconsolidated fine-grained soils

If the soil is overconsolidated, we have to consider two cases depending on the magnitude of $\Delta\sigma_z$. We will approximate the curve in the σ'_z (log scale) versus e space as two straight lines, as shown in Figure 7.18. In case 1, the increase in $\Delta\sigma_z$ is such that $\sigma'_{fin} = \sigma'_{zo} + \Delta\sigma_z$ is less than σ'_{zc} (Figure 7.18a). In this case, consolidation occurs along the URL and

$$\rho_{pc} = \frac{H_o}{1+e_o} C_r \log\frac{\sigma'_{fin}}{\sigma'_{zo}}; \quad \sigma'_{fin} < \sigma'_{zc} \tag{7.29}$$

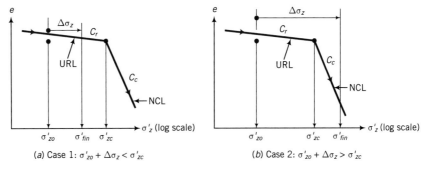

(a) Case 1: $\sigma'_{zo} + \Delta\sigma_z < \sigma'_{zc}$ (b) Case 2: $\sigma'_{zo} + \Delta\sigma_z > \sigma'_{zc}$

Figure 7.18 Two cases to consider for calculating settlement of overconsolidated fine-grained soils.

In case 2, the increase in $\Delta\sigma_z$ is such that $\sigma'_{fin} = \sigma'_{zo} + \Delta\sigma_z$ is greater than σ'_{zc} (Figure 7.18b). In this case, we have to consider two components of settlement—one along the URL and the other along the NCL. The equation to use in case 2 is

$$\rho_{pc} = \frac{H_o}{1+e_o}\left(C_r \log\frac{\sigma'_{zc}}{\sigma'_{zo}} + C_c \log\frac{\sigma'_{fin}}{\sigma'_{zc}}\right), \quad \sigma'_{fin} > \sigma'_{zc} \tag{7.30}$$

or

$$\rho_{pc} = \frac{H_o}{1+e_o}\left[C_r \log(OCR) + C_c \log\frac{\sigma'_{fin}}{\sigma'_{zc}}\right], \quad \sigma'_{fin} > \sigma'_{zc} \tag{7.31}$$

7.8.4   Procedure to calculate primary consolidation settlement

The procedure to calculate primary consolidation settlement is as follows:

1.   Calculate the current vertical effective stress (σ'_{zo}) and the current void ratio (e_o) at the center of the soil layer for which settlement is required.
2.   Calculate the applied vertical total stress increase ($\Delta\sigma_z$) at the center of the soil layer using the appropriate method in Chapter 6.
3.   Calculate the final vertical effective stress $\sigma'_{fin} = \sigma'_{zo} + \Delta\sigma_z$.
4.   Calculate the primary consolidation settlement.

   (a)   If the soil is normally consolidated ($OCR = 1$), the primary consolidation settlement is

$$\rho_{pc} = \frac{H_o}{1+e_o}C_c \log\frac{\sigma'_{fin}}{\sigma'_{zo}}$$

   (b)   If the soil is overconsolidated and $\sigma'_{fin} < \sigma'_{zc}$, the primary consolidation settlement is

$$\rho_{pc} = \frac{H_o}{1+e_o}C_r \log\frac{\sigma'_{fin}}{\sigma'_{zo}}$$

   (c)   If the soil is overconsolidated and $\sigma'_{fin} > \sigma'_{zc}$, the primary consolidation settlement is

$$\rho_{pc} = \frac{H_o}{1+e_o}\left[C_r \log(OCR) + C_c \log\frac{\sigma'_{fin}}{\sigma'_{zc}}\right]$$

where H_o is the thickness of the soil layer.

You can also calculate the primary consolidation settlement using m_v. However, unlike C_c, which is assumed to be constant, m_v varies with stress levels. You should compute an average value of m_v over the stress range σ'_{zo} to σ'_{fin}. To reduce the effects of nonlinearity, the vertical effective stress difference should not exceed 100 kPa in calculating m_v or m_{vr}. The primary consolidation settlement, using m_v, is

$$\rho_{pc} = H_o m_v \Delta\sigma_z \tag{7.32}$$

The advantage of using Equation (7.32) is that m_v is readily determined from displacement data in consolidation tests; you do not have to calculate void ratio changes from the test data as required to determine C_c.

7.9 SECONDARY COMPRESSION

The secondary consolidation settlement is

$$\rho_{sc} = \frac{H_o}{(1+e_p)} C_\alpha \log\left(\frac{t}{t_p}\right) \tag{7.33}$$

7.10 SETTLEMENT OF THICK SOIL LAYERS

For better accuracy, when dealing with thick layers ($H_o > 1$ m), you should divide the soil layer into sublayers (about two to five sublayers) and find the settlement for each sublayer. Add up the settlement of each sublayer to find the total primary consolidation settlement. You must remember that the value of H_o in the primary consolidation equations is the thickness of the sublayer. An alternative method is to use a harmonic mean value of the vertical stress increase for the sublayers in the equations for primary consolidation settlement. The harmonic mean stress increase is

$$\Delta\sigma_z = \frac{n(\Delta\sigma_z)_1 + (n-1)(\Delta\sigma_z)_2 + (n-2)(\Delta\sigma_z)_3 + \cdots + (\Delta\sigma_z)_n}{n + (n-1) + (n-2) + \cdots + 1} \tag{7.34}$$

where n is the number of sublayers and the subscripts 1, 2, ... n, mean the first (top) layer, the second layer, and so on. You can also use the layer thickness rather than n for each layer in Equation (7.34). The advantage of using the harmonic mean is that the settlement is skewed in favor of the upper part of the soil layer. You may recall from Chapter 6 that the increase in vertical total stress from surface loads decreases with depth. Therefore, the primary consolidation settlement of the upper portion of the soil layer can be expected to be more than the lower portion because the upper portion of the soil layer is subjected to higher vertical stress increases.

> ## EXAMPLE 7.8 *Consolidation Settlement of a Normally Consolidated Clay*
>
> The soil profile at a site for a proposed office building consists of a layer of fine sand 10 m thick above a layer of soft, normally consolidated clay 1 m thick (Figure E7.8). Below the soft clay is a deposit of coarse sand. The groundwater table was observed at 3 m below ground level. The void ratio of the sand is 0.76 and the water content of the clay is 34%. The building will impose a vertical stress increase of 80 kPa at the middle of the clay layer. Estimate the primary consolidation settlement of the clay. Assume the soil above the water table to be saturated, $C_c = 0.3$, and $G_s = 2.7$.
>
> **Strategy** In this problem, you are given the stratigraphy, groundwater level, vertical total stress increase, and the following soil parameters and soil condition:
>
> $$e_o \text{ (for sand)} = 0.76$$
>
> $$w \text{ (for clay)} = 34\%, \quad H_o = 1\text{ m}, \quad \Delta\sigma_z = 80\text{ kPa}, \quad C_c = 0.3, \quad G_s = 2.7$$
>
> Since you are given a normally consolidated clay, the primary consolidation settlement is found from Equation (7.28).

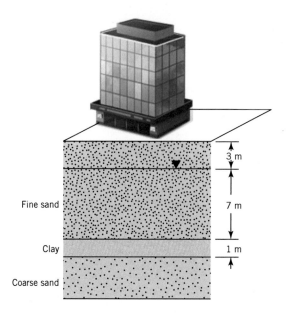

Figure E7.8

Solution 7.8

Step 1: Calculate σ'_{zo} and e_o at the center of the clay layer.

$$\text{Sand: } \gamma_{sat} = \left(\frac{G_s + e}{1 + e}\right)\gamma_w = \left(\frac{2.7 + 0.76}{1 + 0.76}\right)9.81 = 19.3 \text{ kN/m}^3$$

$$\text{Clay: } e_o = wG_s = 2.7 \times 0.34 = 0.918$$

$$\gamma_{sat} = \left(\frac{G_s + e}{1 + e}\right)\gamma_w = \left(\frac{2.7 + 0.918}{1 + 0.918}\right)9.81 = 18.5 \text{ kN/m}^3$$

Porewater pressure at center of clay is $u_o = 7.5 \times 9.81 = 73.6 \text{ kPa}$

The vertical effective stress at the mid-depth of the clay layer is

$$\sigma_{zo} = (19.3 \times 10) + (18.5 \times 0.5) = 202.3 \text{ kPa}$$

$$\sigma'_{zo} = \sigma_{zo} - u_o = 202.3 - 73.6 = 128.7 \text{ kPa}$$

Step 2: Calculate the increase of stress at the mid-depth of the clay layer.

You do not need to calculate $\Delta\sigma_z$ for this problem. It is given as $\Delta\sigma_z = 80 \text{ kPa}$.

Step 3: Calculate σ'_{fin}.

$$\sigma'_{fin} = \sigma'_{zo} + \Delta\sigma_z = 128.7 + 80 = 208.7 \text{ kPa}$$

Step 4: Calculate the primary consolidation settlement.

$$\rho_{pc} = \frac{H_o}{1 + e_o}C_c \log\frac{\sigma'_{fin}}{\sigma'_{zo}} = \frac{1}{1 + 0.918} \times 0.3 \log\frac{208.7}{128.7} = 0.0328 \text{ m} \approx 33 \text{ mm}$$

EXAMPLE 7.9 *Consolidation Settlement of an Overconsolidated Clay*

Assume the same soil stratigraphy as in Example 7.8. But now the clay is overconsolidated with an $OCR = 2.5$, $w = 38\%$, and $C_r = 0.05$. All other soil values given in Example 7.8 remain unchanged. Determine the primary consolidation settlement of the clay.

Strategy Since the soil is overconsolidated, you will have to check whether σ'_{zc} is less than or greater than the sum of the current vertical effective stress and the applied vertical total stress increase at the center of the clay. This check will determine the appropriate equation to use. In this problem, the unit weight of the sand is unchanged but the clay has changed.

Solution 7.9

Step 1: Calculate σ'_{zo} and e_o at mid-depth of the clay layer.

$$e_o = wG_s = 0.38 \times 2.7 = 1.03$$

Clay: $\gamma_{sat} = \left(\dfrac{G_s + e}{1 + e}\right)\gamma_w = \left(\dfrac{2.7 + 1.03}{1 + 1.03}\right)9.81 = 18 \text{ kN/m}^3$

$$\sigma_{zo} = (19.3 \times 10) + (18 \times 0.5) = 202 \text{ kPa}$$

$$\sigma'_{zo} = 202 - 73.6 = 128.4 \text{ kPa}$$

(Note that the increase in vertical effective stress from the unit weight change in this overconsolidated clay is very small.)

Step 2: Calculate the past maximum vertical effective stress.

$$\sigma'_{zc} = \sigma'_{zo} \times OCR = 128.4 \times 2.5 = 321 \text{ kPa}$$

Step 3: Calculate σ'_{fin}.

$$\sigma'_{fin} = \sigma'_{zo} + \Delta\sigma_z = 128.4 + 80 = 208.4 \text{ kPa}$$

Step 4: Check if σ'_{fin} is less than or greater than σ'_{zc}.

$$(\sigma'_{fin} = 208.4 \text{ kPa}) < (\sigma'_{zc} = 321 \text{ kPa})$$

Therefore, use Equation (7.29) to calculate the primary consolidation settlement.

Step 5: Calculate the total primary consolidation settlement.

$$\rho_{pc} = \frac{H_o}{1 + e_o}\left[C_r \log \frac{\sigma'_{fin}}{\sigma'_{zo}}\right]$$

$$= \frac{1}{1 + 1.03} \times \left(0.05 \log \frac{208.4}{128.4}\right) = 0.00486 \text{ m} = 5 \text{ mm}$$

Step 6: Check reasonableness of answer.

Since the soil is overconsolidated, the settlement will be smaller than the same soil in a normally consolidated state.

The estimated settlement is about one-sixth the settlement for the normally consolidated soil (see Example 7.8). The changes in final and initial vertical effective stresses are small. So the settlement ratio of the overconsolidated soil to the normally consolidated soil will be approximately C_r/C_c (= 0.05/0.3 = 1/6). The answer is reasonable.

7.11 ONE-DIMENSIONAL CONSOLIDATION THEORY

Terzaghi (1925) developed the theory of one-dimensional consolidation. We have used this theory throughout this chapter to understand the one-dimensional consolidation of fine-grained soils. The key assumptions of this theory and the key implications are as follows:

- An isotropic, homogeneous, saturated soil. In the absence of dissolved gases, soils under the groundwater level are generally saturated. Soils are anisotropic materials (materials that have different properties in different directions and provide different resistances to flow and load in these directions) and are rarely homogeneous.
- Incompressible soil particles and the water. The soil grains and the water do not compress or enlarge.
- Vertical flow of water. The direction of flow of water is only vertical. In real soils, the flow of water through them is not the same in all directions because soils are anisotropic materials.
- Validity of Darcy's law. Recall that Darcy's law states that the flow velocity is proportional to the hydraulic gradient.
- Small strains. The strains (change in length divided by the original length) in a given direction are infinitesimal ($\approx < 0.001\%$ for practical applications). The assumption of one-dimensional consolidation leads to zero lateral strains, that is, $\varepsilon_x = \varepsilon_y = 0$.

The governing partial differential equation for one-dimensional consolidation (see derivation in Appendix A) is

$$\frac{\partial u}{\partial t} = C_v \frac{\partial^2 u}{\partial z^2} \tag{7.35}$$

The solution of this equation gives the variation of excess porewater pressure with time and depth for a given set of boundary conditions. In the case of a uniform distribution of initial excess porewater pressure in which double drainage occurs, the analytic solution is

$$\frac{\Delta u_z}{\Delta u_o} = \frac{4}{\pi} \sum_{m=1}^{\infty} \left\{ \frac{(-1)^{m-1}}{2m-1} \cos\left[(2m-1)\frac{\pi}{2}\left(\frac{H_{dr}-z}{H_{dr}}\right)\right] \exp\left[-(2m-1)^2 \frac{\pi^2}{4} T_v\right] \right\} \tag{7.36}$$

where m is a positive integer, Δu_z is the excess porewater pressure at any time t, at a depth z, Δu_o is the initial excess porewater pressure (excess porewater pressure at time $t = 0$), H_{dr} is the length of the drainage path, and T_v is time factor.

The degree of consolidation at a depth z (U_z) is the amount of consolidation that has occurred at a given time. This is expressed as

$$U_z = 1 - \frac{\Delta u_z}{\Delta u_o} = \frac{\Delta \sigma_z'}{\Delta u_o} \tag{7.37}$$

where $\Delta \sigma_z'$ is the change in effective stress at depth z and time t from the dissipation of excess porewater pressure.

A plot of Equation (7.36) giving the variation of degree of consolidation, with depth ratio, z/H_{dr}, where z is the depth from the top drainage surface of the soil for different times (different time factor) is shown in Figure 7.19.

At $T_v = C_v t / H_{dr}^2 \approx 2$, the soil has reached approximately 100% consolidation. So we can make at estimate for the time to reach 100% primary consolidation settlement from

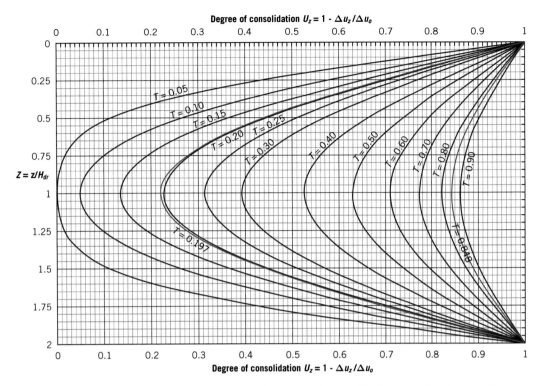

Figure 7.19 Solution of the governing one-dimensional consolidation equation for a uniform initial excess porewater pressure, Δu_o, and double drainage.

$$t \approx \frac{2H_{dr}^2}{C_v} \tag{7.38}$$

For total primary consolidation settlement that is small relative to the original height (about 1% and less),

$$t \approx \frac{H_o^2}{2C_v} \tag{7.39}$$

Key points

1. The one-dimensional consolidation equation allows us to predict the changes in excess porewater pressure at various depths within the soil with time.
2. We need to know the excess porewater pressure at a desired time because we want to determine the vertical effective stress to calculate the primary consolidation settlement.

What's next … Several expressions are available linking consolidation parameters to simple, less time-consuming soil tests such as the Atterberg limits and water content tests. In the next section, some of these expressions are presented.

7.12 TYPICAL VALUES OF CONSOLIDATION SETTLEMENT PARAMETERS AND EMPIRICAL RELATIONSHIPS

Some relationships between simple soil tests and consolidation settlement parameters are given below in Table 7.3, Table 7.4, and Table 7.5. You should be cautious in using these relationships because they may not be applicable to your soil type.

Table 7.3 Typical range of values of C_c and C_r.

C_c = 0.1 to 0.8
C_r = 0.015 to 0.35; also, $C_r \approx C_c/5$ to $C_c/10$
C_α/C_c = 0.03 to 0.08

Table 7.4 Some empirical relationships for C_c and C_r.

Empirical relationships	Reference
$C_c = 0.009(LL - 10)$	Terzaghi and Peck, 1948
$C_c = 1.35PI$ (remolded clays)	Schofield and Wroth, 1968
$C_c = 0.40(e_o - 0.25)$	Azzouz et al., 1976
$C_c = 0.01(w - 5)$	Azzouz et al., 1976
$C_c = 0.37(e_o + 0.003LL - 0.34)$	Azzouz et al., 1976
$C_r = 0.15(e_o + 0.007)$	Azzouz et al., 1976
$C_r = 0.003(w + 7)$	Azzouz et al., 1976
$C_r = 0.126(e_o + 0.003LL - 006)$	Azzouz et al., 1976
$C_r = 0.000463LL\ G_s$	Nagaraj and Murthy, 1985

Note: w is the natural water content (%), LL is the liquid limit (%), e_o is the initial void ratio, and PI is the plasticity index.

Table 7.5 Typical values of C_v.

Soil	c_v (cm²/s × 10⁻⁴)	c_v (m²/yr)
Boston blue clay (CL)	40 ± 20	12 ± 6
Organic silt (OH)	2–10	0.6–3
Glacial lake clays (CL)	6.5–8.7	2.0–2.7
Chicago silty clay (CL)	8.5	2.7
Swedish medium sensitive clays (CL-CH)		
1. laboratory	0.4–0.7	0.1–0.2
2. field	0.7–3.0	0.2–1.0
San Francisco Bay mud (CL)	2–4	0.6–1.2
Mexico City clay (MH)	0.9–1.5	0.3–0.5

Source: Modified from Carter and Bentley (1991).

7.13 MONITORING SOIL SETTLEMENT

Settlements calculated from the one-dimensional consolidation theory and soil parameters from the one-dimensional consolidation test are estimates. Major uncertainties come from the limitations of the theory (see the assumptions in Section 7.11), the lack of knowledge of the actual stress transferred to the soil, the neglect of shear stresses, the inaccuracy of the settlement parameters from lab tests due to sampling disturbances, the simplification of the stratigraphy and drainage boundaries, and secondary compression. It is good practice to measure the actual settlement using field instruments.

There are many instruments to measure soil settlements. They range from conventional land surveying methods to remote sensing. Remote sensing using satellites is becoming popular because they can measure settlement over large areas. One remote sensing technique using satellites is called interferometric synthetic aperture radar (InSAR). Electromagnetic waves are beamed to a swath of ground from radar mounted on the side of a satellite, and the returning signals are recorded for two orbital periods. By comparing the differences in phase of the returning radar signals, the relative ground displacement can be calculated. Another remote sensing system is the global positioning system (GPS), which is a navigation system based on signals transmitted from 24 satellites orbiting the earth twice a day. Table 7.6 summarizes a few of the land-based settlement devices used in practice.

In general, more than one type of instrumentation is used in monitoring settlement.

Table 7.6 Some settlement devices and types of settlement measured.

Type of device	Type of settlement
■ Surface mounted monuments: stakes placed at fixed points on the earth's surface. A level or a total station is used to determine the initial elevation and changes in elevation with time	Surface vertical and horizontal settlements
■ Settlement plate: metal plate with a riser placed on the earth's surface. A level or a total station is used to determine the initial elevation and changes in elevation with time	Vertical settlement at a fixed depth
■ Inclinometer: a probe mounted at the bottom of a casing placed in a borehole. The probe is pulled up to measure the initial position (profile) of the casing and then repeated periodically to measure the displaced profile.	Horizontal settlement with depth (you can also measure the tilt)
■ Extensometer (wire, tape, rod, magnetic): measures distance between two fixed points.	Vertical settlement with depth

7.14 SUMMARY

The estimation of the settlement of soils is important for the design of geosystems. Simplifying assumptions, such as soil is an elastic material, are used in making estimates of soil settlement. The settlement of free draining coarse-grained soil occurs during construction or soon afterward. The settlement of fine-grained soils can occur over decades due to soil consolidation.

Consolidation settlement of a soil is a time-dependent process, which in turn depends on the hydraulic conductivity and thickness of the soil, and the drainage conditions. When an increment of vertical total stress is applied to a soil, the instantaneous (initial) excess porewater pressure is equal to the vertical total stress increment. With time, the initial excess porewater pressure decreases, the vertical effective stress increases by the amount of decrease of the initial excess porewater pressure, and settlement increases. The consolidation settlement is made up of two parts: the early time response called primary consolidation and a later time response called secondary compression.

Soils retain a memory of the past maximum effective stress, which may be erased by loading to a higher stress level. If the current vertical effective stress on a soil was never exceeded in the past (a normally consolidated soil), it would behave elastoplastically when stressed. If the current vertical effective stress on a soil was exceeded in the past (an over-consolidated soil), it would behave elastically (approximately) for stresses less than its past maximum vertical effective stress.

7.14.1 Practical example

EXAMPLE 7.10 *Settlement Due to a Tank Foundation*

A representative stratigraphy at a site for a proposed grain storage tank, 4 m in diameter and 15 m high, is shown in Figure E7.10a. The groundwater is at the top of the clay layer. The tank is located on a circular concrete slab 5 m in diameter that serves as the foundation transmitting the loads to the soil. The weight of the tank full to capacity and of the concrete foundation is 3200 kN. Local code regulations require that the minimum depth of embedment of the foundation be 1 m from the finished surface elevation. Estimated Young's moduli are as follows: poorly graded sand with silt (SP-SM), $E'_{sec} = 20,000$ kPa; poorly graded sand (SP), $E'_{sec} = 15,000$ kPa; lean clay, $E'_{sec} = 40,000$ kPa; well-graded gravel with sand (GW), $E'_{sec} = 45,000$ kPa. One-dimensional consolidation test on the clay (CL) gave $C_c = 0.28$, $C_r = 0.06$, $C_v = 0.05$ m²/day and $OCR = 8$. The specific gravity of clay is 2.7. Assume $\nu' = 0.35$ and neglect the effects (e.g., uplift) of soil excavation.

(a) Estimate the total settlement, neglecting the settlement of the GW soil layer.

(b) Estimate how long it would take for 50% and 100% of primary consolidation settlement in the clay to occur.

(c) If the tank were fully unloaded, and neglecting the settlement of the dead weight of the foundation and the tank, how much settlement would be recoverable.

Strategy You need to calculate the settlement of the sand and the clay to get the total settlement. Since the soil profile consists of layered soil types of finite soil thickness, the increase in vertical total stress using Boussinesq's solution is generally not appropriate. However, we will use it to get a first approximation of the settlement.

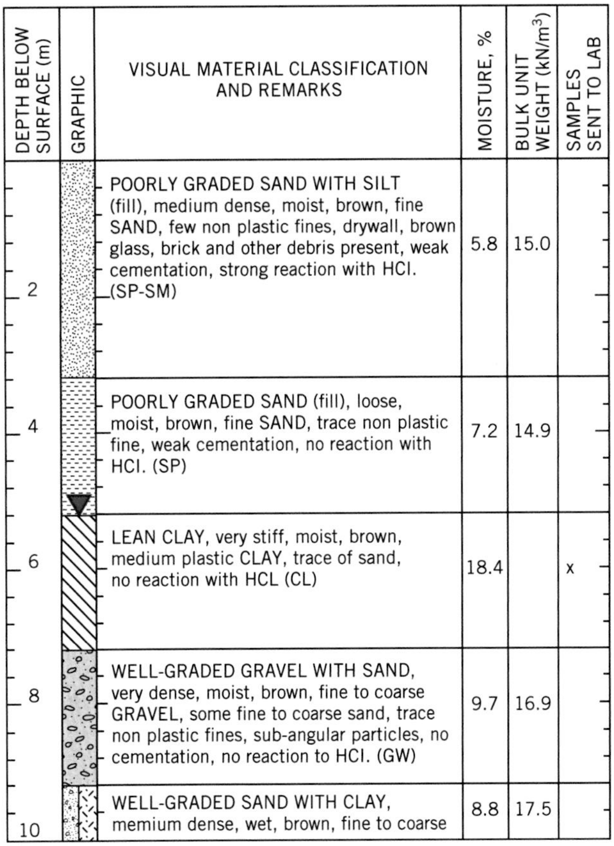

DEPTH BELOW SURFACE (m)	GRAPHIC	VISUAL MATERIAL CLASSIFICATION AND REMARKS	MOISTURE, %	BULK UNIT WEIGHT (kN/m³)	SAMPLES SENT TO LAB
2		POORLY GRADED SAND WITH SILT (fill), medium dense, moist, brown, fine SAND, few non plastic fines, drywall, brown glass, brick and other debris present, weak cementation, strong reaction with HCl. (SP-SM)	5.8	15.0	
4		POORLY GRADED SAND (fill), loose, moist, brown, fine SAND, trace non plastic fine, weak cementation, no reaction with HCl. (SP)	7.2	14.9	
6		LEAN CLAY, very stiff, moist, brown, medium plastic CLAY, trace of sand, no reaction with HCL (CL)	18.4		x
8		WELL-GRADED GRAVEL WITH SAND, very dense, moist, brown, fine to coarse GRAVEL, some fine to coarse sand, trace non plastic fines, sub-angular particles, no cementation, no reaction to HCl. (GW)	9.7	16.9	
10		WELL-GRADED SAND WITH CLAY, memium dense, wet, brown, fine to coarse	8.8	17.5	

Figure E7.10a

Solution 7.10

Step 1: Determine the initial state.

Make a sketch of the tank resting on the finished elevation (Figure E7.10b).

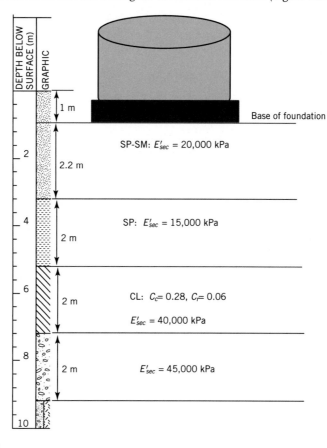

Figure E7.10b

Initial void ratio of clay: $e_o = wG_s = 0.184 \times 2.7 = 0.5$

Clay: $\gamma_{sat} = \dfrac{G_s + e_o}{1 + e_o}\gamma_w = \left(\dfrac{2.7 + 0.5}{1 + 0.5}\right)9.81 = 20.9 \text{ kN/m}^3$

Center of clay layer: $u_o = 1 \times 9.81 = 9.81 \text{ kPa}$; use 9.8 kPa

$$\sigma_{zo} = [15 \times (1 + 1.2)] + (14.9 \times 2) + \left(20.9 \times \dfrac{2}{2}\right) = 98.7 \text{ kPa}$$

$$\sigma'_{zo} = 98.7 - 9.8 = 88.7 \text{ kPa} \approx 89 \text{ kPa}$$

Step 2: Determine the elastic settlement of coarse-grained soils (SP-SM and SP).

SP-SM layer

$$q_s = \dfrac{P_z}{A} = \dfrac{3200}{(\pi \times 4^2)/4} = 255 \text{ kPa}$$

Center of the circular area: $I_{ci} = 1$,

$$\rho_e = \frac{q_s D (1-v)}{E'_{sec}} I_{ci} = \frac{255 \times 4 \times (1-0.35)}{20 \times 10} \times 1 = 44.5 \times 10^3 \text{ m} = 45 \text{ mm}$$

SP layer

We now have to find the stress at the top of the SP layer under the center of the tank.

Under center: $\Delta\sigma_z = q_s I_c$; $I_c = \left[1 - \left(\frac{1}{1+(r_o/z)^2}\right)^{3/2}\right] = \left[1 - \left(\frac{1}{1+(2/2.2)^2}\right)^{3/2}\right] = 0.62$

$$\Delta\sigma_z = q_s I_c = 255 \times 0.692 = 158 \text{ kPa}$$

$$\rho_e = \frac{\Delta\sigma_z D (1-v^2)}{E'_{sec}} I_{ci} = \frac{158 \times 4 \times (1-0.35^2)}{15 \times 10^3} \times 1 = 37 \times 10^{-3} \text{ m} = 37 \text{ mm}$$

Step 3: Calculate the settlement of the clay.

Short term (when loading is applied): For elastic compression,

Under center: $\Delta\sigma_z = q_s I_c$; $I_c = \left[1 - \left(\frac{1}{1+(r_o/z)^2}\right)^{3/2}\right] = \left[1 - \left(\frac{1}{1+(2/4.2)^2}\right)^{3/2}\right] = 0.42$

$$\Delta\sigma_z = q_s I_c = 255 \times 0.42 = 107 \text{ kPa}$$

$$\rho_e = \frac{\Delta\sigma_z D (1-v^2)}{E'_{sec}} I_c = \frac{107 \times 4 \times (1-0.35^2)}{800 \times 10^3} \times 1 = 9 \times 10^{-3} \text{ m} = 9 \text{ mm}$$

Long term: For consolidation,

$$OCR = \frac{\sigma'_{zc}}{\sigma'_{zo}} = 8; \quad \sigma'_{zc} = 8 \times 89 = 712 \text{ kPa}$$

Calculate the vertical stress increase at the center of the clay layer.

Under center: $\Delta\sigma_z = q_s I_c$; $I_c = \left[1 - \left(\frac{1}{1+(r_o/z)^2}\right)^{3/2}\right] = \left[1 - \left(\frac{1}{1+(2/5.2)^2}\right)^{3/2}\right] = 0.39$

$$\Delta\sigma_z = q_s I_c = 255 \times 0.39 = 99.5 \text{ kPa (Use 100 kPa)}$$

$$\sigma'_{fin} = \sigma'_{zo} + \Delta\sigma_z = 89 + 100 = 189 \text{ kPa} < \sigma'_{zc} \ (= 712 \text{ kPa})$$

$$\rho_{pc} = \frac{H_o}{1+e_o} C_r \log\left(\frac{\log\sigma'_{fin}}{\sigma'_{zo}}\right) = \frac{2}{1+0.5} \times 0.06 \times \log\left(\frac{189}{89}\right) = 0.026 \text{ m} = 26 \text{ mm}$$

Step 4: Calculate total settlement.

$$\rho_t = 45 + 37 + 9 + 26 = 117 \text{ mm}$$

Step 5: Calculate times for 50% and 100% primary consolidation clay settlement to occur.

$$C_v = 0.05 \text{ m}^2/\text{yr}$$

Since coarse-grained soils are above and below the clay layer, it is reasonable to assume double drainage

For 50% consolidation, $T_v = 0.197$,

$$t = \frac{T_v H_{dr}^2}{C_v} = \frac{0.197 \times \left(\frac{2}{2}\right)^2}{0.05} = 3.94 \text{ years (use 4 years)}$$

For 100% consolidation,

$$t \approx \frac{H_o^2}{2C_v} \approx \frac{2^2}{2 \times 0.05} = 40 \text{ years}$$

Step 6: Estimate the recoverable settlement on unloading.

The estimation of the settlement of the SP-SM and SP layers is based on the assumption that they behave elastically. The consolidation settlement of the clay occurs on the recompression line, and according to our assumption, the settlement along this line is recoverable. Therefore, we will recover about 117 mm as soon as the loading is removed.

EXERCISES

For all problems, assume $G_s = 2.7$ unless otherwise stated.

Concept understanding

7.1 What is differential soil settlement and how can it affect a structure?

7.2 The differential settlement between the foundations of two columns that are part of the supporting system for a structure is 5 mm. If the distance between the columns is 4 m, determine the angular distortion.

7.3 A building foundation is constructed on a 6 m thick layer of fine sand with a fines content of 25%. Would you expect the settlement to be completed over a construction period of 14 days? Explain your answer.

7.4 What is the difference between consolidation and compaction?

7.5 A clay soil is 80% saturated. Would Terzaghi's consolidation theory be applicable to this soil? Justify your answer.

7.6 In a one-dimensional consolidation test on a saturated soil, the vertical stress applied is 100 kPa. If the at-rest lateral coefficient is 1, what is the value of the instantaneous excess porewater pressure on application of the vertical stress?

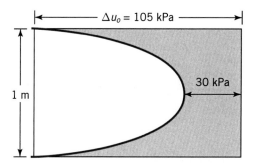

Figure P7.7

7.7 Figure P7.7 shows the excess porewater pressure distribution in a 1 m thick clay at one year after the construction of building. (a) What is the value of the initial vertical stress increase? (b) What is value of the excess porewater pressure at the center of the clay. (c) How much of the initial vertical stress has been transferred to the clay particles (i.e., change in vertical effective stress)?

Problem solving

7.8 A square foundation 1 m × 1 m rests on a surface of a sand classified as SP. If the parameters for the sand is $E'_{sec} = 30,000$ kPa and $\nu = 0.35$, and the foundation must support a load (including its self-weight) of 50 kN, estimate the settlement under the center of the foundation.

7.9 The initial height of a saturated clay sample in a one-dimensional consolidation test is 20 mm. When a vertical load of 40 kPa was applied and the all the excess porewater pressure drained from the sample, the height of the sample was reduced to 19 mm. The initial water content of the sample is 60% and the specific gravity is 2.65. (a) Determine the vertical strain. (b) Determine the change in void ratio.

7.10 Figure P7.10 shows the results of a one-dimensional consolidation test reported by a commercial testing lab on a lean clay.

Sample diameter = 50 mm

LL = 48 %

PL = 38 %

Initial sample height = 19.8 mm

G_s = 2.74

Initial vertical effective stress of the sample in the field = 80 kPa

Initial saturation = 99.6%

Water content at start of test = 43.7%

Water content at end of test = 22.0%

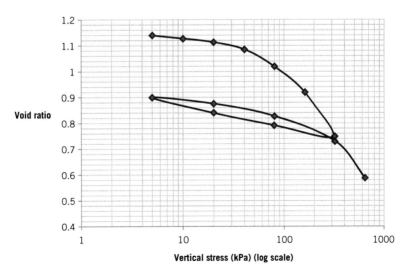

Figure P7.10

You can re-plot Figure P7.10 using the data in the following table. The initial void ratio was calculated from the initial water content.

Loading						
Vertical stress (kPa)	5	10	20	40	80	160	320
Void ratio	1.14	1.128	1.114	1.086	1.02	0.92	0.75
Unloading						
Vertical stress (kPa)	80	20	5			
Void ratio	0.792	0.84	0.9			
Reloading						
Vertical stress (kPa)	20	80	320	640		
Void ratio	0.876	0.828	0.732	0.588		

(a) Determine C_c and C_r. (b) Determine σ'_{zc} using both the Casagrande method and the Brazilian method. (c) The initial vertical effective stress (overburden pressure) is 80 kPa. Calculate the value of OCR. (d) After the test the water content was 22.0 % and the degree of saturation was 100.3%. Calculate the void ratio and dry unit weight at the end of the test. (e) If the void ratio from (d) is different from the void ratio reported, explain the reason or reasons for this difference. (f) What could be the reason for the degree of saturation being slightly over 100%?

7.11 The results of vertical effective stress versus vertical strain (%) for a one-dimensional consolidation test on a silty clay are given in the table below.

Loading						
Vertical stress (kPa)	5	10	20	40	80	160	320
Vertical strain (%)	1.5	2.2	3.3	5	7.4	10	14.2
Unloading						
Vertical stress (kPa)	160	80	40	20	10	5	
Vertical strain (%)	13.8	13	12.2	11.5	11	10.2	
Reloading						
Vertical stress (kPa)	10	20	40	80	160	320	640
Vertical strain (%)	10.6	11.2	11.6	12.3	13.9	14.6	19

(a) Determine m_v and m_{vr} for a vertical effective stress range from 100 kPa to 150 kPa. (b) Calculate the constrained Young's modulus E'_c for vertical effective stress range from 100 kPa to 150 kPa.

7.12 A sample of saturated, normally consolidated clay of height 20 mm and 50 mm diameter was tested in a consolidometer. At the end of the test, the total change in height was 2 mm and the water content was determined as 38%. (a) Calculate the initial void ratio. (b) If the total change in height of the sample after a loading of 80 kPa during primary consolidation were 0.6 mm, calculate the corresponding void ratio.

7.13 A sample of saturated clay of height 20 mm was tested in a one-dimensional consolidometer. Loading and unloading of the sample were carried out. The thickness H_f of the sample at the end of each stress increment/decrement is shown in the table below. At the end of the test, the water content of the sample was determined as 45%.

σ'_z (kPa)	20	40	80	1600	800	400
H_f (mm)	19.64	19.12	18.08	16.78	16.91	17.04

(a) Plot the results as void ratio versus σ'_z (log scale). (b) Determine C_c and C_r. (c) Determine σ'_{zc} using Casagrande's method and the strain energy method.

7.14 The table below shows data recorded during a consolidation test on silty clay sample of diameter 50 mm for an increment of vertical stress of 20 kPa. At the start of the loading, the sample height was 19.94 mm.

Time (min)	0	0.25	1	4	9	16	36	64	100	1440
Total change in height (mm)	0	0.295	0.345	0.480	0.600	0.718	0.885	0.935	0.953	0.963

Determine C_v using the root time and the log time methods.

7.15 A consolidation test on a saturated clay soil gave the following results: $C_c = 0.2$, $C_r = 0.04$, and $OCR = 4.5$. The existing vertical effective stress in the field is 120 kPa. A building foundation will increase the vertical total stress at the center of the clay by 80 kPa. The thickness of the clay layer is 1.5 m and its water content is 36%. (a) Calculate the primary consolidation settlement. (b) What would be the difference in settlement if OCR were 1.5 instead of 4.5?

7.16 The liquid and plastic limits of a clay are 48% and 23%, respectively. The specific gravity is 2.7. Estimate C_c and C_r.

Critical thinking and decision making

7.17 Two adjacent bridge piers rest on saturated clay layers of different thickness but with the same consolidation parameters. Pier 1 imposes a vertical stress increment of 150 kPa to a 2 m thick layer, while pier 2 imposes the same stress to a 3 m thick layer. (a) What is the differential settlement between the two piers if $m_v = 2 \times 10^{-5} \, \text{m}^2/\text{kN}$? (b) If the settlement of the two piers were to be the same, what should be the ratio of the areas of the bases of piers 1 and 2.

7.18 A covered steel (unit weight = 80 kN/m³) tank, 15 m inside diameter × 10 m high and with 10 mm wall thickness, is filled with liquid (unit weight = 9 kN/m³) up to a height of 9.9 m. The tank sits on a concrete (unit weight = 25 kN/m³) foundation of diameter 15.5 m and thickness of 0.75 m. The foundation bottom is 0.75 m below the finished surface elevation. The soil consists of 4.75 m thick soft, normally consolidated marine clay above a very thick layer of gravel (>10 m thick). The geotechnical data of the clay are $C_c = 0.32$, $C_r = 0.08$, $C_v = 0.08 \, \text{m}^2/\text{year}$, and $w = 48\%$. The groundwater level is 0.75 m below the surface. Assume that the foundation is flexible and that the soil above the groundwater level is saturated. (a) Make a sketch of the soil profile with the tank resting on its foundation. (b) Calculate the primary consolidation settlement under the center of the completely filled tank. (c) Calculate the times for 50% and 100% consolidation to occur. (d) Calculate the primary consolidation settlement if the tank was loaded to half its capacity and kept there for 2 years. (e) Calculate the rebound if the tank is now completely emptied.

7.19 Figure P7.19 shows a representative soil profile at a site for a proposed office building 30 m × 60 m. The finished elevation is 1.2 m below the existing ground surface (top of the poorly graded sand with silt). Test data and other pertinent information are shown in Figure P7.19. The base of the concrete slab for the foundation of the building is 0.8 m below the finished elevation. The average vertical surface stress from the building is 150 kPa. (a) Calculate the total settlement over a depth of 5 m below the bottom of the concrete slab. You should estimate the Young's modulus using the N values and Table 7.1. Assume that the rod length for the SPT test is 0.3 m greater than the depth at which the SPT test was done. Use $C_E = 1$ for the hammer. Assume that Poisson's ratio is 0.35 for all soils. Groundwater was not observed up to a depth of 10 m.

7.20 A building of the same size (30 m × 60 m) and applied vertical surface stress (150 kPa) as in P7.19 is located in an area where the representative soil profile is shown in Figure P7.20. In this case, a 1.2 m clay layer was identified starting 4 m from the surface. The groundwater level is at the top of the clay. Consolidation test on a sample of the clay gave $C_c = 0.46$, $C_r = 0.1$, $OCR = 1$, and $C_v = 0.004 \, \text{m}^2/\text{yr}$. The clay's water content is 60% and its specific gravity is 2.65. The averaged values of q_c from cone penetrometer test are: silty sand with gravel, 5.2 MPa; poorly graded sand with silt, 4.4 MPa. The base of the concrete slab for the foundation of the building is 0.75 m below the current surface elevation. Assume that Poisson's ratio is 0.35 for all soils. Neglect the soil above the bottom of the foundation. (a) Calculate the total settlement up to the bottom of the lean clay layer below the bottom of the concrete slab. Young's modulus of the clay is 15,000 kPa. You should estimate Young's modulus for the coarse-grained soils using the cone penetration values and Table 7.1. (b) What percentage of the total settlement is long term settlement? (c) If the design life of the building is 30 years, how much settlement would occur in 15 years and at the end of the design life of the building?

DEPTH BELOW SURFACE (m)	GRAPHIC	VISUAL MATERIAL CLASSIFICATION AND REMARKS	MOISTURE, %	DRY UNIT WEIGHT (kN/m^3)	N
		SILTY SAND WITH GRAVEL, dry, brown, fine to coarse SAND, little non plastic fines, little fine to coarse gravel, max. particle size 50 mm. (SM)			
2 4		POORLY GRADED SAND WITH SILT, medium dense, dry, brown, fine to medium SAND, few non plastic fines, no cementation, no reaction with HCl. (SP-SM)	10.7	14.5	25
6 8		WELL-GRADED SAND, medium dense, dry, brown, fine to coarse SAND, trace non plastic fines, rounded particles, no cementation, no reaction with HCl. (SW) No recovery.	11.2	15.5	19
10		WELL-GRADED GRAVEL WITH SAND, dense, moist, brown, fine to coarse			

Figure P7.19

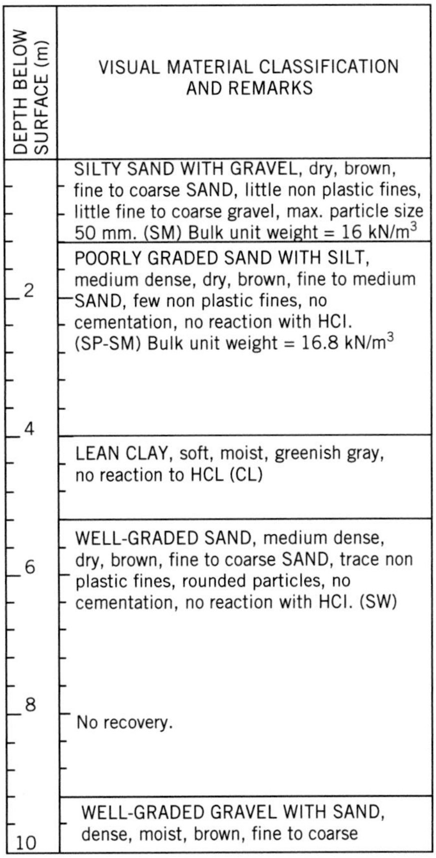

DEPTH BELOW SURFACE (m)	VISUAL MATERIAL CLASSIFICATION AND REMARKS
	SILTY SAND WITH GRAVEL, dry, brown, fine to coarse SAND, little non plastic fines, little fine to coarse gravel, max. particle size 50 mm. (SM) Bulk unit weight = 16 kN/m³
2	POORLY GRADED SAND WITH SILT, medium dense, dry, brown, fine to medium SAND, few non plastic fines, no cementation, no reaction with HCl. (SP-SM) Bulk unit weight = 16.8 kN/m³
4	LEAN CLAY, soft, moist, greenish gray, no reaction to HCL (CL)
6	WELL-GRADED SAND, medium dense, dry, brown, fine to coarse SAND, trace non plastic fines, rounded particles, no cementation, no reaction with HCl. (SW)
8	No recovery.
10	WELL-GRADED GRAVEL WITH SAND, dense, moist, brown, fine to coarse

Figure P7.20

Chapter 8
Soil Strength

8.1 INTRODUCTION

The safety of any geotechnical structure is dependent on the strength of the soil. The term "strength of soil" normally refers to the shearing strength or shear strength. If the soil fails, a structure founded on or within it can collapse, endangering lives and causing economic damage. In this chapter, we will define, describe, and determine the shear strength of soils.

Learning outcomes

When you complete this chapter, you should be able to do the following:

- Understand the concept of shear strength of soils.
- Understand typical stress–strain behavior of soils.
- Understand the differences between drained and undrained shear strength.
- Interpret laboratory and field test results to obtain shear strength parameters.
- Determine the type of shear test that best simulates field conditions.

8.2 DEFINITIONS OF KEY TERMS

Shear strength of a soil (τ_f) is the maximum internal shear resistance to applied shearing forces.

Effective friction angle (ϕ') is a measure of the shear strength of soils due to friction. It is also called angle of shearing resistance.

Cementation (c_{cm}) is a measure of the shear strength (can also be interpreted as bond strength) of a soil from forces that cement the particles.

Soil tension (c_t) is a measure of the apparent shear strength of a soil from soil suction (negative porewater pressures or capillary stresses).

Cohesion (c_o) is a measure of the resistance due to intermolecular forces.

Undrained shear strength (s_u) is the shear strength of a soil when sheared at constant volume.

Apparent cohesion (C) is the apparent shear strength at zero normal effective stress.

Soil Mechanics Fundamentals, First Edition. Muni Budhu.
© 2015 John Wiley & Sons, Ltd. Published 2015 by John Wiley & Sons, Ltd.
Companion website: www.wiley.com\go\budhu\soilmechanicsfundamentals

Critical state is a stress state (failure stress state) reached in a soil when continuous shearing occurs at constant shear stress to normal effective stress ratio and constant volume.

Dilation is a measure of the change in volume of a soil when the soil is distorted by shearing.

8.3 BASIC CONCEPT

The shear strength of a soil is the maximum internal resistance to applied shearing forces. A soil mass will distort when shear forces are applied to it. A measure of this distortion is the shear strain. The shear strain or engineering strain is $\gamma_{xz} = \varepsilon_1 - \varepsilon_3$ (ε_1 is the major principal strain and ε_3 is the minor principal strain), and the shear stress, τ, is the shear force divided by the area of the plane that it acts on.

When a soil is sheared it mobilizes its fabric (structural arrangement of its particles or grains) to effectively resist the imposed shear stresses. Failure will occur along a path (or swath) within the soil fabric that offers the least resistance. Failure has many connotations in engineering. Failure in our context is a condition in which the soil fabric cannot resist further shearing stresses. Along the path of least resistance, the soil reaches a critical density (unit weight) that remains constant under continuous shearing.

The initial state (initial stresses, initial unit weight or density, and overconsolidation ratio) and the shearing stresses (magnitude and direction) or shear strains have significant influence on the soil fabric that is mobilized during shearing. Thus, dense sand mobilizes a different soil fabric than loose sand to resist the same shear stress. Unlike a three-dimensional truss where the members are held together at joints, the joints in soils are grain or particle contacts. It is these grain to grain contacts that provide the shearing resistance. The prevailing theory is that failure occurs when the average frictional resistance of grains along the path (or swath) of lease resistance is exceeded. We will explore this theory and its applications later in this chapter.

> *What's next* ... In the next section, typical soil responses to shearing are considered.

8.4 TYPICAL RESPONSE OF SOILS TO SHEARING FORCES

We are going to summarize the important features of the responses of two groups of uncemented soils—type I and type II—when subjected to a constant vertical (normal) effective stress, σ_n' and increasing shear strain, γ_{zx}. These are artificial groups created here to distinguish two basic types of soil responses observed during shearing. We will consider the shear stress versus the shear strain, the vertical strain, which is the change in height, ΔH_o, divided by the original height, H_o, that is, $\epsilon_z = \Delta H_o/H_o$, versus the shear strain, and the void ratio, e, versus the shear strain responses, as illustrated in Figure 8.1. Compression is taken as positive; expansion as negative. We will assume that the lateral strains are zero, so that the change in vertical strain is the same as the change in volume. Type I soils represent mostly loose sands and normally consolidated and lightly overconsolidated clays ($OCR \leq 2$). Type II soils represent mostly dense sands and overconsolidated clays ($OCR > 2$).

Type I soils—*loose sands, normally consolidated and lightly overconsolidated clays* ($OCR \leq 2$)—are observed to:

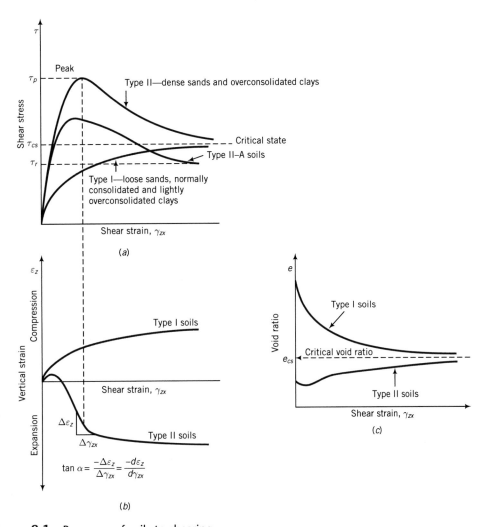

Figure 8.1 Response of soils to shearing.

- Show gradual increase in shear stresses as the shear strain increases (strain-hardens) until an approximately constant shear stress, which we will call the critical state shear stress, τ_{cs}, is attained (Figure 8.1a).
- Compress, meaning they become denser until no further change in volume occurs (Figure 8.1b) or until a constant void ratio, which we will call the critical void ratio, e_{cs}, is reached (Figure 8.1c).

Type II soils—*dense sands and heavily overconsolidated clays* (OCR > 2)—are observed to:

- Show a rapid increase in shear stress, reaching a peak value, τ_p, at low shear strains (compared to type I soils) and then show a decrease in shear stress with increasing shear strain (strain-softens), ultimately attaining a critical state shear stress (Figure 8.1a). The strain-softening response generally results from localized failure zones called shear bands. These shear bands are soil pockets that have loosened and reached the critical state shear stress. Between the shear bands are denser soils that gradually loosen as shearing continues. The shear bands are synonymous with the swath within the soil fabric that offers the least resistance as presented in Section 8.3.

When a shear band develops in some types of overconsolidated clays, the particles become oriented parallel to the direction of the shear band, causing the final shear stress of these clays to decrease below the critical state shear stress. We will call this type of soil type II-A, and the final shear stress attained the residual shear stress, τ_r. Type II-A soils have often been observed in slopes with clay-rich soils that have failed in the past. The prior movement of these slopes polished the soil particles reducing the frictional resistance. Type I soils at very low normal effective stress can also exhibit a peak shear stress during shearing.

- Compress initially (attributed to particle adjustment) and then expand, that is, they become looser (Figure 8.1b, c) until a critical void ratio (the same void ratio as in type I soils) is attained.

The volume expansion is called dilation, which is a measure of the increase in volume of the soil with respect to the change in shear strain. Dilation depends essentially on the structural arrangement of the soil particles and the applied stress path (a representation of stress that includes magnitude and direction, and type such as compression, shear, or both). The structural arrangements of the soil particles depend on the shape, size, and distribution of the particles within the soil, depositional history, and prior loadings. The structural arrangements of soil particles influence the soil's porosity and therefore its density and permeability. Dilation can be seen in action at a beach. If you place your foot on beach sand just following a receding wave, you will notice that the initially wet, saturated sand around your foot momentarily appears to be dry (whitish color). This results from the sand mass around your foot dilating and sucking water up into the voids. The water is released, showing up as surface water, when you lift your foot up.

The critical state shear stress is reached for all soils when no further volume change occurs under continued shearing. We will use the term *critical state* (synonymous with failure as defined in Section 3) to define the stress state reached by a soil when no further change in shear stress and volume occurs under continuous shearing at a constant normal effective stress. For some dense coarse-grained soils and heavily overconsolidated clays, the critical state is not always discernible from laboratory shear tests because the formation of shear bands leads to some parts of the soil mass reaching critical state while other parts have not (the soil mass is not deforming as a single unit). You would have to shear the soil mass to very large strains ($\gg 10\%$) for it to fully develop critical state.

8.4.1 Effects of increasing the normal effective stress

The effects of increasing the normal effective stress are as follows:

1. *Type I soils:* The amount of compression and the magnitude of the critical state shear stress increase (Figure 8.2a, b). In a plot of normal effective stresses versus the critical state shear stresses, an approximate straight line from the origin, OA in Figure 8.2c, is normally observed. The angle between OA and the σ'_n axis is the critical state angle of shearing resistance, ϕ'_{cs}. We will call the angle of shearing resistance the friction angle. The critical state friction angle is a constant for a given soil and is a fundamental soil property (property does not vary with loading conditions).

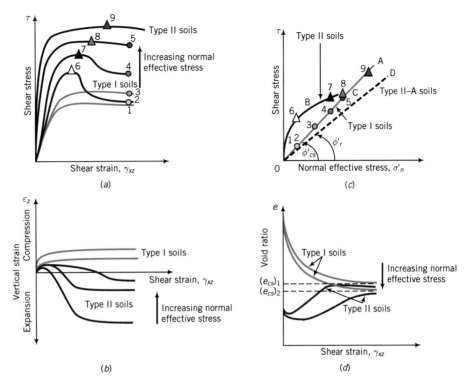

Figure 8.2 Effects of increasing normal effective stresses on the response of soils.

2. *Type II soils:* The peak shear stress tends to disappear, the critical state shear stress increases, and the change in volume expansion decreases (Figure 8.2a, b). In a plot of normal effective stresses versus the peak shear stresses, a curve (OBCA, Figure 8.2c) is normally observed. At large normal effective stresses, the peak shear stress is suppressed, and only a critical state shear stress is observed and appears as a point (point 9) located on OA (Figure 8.2c). The angle between the normal effective stress axis and a line from the origin to each peak shear stress gives the peak friction angle, ϕ'_p, at the corresponding normal effective stress.

3. *Type II-A soils:* In a plot of normal effective stresses versus the residual shear stresses, we normally get a line OD below OA (Figure 8.2c). The angle between OD and the σ'_n axis gives the residual friction angle, ϕ'_r.

As the normal effective stress increases, the critical void ratio decreases (Figure 8.2d). Thus, the critical void ratio is dependent on the magnitude of the normal effective stress and is not a fundamental soil parameter.

8.4.2 Effects of overconsolidation ratio, relative density, and unit weight ratio

The initial state of the soil dictates the response of the soil to shearing forces. For example, two overconsolidated homogeneous fine-grained soils with different overconsolidation ratios but the same mineralogical composition would exhibit different peak shear stresses and volume expansion, as shown in Figure 8.3. The higher overconsolidated soil generally tends to give a higher peak shear strength and greater volume expansion. The effects of relative

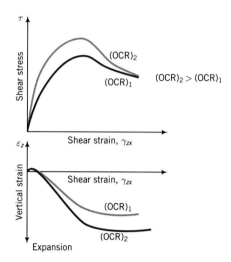

Figure 8.3 Effects of *OCR* on peak strength and volume expansion.

density on the response of coarse-grained soils are similar to the effects of overconsolidation ratios on fine-grained soils.

For the same void ratio, a soil may have different soil fabrics. It is the characteristic of the soil fabric rather than the void ratio that is crucial for strength, stability and settlement of soils. We do not know the characteristic of the soil fabric, so we resort to void ratio to assist us in interpreting soil behavior. The unit weight ratio, R_d, can also be used instead of void ratio as illustrated in Figure 8.4.

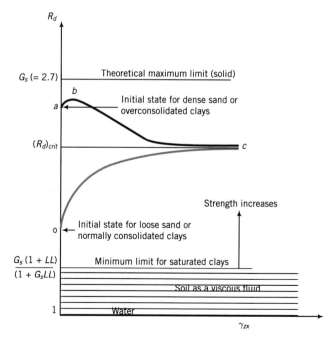

Figure 8.4 Effects of unit weight ratio.

The unit weight ratio for a saturated soil can be expanded as $R_d = [G_s/(1 + e)] + [e/(1 + e)]$. The term $[G_s/(1 + e)]$ is the solid fraction contribution; the term $[e/(1 + e)]$ is void fraction, which is the porosity. Since the shearing resistance of the void fraction (water and air) is zero, all the shear resistance is provided by the solid fraction. Higher values of the solid fraction, that is, lower values of e (denser soil), will result in higher shearing resistance. For loose sand or normally consolidated clays, the shear resistance (strength) will continuously increase as the soil mobilized a soil fabric with a denser configuration until a critical unit weight ratio, $(R_d)_{crit}$, is achieved. For dense sand or overconsolidated clays, a denser soil fabric is initially mobilized compatible with the confining stresses (mean effective stress) applied. Thereafter, shear strains or shear stresses forces the soil fabric to loosen. During this loosening process, the shearing resistance increases until the soil mobilizes a fabric that provides the greatest resistance under the confining stress. Afterwards, the shear resistance decreases until $(R_d)_{crit}$ is achieved.

8.4.3 Effects of drainage of excess porewater pressure

In geotechnical practice, we consider drained (long-term loading, effective stress analysis) and undrained conditions (short-term loading, total stress analysis) in evaluating the stability of soil structures. Drained condition occurs when the excess porewater pressure developed during loading of a soil dissipates, that is, $\Delta u = 0$. The volume of the soil then changes with loading. Undrained condition occurs when the excess porewater pressure cannot drain, at least quickly, from the soil; that is, $\Delta u \neq 0$. The volume of the soil does not change during undrained loading. The existence of either condition—drained or undrained—depends on the soil type, the geological features (e.g., fissures, sand layers in clays), and the rate of loading.

In reality, neither condition is true. They are limiting theoretical conditions that set up the bounds within which the true condition lies.

1. A soil with a tendency to compress during drained loading will exhibit an increase in excess porewater pressure (positive excess porewater pressure, Figure 8.5) under undrained condition, resulting in a decrease in effective stress.

2. A soil that expands during drained loading will exhibit a decrease in excess porewater pressure (negative excess porewater pressure, Figure 8.5) under undrained condition, resulting in an increase in effective stress. These changes in excess porewater pressure occur because the void ratio does not change during undrained loading; that is, the volume of the soil remains constant.

3. During the life of a geotechnical structure, called the long-term condition, the excess porewater pressure developed by a loading dissipates, and drained condition applies. Clays usually take many years to dissipate the excess porewater pressures.

4. During construction and shortly after, called the short-term condition, soils with low hydraulic conductivity (fine-grained soils) do not have sufficient time for the excess porewater pressure to dissipate, and undrained condition applies.

5. The hydraulic conductivity of clean coarse-grained soils (<5% fines) is sufficiently large that under static loading conditions the excess porewater pressure dissipates quickly. Consequently, undrained condition does not apply to clean, coarse-grained soils under static loading, but only to fine-grained soils and to mixtures of coarse- and fine-grained soils. Coarse-grained soils with fines > 35% are likely to behave like fine-grained soils. Coarse-grained soils with fines content between 5% and 35%, are likely to behave somewhere between fine-grained soils and clean, coarse-grained soils. Dynamic loading,

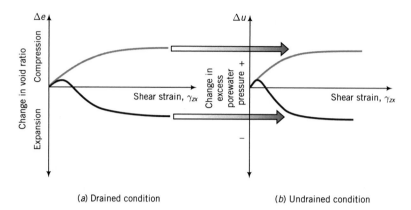

(a) Drained condition   (b) Undrained condition

Figure 8.5 Effects of drained and undrained conditions on volume and excess porewater pressure changes.

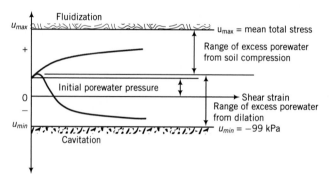

Figure 8.6 Porewater pressure limits.

such as during an earthquake, is imposed so quickly that even clean coarse-grained soils do not have sufficient time for the excess porewater pressure to dissipate, and undrained condition applies.

6. The possible maximum porewater cannot exceed the mean principal total stress (1/3 the sum of vertical and lateral principal total stresses). Recall from your Mechanics of Materials course that principal stress occurs on planes of zero shear stress. For isotropic loading (same normal loading in all directions), the mean principal total stress is equal to the vertical total stress. When the porewater pressure becomes equal to the mean principal total stress, a state of fluidization occurs. The soil behaves like a viscous fluid. The possible minimum porewater pressure cannot be less than a gauge pressure of -99 kPa (vapor pressure of water at 20°C). At this minimum pressure, cavitation (cavities or bubbles are formed within the water in the void spaces and implode) can occur (Figure 8.6).

A summary of the essential differences between drained and undrained conditions is shown in Table 8.1.

8.4.4 Effects of cohesion

Cohesion, c_o, represents the action of intermolecular forces on the shear strength of soils. These forces do not contribute significant shearing resistance for practical consideration. In

Table 8.1 Differences between drained and undrained loading.

Condition	Drained	Undrained
Excess porewater pressure	$\sim$0	Not zero; could be positive or negative
Volume change	Compression Expansion	Positive excess porewater pressure Negative excess porewater pressure
Consolidation	Yes, fine-grained soils	No
Compression	Yes	Yes, but lateral expansion must occur so that the volume change is zero.
Analysis	Effective stress	Total stress
Design strength parameters	ϕ'_{cs} (or ϕ'_p or ϕ'_r)	s_u

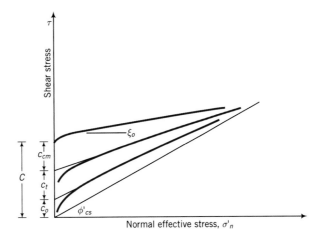

Figure 8.7 Peak shear stress envelope for soils resulting from cohesion, soil tension, and cementation.

a plot of normal effective stresses versus the peak shear stresses using shear test data, an intercept shear stress, c_o, would be observed (Figure 8.7) when a best-fit straight line is used as the trend line.

8.4.5 Effects of soil tension and saturation

Soil tension is the result of surface tension of water on soil particles in unsaturated soils. A suction pressure (negative porewater pressure from capillary stresses) is created that pulls the soil particles together. Recall that the effective stress is equal to total stress minus porewater pressure. Thus, if the porewater pressure is negative, the normal effective stress increases (total normal stress is constant). For soil as a frictional material, this increase in normal effective stress leads to a gain in shearing resistance. The intergranular friction angle or critical state friction angle does not change.

If the soil becomes saturated, the soil tension reduces to zero. Thus, any gain in shear strength from soil tension is only temporary. It can be described as an apparent shear strength, c_t. In practice, you should not rely on this gain in shear strength, especially for long-term loading. However, with experience, you will learn to utilize this gain in shearing strength for short-term geotechnical applications such as in constructing an excavation in overconsolidated soils without bracing.

Unsaturated soils generally behave like type II soils because negative excess porewater pressure increases the normal effective stress and, consequently, the shearing resistance. In a plot of normal effective stresses versus the peak shear stresses using shear test data, an intercept shear stress, c_t, would be observed (Figure 8.7).

8.4.6 Effects of cementation

Nearly all natural soils have some degree of cementation, wherein the soil particles are chemically bonded. Salts such as calcium carbonate ($CaCO_3$) are the main natural compounds for cementing soil particles. The degree of cementation can vary widely, from very weak bond strength (soil crumbles under finger pressure) to the bond strength of weak rocks.

Cemented soils possess shear strength even when the normal effective stress is zero. They behave much like type II soils except that they have an initial shear strength, c_{cm}, under zero normal effective stress. In this textbook, we will call this initial shear strength the cementation strength. In a plot of normal effective stresses versus the peak shear stresses using shear test data, an intercept shear stress, c_{cm}, would be observed (Figure 8.7). The slope angle, ξ_o, of the best-fit straight line from peak shear test data is the apparent friction angle (Figure 8.7).

The shear strength from cementation is mobilized at small shear strain levels (generally $< 0.001\%$). In most geotechnical structures, the soil mass is subjected to much larger shear strain levels. You need to be cautious in utilizing c_{cm} in design because at large shear strains, any shear strength due to cementation in the soil will be destroyed. Also, the cementation of natural soils is generally nonuniform. Thus, over the footprint of your structure, the shear strength from cementation will vary.

Key points

1. Type I soils—loose sands and normally consolidated and lightly overconsolidated clays—strain-harden to a critical state shear stress and compress toward a critical void ratio.
2. Type II soils—dense sands and overconsolidated clays—reach a peak shear stress, strain-soften to a critical state shear stress, and expand toward a critical void ratio after an initial compression at low shear strains.
3. The peak shear stress of type II soils is suppressed and the volume expansion decreases when the normal effective stress is large.
4. Just before peak shear stress is attained in type II soils, shear bands develop. Shear bands are loose pockets or bands of soil masses that have reached the critical state shear stress. With continued shearing, denser soil masses adjacent to shear bands gradually become looser.
5. All soils reach a critical state, irrespective of their initial state, at which continuous shearing occurs without changes in shear stress and volume for a given normal effective stress.
6. The critical state shear stress and the critical void ratio depend on the normal effective stress.

7. Higher normal effective stresses result in higher critical state shear stresses and lower critical void ratios.
8. The critical void ratio is not a fundamental soil property.
9. At large strains, the particles of some overconsolidated clays become oriented parallel to the direction of shear bands, and the final shear stress attained is lower than the critical state shear stress.
10. Higher overconsolidation ratios of homogeneous soils result in higher peak shear stresses and greater volume expansion.
11. Volume changes that occur under drained condition are suppressed under undrained condition. The result of this suppression is that a soil with a compression tendency under drained condition will respond with positive excess porewater pressures during undrained condition, and a soil with an expansion tendency during drained condition will respond with negative excess porewater pressures during undrained condition.
12. Cohesion, defined as the shearing resistance from intermolecular forces, is generally small for consideration in geotechnical application.
13. Soil tension resulting from surface tension of water on soil particles in unsaturated soils creates an apparent shear resistance that disappears when the soil is saturated. You need to be cautious in utilizing this additional shearing resistance in certain geotechnical applications such as shallow excavations.
14. Cementation—the chemical bonding of soil particles—is present to some degree in all natural soils. It imparts shear strength to the soil at zero normal effective stress. The shear strain at which this shear strength is mobilized is very small. You should be cautious in using this shear strength in designing geotechnical systems because in most of these systems the shear strain mobilized is larger than that required to mobilize the shear strength due to cementation.
15. The critical state friction angle is a fundamental soil parameter. It does not change with changes in loading conditions. The peak friction angle, the dilation angle, and the critical void ratio are not fundamental soil parameters. They change with changes in the shape, size, and distribution of the soil particles, the magnitude of the normal effective stress, the initial porosity (or initial void ratio or initial relative density or initial confining pressure), the applied stress path and the strength of the particles.

What's next ... You should now have a general idea of the responses of soils to shearing forces. How do we interpret these responses using mechanical models? In the next section, three models are considered.

8.5 THREE MODELS FOR INTERPRETING THE SHEAR STRENGTH OF SOILS

There are several soil failure criteria that are used to interpret the shear strength of soils. We will present three such criteria in this textbook. The details of these failure criteria and some others are described in Muni Budhu *Soil Mechanics and Foundations* (3rd ed., Wiley, 2011). Only the essential details of the three criteria are presented here. These criteria will be used to interpret the results from laboratory soil strength tests.

The subscript f will be used in this textbook as a generic notation of failure. For these models, failure is associated with either peak stress state or critical state. We will differentiate these states in discussing the use of these models for soils.

8.5.1 Coulomb's failure criterion

Coulomb's failure criterion states that the shear strength of a soil is proportional to normal effective stress acting on the failure (or slip) plane. Coulomb's failure criterion for saturated, uncemented soils at critical state (Figure 8.8) is expressed as

$$\tau_{cs} = (\sigma'_n)_{cs} \tan\phi'_{cs} \tag{8.1}$$

where $\tau_{cs}(= T_{cs}/A)$, with T_{cs} as the shear force at critical state and A as the area of the plane parallel to T_{cs}), is the shear stress at critical state, $(\sigma'_n)_{cs}$ is the normal effective stress at critical state, and ϕ'_{cs} is the critical state friction angle. The critical state friction angle is a fundamental soil parameter and, for a given soil, does not change with loading conditions or the initial stress state (a state of stress representing all stresses on all planes. Mohr's circle (see Appendix B) that you have learned in a Mechanics of Materials course is one approach to finding stress state). Since the area on which the shear force and the normal force act is the same, we can re-write Equation (8.1) as $T_{cs} = P_{cs} \tan\phi'_{cs}$ where P is the normal force.

Coulomb's failure criterion for uncemented soils (Figure 8.7) at peak stress state is expressed as

$$\tau_p = (\sigma'_n)_p \tan\phi'_p = (\sigma'_n)_p \tan(\phi'_{cs} + \alpha_p) \tag{8.2}$$

where $\tau_p(= T_p/A)$, with T_p as the shear force at peak state and A as the area of the plane parallel to T_p), is the shear stress at peak state, $(\sigma'_n)_p$ is the normal effective stress on the plane on which slip is initiated, ϕ'_p is the effective friction angle at peak state, which we will call the peak friction angle, and α_p is the peak dilation angle. The peak friction angle is not a fundamental soil parameter. The dilation angle α is also not a fundamental soil parameter but changes with loading conditions and the initial stress state as discussed earlier (Section 4 of this chapter). At critical state, $\alpha = 0$ and Equation (8.2) reduces to Equation (8.1). If a soil mass is constrained in the lateral directions, the dilation angle is represented (Figure 8.1b) as

$$\alpha = \tan^{-1}\left(\frac{-\Delta H_o}{\Delta x}\right) \tag{8.3}$$

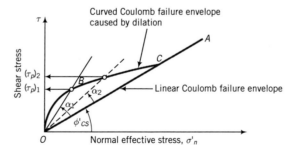

Figure 8.8 Coulomb's failure criterion.

where ΔH_o and Δx are the change in height and the change in horizontal displacement respectively; ΔH_o is negative for expansion. The dilation angle for a soil that tends to expand will be positive. From Equation (8.2), the peak shear stress increases as the dilation angle increases. Thus, the peak friction angle is dependent on the ability of the soil to dilate.

In the case of an unsaturated soil with some degree of cementation and cohesion, the Coulomb's frictional law can be written as

$$\tau_p = C + (\sigma'_n)_p \tan(\xi_o) \tag{8.4}$$

where $C = c_o + c_t + c_{cm}$ is the apparent shear strength at zero normal effective stress (C is the sum of the cohesion, soil tension and cementation strength, Figure 8.7), and ξ_o is the apparent friction angle. If c_o or c_t is small, then $C = c_{cm}$ is the cementation strength. If the soil is uncemented but unsaturated, then $c_{cm} = 0$ and $C = c_o + c_t$ or $C = c_t$ if c_o is neglected. Neither C (c_{cm} or c_o or c_t) nor ξ_o is a fundamental soil parameter. Also, adding C or any one of its components (c_{cm} or c_o or c_t) to the apparent frictional strength $\left[(\sigma'_n)_p \tan \xi_o\right]$ is not strictly correct because these components are not mobilized at the same shear strains. Equation (8.4) represents a linear fit to shear test data to estimate the shear resistance at different normal effective stresses for practical applications.

Coulomb's model applies strictly to soil failures that occur along a presumptive or known slip plane, such as a joint, the interface of two soils, or the interface between a structure and a soil. Stratified soil deposits such as overconsolidated varved clays (regular layered soils that depict seasonal variations in deposition) and fissured clays are likely candidates for failure analysis using Coulomb's model, especially if the direction of shearing is parallel to the direction of the bedding plane.

Key points

1. Coulomb's frictional law for the peak shear stress is $\tau_p = (\sigma'_n)_p \tan(\phi'_{cs} + \alpha_p)$, where τ_p is the shear stress when slip is initiated, $(\sigma'_n)_p$ is the normal effective stress on the slip plane, ϕ'_{cs} is the critical state friction angle, and α_p is the peak dilation angle.
2. Dilation increases the shear strength of the soil and causes Coulomb's failure envelope to curve.
3. Large normal effective stresses tend to suppress dilation.
4. At the critical state, the dilation angle is zero, and $\tau_{cs} = (\sigma'_n)_{cs} \tan \phi'_{cs}$.
5. For unsaturated soils with cementation and cohesion, Coulomb's frictional law is $\tau_p = C + (\sigma'_n)_p \tan(\xi_o)$, where $C = c_o + c_t + c_{cm}$ is the apparent shear strength at zero normal effective stress and ξ_o is the apparent friction angle.
6. Information on the deformation of the soil is not included in the interpretation of soil strength using Coulomb's failure criterion.

8.5.2 Mohr–Coulomb failure criterion

Coulomb's frictional law for finding the shear strength of soils requires that we know the friction angle and the normal effective stress on the slip plane. Both of these components are not readily known because soils are usually subjected to a variety of stresses. By combining Mohr's circle for finding stress states (see Appendix B) with Coulomb's frictional law, we can develop a generalized failure criterion. The Mohr–Coulomb (MC) failure criterion defines failure when the maximum principal effective stress ratio, called maximum effective stress obliquity, $\left(\sigma_1'/\sigma_3'\right)_{max}$, is achieved and not when the maximum shear stress, $[(\sigma_1' - \sigma_3')/2]_{max}$, is achieved (Figure 8.9). The MC failure criterion is expressed as:

8.5.2.1 Saturated or clean, dry, uncemented soils at critical state (Figure 8.9)

$$\sin\phi_{cs}' = \left(\frac{\sigma_1' - \sigma_3'}{\sigma_1' + \sigma_3'}\right)_{cs} \tag{8.5}$$

or

$$\tau_{cs} = \frac{\sigma_1' - \sigma_3'}{2}\cos\phi_{cs}' \tag{8.6}$$

8.5.2.2 Saturated or clean, dry, uncemented soils at peak state

$$\sin\phi_p' = \left(\frac{\sigma_1' - \sigma_3'}{\sigma_1' + \sigma_3'}\right)_p \tag{8.7}$$

or

$$\tau_p = \left(\frac{\sigma_1' - \sigma_3'}{2}\right)_p \cos\phi_p' = \left(\frac{\sigma_1' - \sigma_3'}{2}\right)_p \cos\left(\phi_{cs}' + \alpha_p\right) \tag{8.8}$$

8.5.2.3 Unsaturated, cemented, cohesive soils (Figure 8.10)

$$\sin\xi_o = \frac{\left(\sigma_1' - \sigma_3'\right)}{2C\cot\xi_o + \left(\sigma_1' + \sigma_3'\right)} \tag{8.9}$$

$$\tau_p = C + \frac{1}{2}\tan\xi_o\left[\sigma_1'(1 - \sin\xi_o) + \sigma_3'(1 + \sin\xi_o)\right] \tag{8.10}$$

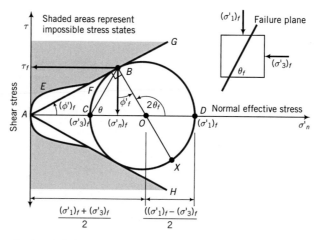

Figure 8.9 Mohr–Coulomb failure criterion. The subscript *f* is replaced by subscript *cs* for critical state and subscript *p* for peak stress.

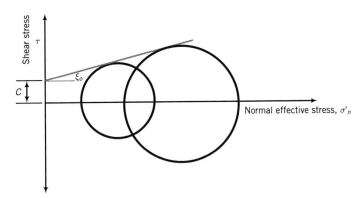

Figure 8.10 The Mohr–Coulomb failure envelope for an unsaturated, cemented soil with cohesion.

In Figure 8.9, the angle $BCO = \theta_f$ represents the inclination of the failure plane (BC) or slip plane to the plane on which the major principal effective stress acts in Mohr's circle. This angle is

$$\text{Critical state:} \quad \theta_{cs} = 45° + \frac{\phi'_{cs}}{2} \tag{8.11}$$

$$\text{Peak state:} \quad \theta_p = 45° + \frac{\phi'_p}{2} \tag{8.12}$$

The MC failure criterion is a limiting stress criterion. Therefore, the failure lines AG and AH (Figure 8.9) are fixed lines in $[\tau, \sigma'_n]$ space. The line AG is the failure line for compression, while the line AH is the failure line for extension (soil elongates; the lateral effective stress is greater than the vertical effective stress). The shear strength in compression and in extension from interpreting soil strength using the MC failure criterion is identical. In reality, this is not so.

The MC failure criterion, like the Coulomb failure criterion, treats the soil above and below the failure plane as rigid bodies. Strains, which are important for geosystems design, are not considered. Also, the MC criterion does not consider the loading history of the soil that is known to influence the shearing responses of soils.

Key points

1. Coupling Mohr's circle with Coulomb's frictional law allows us to define shear failure based on the stress state of the soil.
2. Failure occurs, according to the Mohr–Coulomb failure criterion, when the soil reaches the maximum principal effective stress obliquity, that is, $(\sigma'_1/\sigma'_3)_{max}$.
3. The failure plane or slip plane is inclined at an angle $\theta_f = 45° + (\phi'_f/2)$ to the plane on which the major principal effective stress acts where the subscript f is replaced by cs for critical state and by p for peak state.
4. The maximum shear stress, $\tau_{max} = [(\sigma'_1 - \sigma'_3)/2]_{max}$, is not the failure shear stress.
5. Information on the deformation or the initial stress state of the soil is not included in the interpretation of soil strength using the Mohr–Coulomb failure criterion.

8.5.3 *Tresca's failure criterion*

The shear strength of a fine-grained soil under undrained conditions is called the undrained shear strength, s_u. To interpret the undrained shear strength, we use the Tresca's failure criterion, which states that the shear stress at failure (actually, Tresca's criterion is a yield not a failure criterion, but the latter is accepted for soils) is one-half the principal stress difference. The undrained shear strength, s_u, is the radius of the Mohr total stress circle; that is,

$$s_u = \frac{(\sigma_1)_f - (\sigma_3)_f}{2} = \frac{(\sigma_1')_f - (\sigma_3')_f}{2} \qquad (8.13)$$

as shown in Figure 8.11a. The subscript f in Equation (8.13) is replaced by subscript cs for critical state and subscript p for peak stress. The shear strength under undrained loading depends only on the initial void ratio or the initial water content or initial confining pressure. An increase in initial normal effective stress, sometimes called confining pressure, causes a decrease in the initial void ratio and a larger change in excess porewater pressure when a soil is sheared under undrained conditions. The result is that the Mohr's circle of total stress expands and the undrained shear strength increases (Figure 8.11b). Thus, s_u is not a fundamental soil property. The value of s_u depends on the magnitude of the initial confining pressure or the initial void ratio (or initial water content). Analyses of soil strength and soil stability problems using s_u are called total stress analyses (TSA).

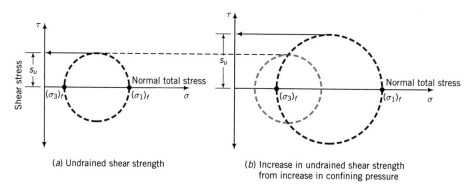

(a) Undrained shear strength

(b) Increase in undrained shear strength from increase in confining pressure

Figure 8.11 Tresca's failure criterion for undrained conditions as represented by Mohr's circles. The subscript f is replaced by subscript cs for critical state and subscript p for peak stress.

Key points

1. For a total stress analysis, which applies to fine-grained soils, the shear strength parameter is the undrained shear strength, s_u.
2. Tresca's failure criterion is used to interpret the undrained shear strength of fine-grained soils.
3. The undrained shear strength depends on the initial void ratio or initial water content or initial confining pressure. It is not a fundamental soil shear strength parameter.
4. Information on the deformation of the soil is not needed to interpret soil strength using Tresca's failure criterion.

Summaries of the three failure criteria are shown in Table 8.2 and Table 8.3

Typical values of ϕ'_{cs}, ϕ'_p, ϕ'_r, and α_p for soils are shown in Table 8.4 and Table 8.5. The ranges of values in these tables and others in this book are not absolute values. They represent ranges of values observed and reported for various soil types. They are intended for guidance. Typical values of s_u are given in Table 8.6.

Table 8.2 Differences among the three failure criteria.

Name	Failure criteria	Soil treated as	Best used for	Test data interpretation[a]
Coulomb	Failure occurs by impending, frictional sliding on a slip plane.	Rigid, frictional material	Layered or fissured overconsolidated soils or a soil where a prefailure plane exists	Direct shear
Mohr–Coulomb	Failure occurs by impending, frictional sliding on the plane of maximum principal effective stress obliquity.	Rigid, frictional material	Long-term (drained condition) strength of overconsolidated fine-grained and dense coarse-grained soils	Triaxial
Tresca	Failure occurs when one-half the maximum principal stress difference is achieved.	Homogeneous solid	Short-term (undrained condition) strength of fine-grained soils	Triaxial

[a]See Section 8.5 for description of these tests.

Table 8.3 Summary of equations for the three failure criteria.

Name	Peak	Critical state
Coulomb	Saturated, uncemented soils: $$\tau_p = (\sigma'_n)_p \tan\phi'_p = (\sigma'_n)_p \tan(\phi'_{cs}+\alpha_p)$$ Unsaturated, cemented soils with cohesion: $$\tau_p = C + (\sigma'_n)_p \tan(\xi_o)$$ $$C = c_o + c_t + c_{cm}$$	$$\tau_{cs} = (\sigma'_n)_{cs}\tan\phi'_{cs}$$
Mohr–Coulomb	Saturated, uncemented soils: $$\sin\phi'_p = \left(\frac{\sigma'_1 - \sigma'_3}{\sigma'_1 + \sigma'_3}\right)_p$$ $$\frac{(\sigma'_3)_p}{(\sigma'_1)_p} = \frac{1-\sin\phi'_p}{1+\sin\phi'_p} = \tan^2\left(45° - \frac{\phi'_p}{2}\right)$$ Unsaturated, cemented soils with cohesion: $$\sin\xi_o = \frac{(\sigma'_1 - \sigma'_3)}{2C\cot\xi_o + (\sigma'_1 + \sigma'_3)}$$ $$C = c_o + c_t + c_{cm}$$ Inclination of the failure plane to the plane on which the major principal effective stress acts. $$\theta_p = 45° + \frac{\phi'_p}{2}$$	$$\sin\phi'_{cs} = \left(\frac{\sigma'_1 - \sigma'_3}{\sigma'_1 + \sigma'_3}\right)_{cs}$$ $$\frac{(\sigma'_3)_{cs}}{(\sigma'_1)_{cs}} = \frac{1-\sin\phi'_{cs}}{1+\sin\phi'_{cs}} = \tan^2\left(45° - \frac{\phi'_{cs}}{2}\right)$$ Inclination of the failure plane to the plane on which the major principal effective stress acts. $$\theta_{cs} = 45° + \frac{\phi'_{cs}}{2}$$
Tresca	$$(s_u)_p = \frac{(\sigma_1 - \sigma_3)_p}{2}$$	$$(s_u)_{cs} = \frac{(\sigma_1 - \sigma_3)_{cs}}{2}$$

Table 8.4 Ranges of friction angles for soils (degrees).

Soil type	ϕ'_{cs}	ϕ'_p	ϕ'_r
Gravel	30–35	35–50	
Mixtures of gravel and sand with fine-grained soils	28–33	30–40	
Sand	27–37[a]	32–50	
Silt or silty sand	24–32	27–35	
Clays	15–30	20–30	5–15

[a]Higher values (32°–37°) in the range are for sands with significant amount of feldspar (Bolton, 1986). Lower values (27°–32°) in the range are for quartz sands.

Table 8.5 Typical ranges of dilation angles for soils.

Soil type	α_p (degrees)
Dense sand	10–15
Loose sand	<10
Normally consolidated clay	0

Table 8.6 Typical values of s_u for saturated fine-grained soils.

Description	s_u (kPa)
Very soft (extremely low)	<10
Soft (low)	10–25
Medium stiff (medium)	25–50
Stiff (high)	50–100
Very stiff (very high)	100–200
Extremely stiff (extremely high)	>200

What's next ... We have studied the responses of soils to loading and have considered three models to interpret these responses. These models allow us to determine strength parameters to be used in the design of geotechnical structures (geostructures). In the next section, the key factors that affect these parameters are discussed.

8.6 FACTORS AFFECTING THE SHEAR STRENGTH PARAMETERS

Knowledge of the shear strength of soils is important for the design and construction, and safety of most geosystems such as foundations, excavations, tunnels, dams, and slopes. Geoengineers need the accurate value of strength parameters in order to design and construct these geosystems. For uncemented soils, the shear strength parameters are the critical state friction angle ϕ'_{cs}, the peak friction angle ϕ'_p, and the undrained shear strengths $(s_u)_{cs}$ and $(s_u)_p$. For cemented soils, the shear strength parameters are the apparent cohesion C (consisting of cohesion c_o, cementation or bond strength c_{cm}, and soil tension c_t), and the apparent friction angle ξ_o. For soils with residual strength, the shear strength parameters are the residual friction angle ϕ'_r and residual undrained shear strength $(s_u)_r$.

The critical state friction angle is a fundamental soil parameter for a given soil. It does not change with the initial stress state (overconsolidation ratio, relative density, initial confining pressure) of the soil or the applied stress path. It is a measure of the average interparticle sliding friction.

The peak friction angle is not a constant for a given soil and is not a fundamental soil property. Its value depends on the soil's capacity to dilate. Thus the peak friction angle is influenced by factors such as the shapes, sizes, gradations, hardness, surface roughness and asperities of the soil particles, the magnitude of the normal effective stress, the initial porosity (or initial void ratio or initial relative density or initial confining pressure), the initial stress state (or overconsolidation ratio), and the applied stress path. The use of the peak friction angle in design depends largely on the geoengineer's experience.

The undrained shear strength is not a constant for a given soil (fine-grained soil) and is not a fundamental soil property. It changes with the initial void ratio (initial porosity or initial water content) or initial effective stress or initial confining pressure. For example, if the void ratio decreases because the soil is subjected to a greater initial effective stress, then the undrained shear strength of the soil theoretically should increase. For a given soil, the undrained shear strength would decrease if the water content increases. The key test information that you should know in utilizing s_u is the initial effective stress or initial confining pressure, the overconsolidation ratio, and whether it was determined at peak or critical state.

The apparent cohesion is comprised of shear strengths from cohesion, soil tension, and cementation. The contribution from either cohesion or soil tension or cementation is difficult to distinguish from test results. The shear strength from cohesion (intermolecular forces) is small for consideration in practical geotechnical projects. Soil tension from unsaturated soils is unreliable. Cementation or bond strength depends on the cementing agent, the uniformity and degree of cementation. The degree of cementation in natural soils is spatially variable and cannot be determined, at least economically. This apparent cohesion is only mobilized at very small shear strains, so it can be utilized only in geotechnical projects where very small soil movements (strains much less than 0.1%) are likely to occur. The apparent cohesion (C) is not equivalent to the undrained shear strength.

The apparent friction angle is not a fundamental soil parameter. It is a convenient practical measure to estimate the shear strength in unsaturated soils with/without cementation at different normal effective stresses.

The residual soil friction depends on the history and degree of loading and the type of soil minerals present. This value of the residual friction can be just a fraction of the critical state friction angle. Particular attention should be paid to this value especially in overconsolidated clays where movements such as in slopes have occurred historically.

What's next ... We have identified the shear strength parameters that are important for the analysis and design of geosystems. A variety of laboratory tests and field tests are used to determine these parameters. We will briefly describe a number of these tests and interpret the results. You may choose to perform some of these tests in the laboratory section of your course. Detailed test procedures are presented in the relevant codes or standards in your country and are not duplicated here.

8.7 LABORATORY TESTS TO DETERMINE SHEAR STRENGTH PARAMETERS

8.7.1 A simple test to determine the critical state friction angle of clean coarse-grained soils

The critical state friction angle ϕ'_{cs} for a clean coarse-grained, dry soil can be found by pouring the soil into a loose heap about 75 mm high on a horizontal surface (usually a large glass plate) using a funnel. The soil at the base of the heap is carefully scraped away until the soil particles are seen moving down the slope's face. Measuring the angle of the heap's slope relative to the horizontal surface gives the angle of repose, which closely approximates ϕ'_{cs}.

8.7.2 Shear box or direct shear test

The direct shear (DS) is a popular test to determine the shear strength parameters because it is simple, easy to perform, and cheap. This test is useful when a soil mass is likely to fail along a thin zone under plane strain conditions (strain in one direction is zero). The sample container for a DS test is either a horizontally split, open metal box (Figure 8.12), called a shear box, or a horizontally split metal cylinder. Soil is placed in the box (or cylinder), and one-half of the box (or cylinder) is moved relative to the other half. Failure is thereby constrained along a thin zone of soil on the horizontal plane (AB). Serrated or grooved metal plates or porous stones are placed at the top and bottom faces of the soil to generate the shearing force.

Vertical forces are applied through a metal platen resting on the top serrated plate or porous stone. Horizontal forces are applied through a motor for displacement control or by weights through a pulley system for load control. Most DS tests are conducted using horizontal displacement control (rate of displacement is between 0.0025 to 1 mm/min) because we can get both the peak shear force and the critical shear force. In load control tests, you cannot get data beyond the maximum or peak shear force.

The horizontal displacement, Δx, the vertical displacement, ΔH_o, the vertical loads (forces), P_z, and the horizontal loads (forces), P_x, are measured. Usually, three or more tests are carried out on a soil sample using three different constant vertical forces. Failure is determined when the soil cannot resist any further increment of horizontal force. The stresses and strains in

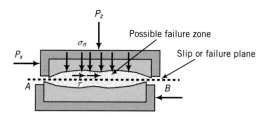

Figure 8.12 Shear box.

the shear box test are difficult to calculate from the forces and displacements measured. The stresses in the thin (dimension unknown) constrained failure zone (Figure 8.12) are not uniformly distributed, and strains cannot be determined.

The DS apparatus cannot prevent drainage, but one can get an estimate of the undrained shear strength of clays by running the shear box test at a fast rate of loading so that the test is completed quickly. Generally, three or more tests are performed on a soil. The soil sample in each test is sheared under a constant vertical force, which is different in each test. The data recorded for each test are the horizontal displacements, the horizontal forces, the vertical displacements, and the constant vertical force under which the test is conducted. From the recorded data, you can find the following strength parameters: τ_p, τ_{cs}, ϕ'_p, ϕ_{cs}, α_p, (and s_u), if the fine-grained soils are tested quickly. Coulomb's failure criterion is used to determine the shear strength. The strength parameters are generally determined by plotting the data, as illustrated in Figure 8.12 for sand.

Only the results of one test at a constant value of P_z are shown in Figure 8.13a, b. The results of $(P_x)_p$ and $(P_x)_{cs}$ plotted against P_z for all tests are shown in Figure 8.13c. If the soil is dilatant, it will exhibit a peak shear force (Figure 8.13a, dense sand) and expand (Figure 8.13b, dense sand), and the failure envelope will be curved (Figure 8.13c, dense sand). The peak shear stress is the peak shear force divided by the cross-sectional area (A) of the test sample; that is,

$$\tau_p = \frac{(P_x)_p}{A} \tag{8.14}$$

The critical state shear stress is

$$\tau_{cs} = \frac{(P_x)_{cs}}{A} \tag{8.15}$$

In a plot of vertical forces versus horizontal forces (Figure 8.13c), the points representing the critical horizontal forces should ideally lie along a straight line through the origin. Experimental results usually show small deviations from this straight line, and a "best-fit" straight line is conventionally drawn. The angle subtended by this straight line and the horizontal axis is ϕ_{cs}. Alternatively,

$$\phi'_{cs} = \tan^{-1}\frac{(P_x)_{cs}}{P_z} \tag{8.16}$$

For dilatant soils, the angle between a line from the origin to each peak horizontal force that does not lie on the best-fit straight line in Figure 8.13c, and the abscissa (normal effective stress axis) represents a value of ϕ'_p at the corresponding vertical force. Recall that ϕ'_p is not constant but varies with the magnitude of the normal effective stress (P_z/A). Usually, the normal effective stress at which ϕ'_p is determined should correspond to the maximum anticipated normal effective stress in the field. The value of ϕ'_p is largest at the lowest value of the applied normal effective stress, as illustrated in Figure 8.13c. You would determine ϕ'_p by drawing a line from the origin to the point representing the peak horizontal force at the desired normal force, and measuring the angle subtended by this line and the horizontal axis (Figure 8.13c). Alternatively,

$$\phi'_p = \tan^{-1}\frac{(P_x)_p}{P_z} \tag{8.17}$$

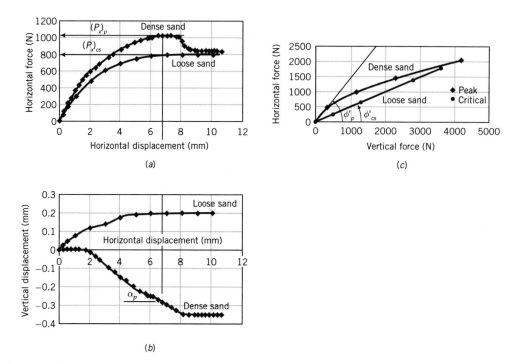

Figure 8.13 Results from a DS using a shear box test on a dense and a loose sand.

You can also determine the peak dilation angle directly for each test from a plot of horizontal displacement versus vertical displacement, as illustrated in Figure 8.13b. The peak dilation angle is

$$\alpha_p = \tan^{-1}\frac{(-\Delta H_o)}{\Delta x} \tag{8.18}$$

We can find α_p from

$$\alpha_p = \phi'_p - \phi'_{cs} \tag{8.19}$$

In general, Equations (8.18) and (8.19) do not give exactly the same result because of nonuniform stresses and strains within the test sample.

EXAMPLE 8.1   *Interpreting Direct Shear Test Results*

A moist, light yellowish-brown clayey sand was taken from a depth of 4 m under a proposed foundation for a building, using a 65 mm diameter sampling tube. It was tested in a direct shear device that accepts a circular sample of the same diameter as the sampling tube. The applied vertical load was 780 N and the horizontal displacement rate was 0.02 mm/min. The shear force versus shear displacement plot is shown in Figure E8.1a.

(a) Is the soil dense or loose sand?

(b) Identify and determine the peak shear force and critical state shear force.

(c) Calculate the peak and critical state friction angles and the peak dilation angle using Coulomb's model.

(d) Calculate the peak dilation angle from the friction angles.

(e) Determine the peak dilation angle from the vertical displacement versus horizontal displacement plot.

Strategy Identify the peak and critical state shear force. The friction angle is the arctangent of the ratio of the shear force to the normal force. The constant shear force at large displacement gives the critical state shear force. From this force, you can calculate the critical state friction angle. Similarly, you can calculate ϕ_p' from the peak shear force.

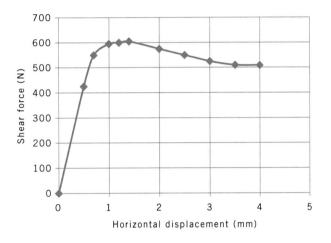

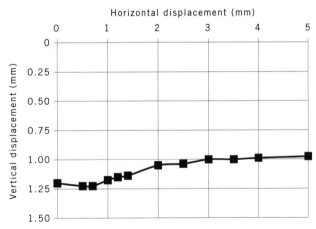

Figure E8.1a

Solution 8.1

Step 1: Determine whether soil is dense or loose.

Since the shear force–shear displacement plot (Figure E8.1a) shows a peak shear force and the soil expands after an initial compression, the soil is likely dense.

Step 2: Identify peak and critical state shear forces.

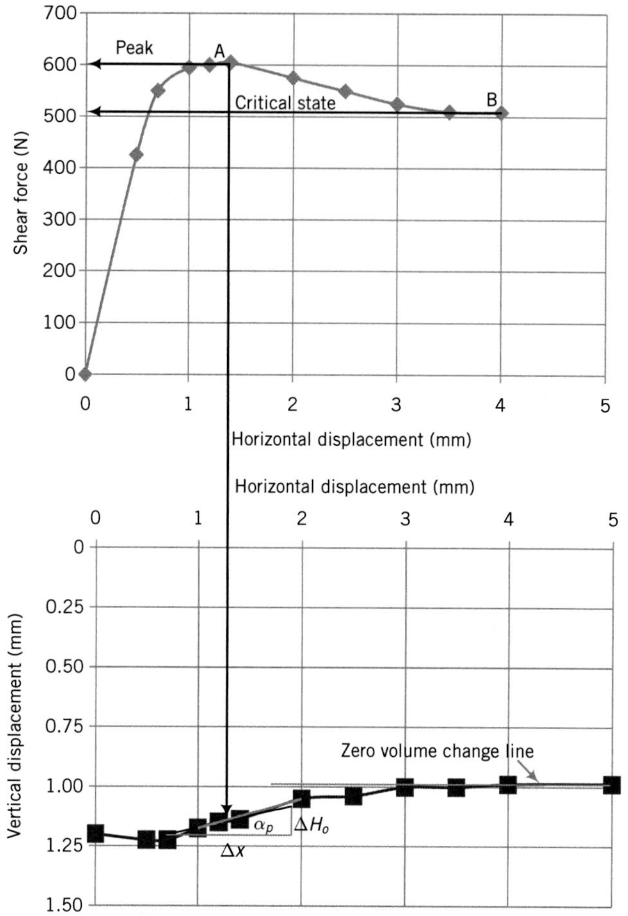

Figure E8.1b

With reference to Figure E8.1b, point A is the peak shear force and point B is the critical state shear force (the shear force and the vertical displacement are approximately constant).

Step 3: Read values of peak and critical state shear forces from the graph in Figure E8.1b.

$$(P_x)_p = 600 \text{ N}$$

$$(P_x)_{cs} = 505 \text{ N}$$

$$P_z = 780 \text{ N (given)}$$

Step 4: Determine friction angles.

$$\phi'_p = \tan^{-1}\left(\frac{(P_x)_p}{P_z}\right) = \tan^{-1}\left(\frac{600}{780}\right) = 37.6°$$

$$\phi'_{cs} = \tan^{-1}\left(\frac{(P_x)_{cs}}{P_z}\right) = \tan^{-1}\left(\frac{505}{780}\right) = 32.9°$$

Step 5: Determine peak dilation angle using the peak and critical state friction angles.
$$\alpha_p = \phi'_p - \phi'_{cs} = 37.6 - 32.9 = 4.7°$$

Step 6: Determine peak dilation angle from the vertical displacement versus horizontal displacement plot.

$$\alpha_p = \tan^{-1} \frac{(-\Delta H_o)}{\Delta x}$$

The maximum slope of the vertical displacement versus horizontal displacement plot corresponding to the peak shear force is

$$\alpha_p = \tan^{-1} \frac{(-\Delta H_o)}{\Delta x} = \tan^{-1} \frac{-(1.08 - 1.16)}{(2 - 1)} = 4.6°$$

This value is very close to that calculated using the peak and critical state friction angles.

Step 7: Check reasonableness of answers.

From Table 8.4 and Table 8.5, the calculated friction angles and peak dilation angles are reasonable. These tables give observed range of values. To determine reasonableness, experience is also required. The peak dilation angle is less than 1/2 of the minimum value for dense sands in Table 8.5. However, the peak dilation angle is strongly influenced by the normal effective stress. Higher normal effective stresses result in lower values of dilation angle. Thus, Table 8.5 is not to be relied on to judge the reasonableness of the results.

EXAMPLE 8.2 *Interpretation of Shear Box Test Data*

Shear box tests data on a dense, dry, clean sand are shown in the table below. The peak or maximum horizontal forces were recorded by the technician performing the tests.

Test number	Vertical force (N)	Horizontal force (N)
Test 1	100	98
Test 2	200	158
Test 3	300	189
Test 4	400	218
Test 5	500	262

Determine the following:

(a) ϕ'_{cs}
(b) ϕ'_p at vertical forces of 200 N and 300 N for sample B
(c) The dilation angle at vertical forces of 200 N and 300 N for sample B

Strategy To obtain the desired values, it is best to plot a graph of vertical force versus horizontal force.

Solution 8.2

Step 1: Plot a graph of the vertical forces versus failure horizontal forces for each sample. See Figure E8.2.

Step 2: Extract ϕ'_{cs}.

The last two points fall on a straight line through the origin. The angle that this line subtends with the horizontal axis gives the critical state friction angle.

$$\phi'_{cs} = \tan^{-1}\left(\frac{264}{500}\right) = 27.8°$$

We have assumed that the volume change is zero for these last two points.

Step 3: Determine ϕ'_p.

The horizontal forces corresponding to vertical forces at 200 N and 300 N do not lie on the straight line corresponding to ϕ'_{cs}. Therefore, each of these forces has ϕ'_p associated with it.

$$\left(\phi'_p\right)_{200\,N} = \tan^{-1}\left(\frac{158}{200}\right) = 38.3°$$

$$\left(\phi'_p\right)_{300\,N} = \tan^{-1}\left(\frac{189}{300}\right) = 32.2°$$

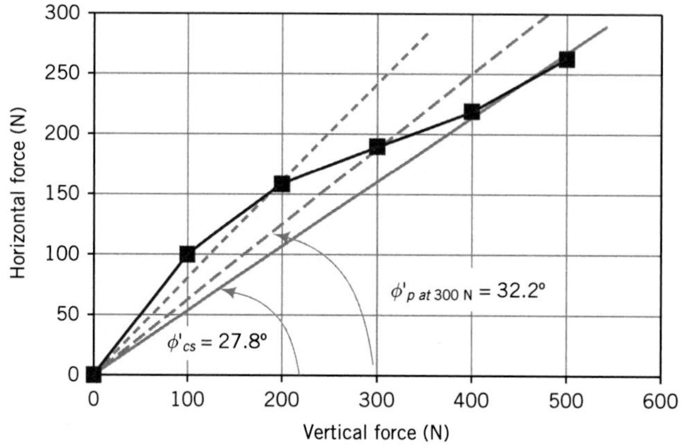

Figure E8.2

Step 4: Determine α_p.

$$\alpha_p = \phi'_p - \phi'_{cs}$$

$$\left(\alpha_p\right)_{200\,N} = 38.3 - 27.8 = 10.5°$$

$$\left(\alpha_p\right)_{300\,N} = 32.2 - 27.8 = 4.4°$$

Note that as the normal force increases, α_p decreases.

Step 5: Check reasonableness of answers.
From Table 8.4 and Table 8.5, the calculated friction angles and peak dilation angles are reasonable. See the comment on dilation angle in Step 7 of Example 8.1.

EXAMPLE 8.3 *Interpretation of Shear Box Test on a Cemented Soil*

Shear box tests results from a saturated, creamish, cemented soil (caliche) are shown in the table below. The shear box sample was 65 mm × 65 mm × 25 mm thick.

Test number	Vertical force (N)	Horizontal force (N)
Test 1	200	196 (peak)
Test 2	400	234 (peak)
Test 3	600	273 (peak)
Test 4	800	298 (peak; maximum value recorded)

(a) Determine the shear strength parameters. (b) Write the shear strength equation.

Strategy To obtain the desired values, it is best to plot a graph of vertical force versus horizontal force and then use the Coulomb's failure criterion for cemented soils.

Solution 8.3

Step 1: Plot a graph of the vertical forces versus failure horizontal forces for each sample.

See Figure E8.3.

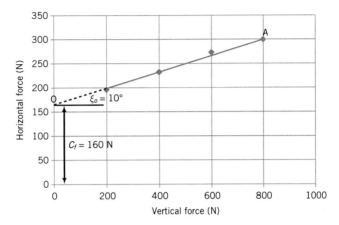

Figure E8.3

Step 2: Draw a best-fit straight line through the data.

The best-fit straight line is shown as OA in Figure E8.3. The dashed part of line OA indicates an assumption that test data points lower than the lowest applied vertical force will be located on it (line OA).

Step 3: Determine the shear strength parameters.

The intercept of line OA gives the apparent cohesion. We will assume that it comprises predominantly the cementation strength $C_f = 160\,\mathrm{N}$, and the cementation shear strength is

$$c_{cm} = \frac{C_f}{A} = \frac{160}{0.065 \times 0.065} = 37{,}870\,\mathrm{Pa} = 37.9\,\mathrm{kPa}$$

The slope of OA is the apparent friction angle ξ_o. By measurement, $\xi_o \approx 10°$.

Note: If the caliche was not saturated then C_f is a combination of the effects of soil suction from the unsaturated state and from cementation.

Step 4: Write the equation for the shear strength.

$$\tau_p = c_{cm} + (\sigma'_n)_f \tan(\xi_o) = 37.9 + (\sigma'_n)_f \tan(10) = 37.9 + 0.176(\sigma'_n)_f \ \mathrm{kPa}$$

Note: $C = c_{cm}$ because the soil is saturated ($c_t = 0$) and we are assuming that cohesion is small ($c_o \approx 0$).

EXAMPLE 8.4 *Predicting the Shear Stress at Failure Using Coulomb Failure Criterion*

The critical state friction angle of a soil is 28°. Determine the critical state shear stress if the normal effective stress is 200 kPa.

Strategy This is a straightforward application of the Coulomb failure criterion.

Solution 8.4

Step 1: Determine the failure shear stress.

$$\tau_{cs} = (\sigma'_n)_f \tan \phi'_{cs}$$

$$\tau_{cs} = 200 \tan 28° = 106.3 \text{ kPa}$$

EXAMPLE 8.5 *Interpreting Shear Strength Parameters for a Compacted Sand with Traces of Clay from Shear Box Test Data*

A sandy clay was compacted using the Proctor test, and a sample was extracted and tested in a direct shear apparatus. The sample diameter was 65 mm and its thickness was 25 mm. The rate of horizontal displacement applied was 0.1 mm/min. Data were recorded every 10 seconds. The table below shows data at every third point for a vertical force of 4000 N. A negative sign denotes vertical expansion.

(a) Plot graphs of (1) horizontal forces versus horizontal displacements and (2) vertical displacements versus horizontal displacements.

(b) Determine (1) the maximum or peak shear stress, (2) the critical state shear stress, (3) the peak dilation angle, (4) ϕ'_p, and (5) ϕ'_{cs}.

Horizontal displacement (mm)	Horizontal force (N)	Vertical displacement (mm)
0.00	0.0	0.000
0.10	184.6	0.000
0.20	353.2	0.000
0.30	559.9	0.000
0.40	793.7	0.000
0.50	953.6	0.004
0.60	1095.1	0.000
0.70	1205.9	0.000
0.80	1279.7	−0.004
0.95	1414.8	−0.012
1.05	1486.3	−0.020
1.30	1700.8	−0.037
1.45	1882.0	−0.049
1.60	2043.0	−0.066
1.75	2081.0	−0.086
1.96	1602.0	−0.115
2.07	1602.0	−0.120
2.20	1603.0	−0.120
2.25	1601.0	−0.120
2.30	1599.0	−0.120
2.35	1610.0	−0.120

Strategy Make plots of the horizontal displacement versus horizontal force and of the horizontal displacement versus vertical displacement. You can then extract the relevant parameters for the calculation of the desired quantities.

Solution 8.5

Step 1: Plot graphs.

See Figure E8.5.

Step 2: Extract the required values.

$$(P_x)_p = 2080 \text{ N}, \quad (P_x)_{cs} = 1600 \text{ N}$$

Step 3: Calculate the required values.

Cross-sectional area of sample: $A = \dfrac{\pi D^2}{4} = \dfrac{\pi \times 0.065^2}{4} = 0.0033 \text{ m}^2$

$$\tau_p = \frac{(P_x)_p}{A} = \frac{2080}{3.3 \times 10^{-3}} = 630,303 \text{ Pa} = 630.3 \text{ kPa}$$

$$\tau_{cs} = \frac{(P_x)_{cs}}{A} = \frac{1600}{3.3 \times 10^{-3}} = 484,848 \text{ Pa} = 484.8 \text{ kPa}$$

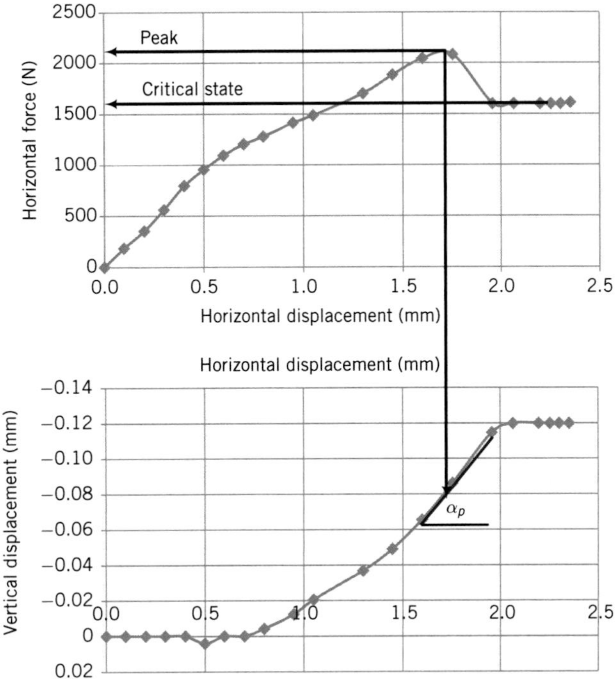

Figure E8.5

Normal effective stress: $\sigma'_n = \left(\dfrac{4000}{3.3 \times 10^{-3}}\right) = 1,212,121 \text{ Pa} = 1212.1 \text{ kPa}$

$$\phi'_p = \tan^{-1}\left(\frac{\tau_p}{\sigma'_n}\right) = \tan^{-1}\left(\frac{630.3}{1212.1}\right) = 27.5°$$

$$\phi'_{cs} = \tan^{-1}\left(\frac{\tau_{cs}}{\sigma'_n}\right) = \tan^{-1}\left(\frac{484.8}{1212.1}\right) = 21.8°$$

$$\alpha_p = \phi'_p - \phi'_{cs} = 27.5 - 21.8 = 5.7°$$

Also, from the vertical displacement versus horizontal displacement plot (Figure E8.5), we get

$$\alpha_p = \tan^{-1}\left(\frac{-\Delta H_o}{\Delta x}\right) = \tan^{-1}\left(\frac{-(-0.11 - 0.07)}{1.95 - 1.6}\right) = 6.5°$$

Step 4: Check reasonableness of answers.

From Table 8.4 and Table 8.5, the calculated friction angles and peak dilation angles are reasonable. See the comment on dilation angle in Step 7 of Example 8.1.

8.7.3 *Conventional triaxial apparatus*

A widely used apparatus to determine the shear strength parameters and the stress–strain behavior of soils is the triaxial apparatus. The name is a misnomer since two, not three, stresses can be controlled. In the triaxial test, a cylindrical sample of soil, usually with a length-to-diameter ratio of 2, is subjected to either controlled increases in axial stresses or axial displacements and radial stresses. The sample is laterally confined by a membrane, and confining stresses (cell pressures, consolidation pressures) are applied by pressurizing water in a chamber (Figure 8.14).

For certain types of tests, loading is applied in two phases. The first phase is the consolidation phase, whereby the soil is consolidated by applying a cell pressure and allowing the excess porewater pressure to drain. For unsaturated soils, a backpressure is applied through one of the porous stone into the soil sample forcing the air out of the voids. The cell pressure must be greater than the back pressure to apply a positive confining pressure. The back pressure is the excess porewater pressure, and the cell pressure is the total stress at this stage of the test. So, if the cell pressure is 50 kPa and the back pressure in 30 kPa, then the resultant cell pressure is 20 kPa. After saturation and consolidation are complete, the excess porewater pressure is zero (usually near to zero) and the cell pressure is 20 kPa. In practice, it is difficult to attain full saturation of the soil sample. In routine testing, values of saturation > 95% are accepted as full saturation. Saturated soils extracted from field become unsaturated because of stress release and the consequent suction pressure that pulls air into the voids.

The second phase of loading involves keeping the cell pressure constant and applying axial stresses by loading the plunger. In the second loading phase, any excess porewater pressure

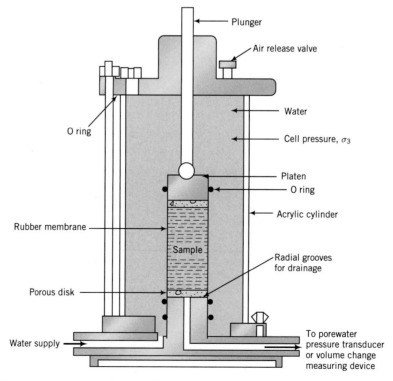

Figure 8.14 Schematic of a triaxial cell.

is either allowed to drain out from the soil (drained condition) or prevented from draining out (undrained condition). If the axial stress is greater than the radial stress, the soil is compressed vertically and the test is called triaxial compression (TC). If the radial stress is greater than the axial stress, the soil is compressed laterally and the test is called triaxial extension (TE).

The applied stresses are principal stresses and the loading condition is axisymmetric. For compression tests, we will denote the radial stresses σ_r as σ_3 and the axial principal total stresses σ_z as σ_1. The average stresses and strains on a soil sample in the triaxial apparatus for compression tests are as follows:

$$\text{Axial stress or deviatoric stress: } \sigma_a = \sigma_1 - \sigma_3 = \frac{P_z}{A} \tag{8.20}$$

$$\text{Cell pressure or chamber pressure or radial stress or minor principal total stress} = \sigma_3 \tag{8.21}$$

$$\text{Axial principal total stress: } \sigma_1 = \frac{P_z}{A} + \sigma_3 \tag{8.22}$$

$$\text{Shear stress: } \tau = \frac{\sigma_1 - \sigma_3}{2} = \frac{P_z}{2A} \tag{8.23}$$

$$\text{Axial strain: } \varepsilon_1 = \varepsilon_a = \frac{\Delta H_o}{H_o} \tag{8.24}$$

$$\text{Radial strain: } \varepsilon_3 = \frac{\Delta r}{r_o} \tag{8.25}$$

$$\text{Volumetric strain: } \varepsilon_p = \frac{\Delta V}{V_o} = \varepsilon_1 + 2\varepsilon_3 \tag{8.26}$$

$$\text{Shear strain: } \gamma = \varepsilon_1 - \varepsilon_3 \tag{8.27}$$

where P_z is the axial load on the plunger, A is the cross-sectional area of the soil sample, r_o is the initial radius of the soil sample, Δr is the change in radius, V_o is the initial volume, ΔV is the change in volume, H_o is the initial height, and ΔH_o is the change in height.

The area of the sample changes during loading, and at any given instance the area is

$$A = \frac{V}{H} = \frac{V_o - \Delta V}{H_o - \Delta H_o} = \frac{V_o[1 - (\Delta V/V_o)]}{H_o[1 - (\Delta H_o/H_o)]} = \frac{A_o(1 - \varepsilon_p)}{1 - \varepsilon_1} = \frac{A_o(1 - \varepsilon_1 - 2\varepsilon_3)}{1 - \varepsilon_1} \tag{8.28}$$

where $A_o (= \pi r_o^2)$ is the initial cross-sectional area and H is the current height of the sample. A set of test data from triaxial drained tests on quart sand is shown in Figure 8.15.

Undrained tests are desired to measure the shear strength of the soil at the void ratio existing in the field or some desirable constant void ratio based on the loading condition. Undrained conditions in the triaxial test refer to external prevention of drainage, so the soil volume as a whole remains constant. However, because the soil does not deform uniformly in the test; some parts of it may be dilating or have a tendency to dilate while other parts are compressing or have a tendency to compress during shear. The combination of dilation and compression is such that the total volume remains constant. Water can migrate from the nondilating (or tendency to compress) parts to the dilating (or tendency to dilate) parts. This internal movement of water (drainage) can significantly influence the stress–strain behavior of soils. For parts that have a tendency to dilate, the principal effective stresses will

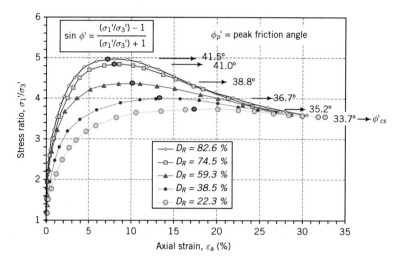

Figure 8.15 Stress–strain results from triaxial drained tests on quart sands. D_R is relative density. (Source: Koerner, 1970.)

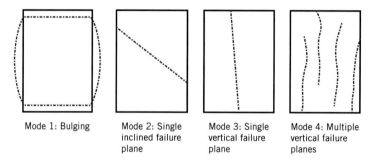

Figure 8.16 Typical failure modes in triaxial tests.

be increasing because the excess porewater pressure is decreasing; the reverse will occur for the nondilating parts (recall from the principle of effective stress [Chapter 6] that, for a saturated soils, if the total stress remains constant and the porewater pressure decreases, then the effective stress will increase). The peak shear strength will not represent the shear strength of the soil at a constant void ratio.

The triaxial apparatus is versatile. We can (1) independently control the applied axial and radial stresses, (2) conduct tests under drained and undrained conditions, and (3) control the applied displacements or stresses. A variety of tests can be conducted in the triaxial apparatus. Only the popular tests used in geotechnical practice are summarized in Table 8.7.

Typical failure modes observed as illustrated in Figure 8.16. Failure mode 1 is mostly associated with soft fine-grained soils and loose coarse-grained soils. Failure modes, 2, 3, and 4 are most associated with stiff fine-grained soils and dense coarse-grained soils. These soils normally exhibit a peak shear stress response to loading. The stresses and strains and, in particular, the porewater pressure are unreliable after the initiation of failure modes 2, 3, and 4. Recall from Figure 8.14 that the porewater pressure are measured at the top and/or bottom boundaries of the sample in the triaxial cell, so changes that are occurring at the failure planes are not captured.

Table 8.7 Summary of common triaxial tests.

Type of test	Soil type	Consolidation	Drainage condition	Duration	Failure criterion	Strength parameter[a]	Practical use	Cost
Unconfined compression (UC)	Fine-grained	No Cell pressure, $\sigma_3 = 0$	Undrained Excess porewater pressure, Δu, greater than or less than zero. Usually not measured	Quick	Tresca	$(s_u)_i = \dfrac{(P_z)_i}{2A} = \dfrac{1}{2}(\sigma_a)_i$ $A = \dfrac{\pi r_o^2}{1 - [(\Delta H_o)_i / H_o]}$	Analysis of short-term stability of slopes, foundations, retaining walls, excavations, and other earthworks. Comparison of the shear strengths of soils from a site to establish soil strength variability quickly; cost-effective. Determination of the stress–strain characteristics under fast loading conditions.	$
Consolidated undrained (CU)	All	Yes Cell pressure, $\sigma_3 > 0$ $\sigma_3' = \sigma_3$	Undrained Excess porewater pressure, Δu, greater than or less than zero.	Quick	MC Tresca	$\sin \phi_i' = \left[\dfrac{P_z/A}{(P_z/A) + 2(\sigma_3 - \Delta u)} \right]_i$ $(s_u)_i = \dfrac{(P_z)_i}{2A} = \dfrac{1}{2}(\sigma_a)_i$ $A = \dfrac{\pi r_o^2}{1 - [(\Delta H_o)_i / H_o]}$	Analysis of both short-term and long-term stability of slopes, foundations, retaining walls, excavations, and other earthworks. Determination of the stress–strain characteristics so that soil stiffness can be calculated.	$$
Consolidated drained (CD)	All but mostly used for coarse-grained soils	Yes Cell pressure, $\sigma_3 > 0$ $\sigma_3' = \sigma_3$	Drained Excess porewater pressure, $\Delta u = 0$.	Long	MC	$\sin \phi_i' = \left[\dfrac{P_z/A}{(P_z/A) + 2\sigma_3'} \right]_i$ $A = \dfrac{\pi r_o^2 \{ 1 - [(\Delta H_o)_i / H_o] - 2[(\Delta V)_i / V_o] \}}{1 - [(\Delta H_o)_i / H_o]}$	Same as for CU except the results apply only to long-term condition.	$$$ (extra cost due to extra time required to perform the test compared with a CU test)
Unconsolidated undrained (UU)	Fine-grained	No Cell pressure, $\sigma_3 > 0$	Undrained excess porewater pressure, Δu, greater than or less than zero.	Quick	Tresca	$(s_u)_i = \dfrac{(P_z)_i}{2A} = \dfrac{1}{2}(\sigma_a)_i$ $A = \dfrac{\pi r_o^2}{1 - [(\Delta H_o)_i / H_o]}$	Same as UC	$

[a] $i = p$ for peak and $i = cs$ for critical state.

EXAMPLE 8.6   *Undrained Shear Strength from a UC Test*

An unconfined compression test was carried out on a saturated clay sample, 38 mm diameter × 76 mm long. The clay was extracted from a borehole in a delta in Southeast Asia and classified as CL. The axial load versus the axial (vertical) displacement is shown in the Figure E8.6. **(a)** Determine the undrained shear strength, and **(b)** describe the consistency of the clay (stiff, soft, etc.)

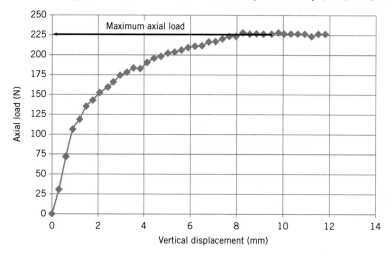

Figure E8.6

Strategy Since the test is a UC test, $\sigma_3 = 0$ and $(\sigma_1)_f$ is the failure axial stress. You can find s_u by calculating one-half the failure axial stress.

Solution 8.6

Step 1: Inspect the plot and determine the peak and critical state axial loads and the corresponding axial displacements.

No peak axial load is discernible. The maximum axial load, taken as the critical state axial load, is 225 N at an axial displacement of about 8.5 mm.

Step 2: Determine the sample area at failure.

Diameter $D_o = 38$ mm; length $H_o = 76$ mm; axial displacement $= \Delta H_o = 8.5$ mm

$$A = \frac{\pi r_o^2}{1 - [(\Delta H_o)/H_o]} = \frac{\pi \times (38/2)^2}{1 - (8.5/76)} = 1277 \text{ mm}^2 = 1277 \times 10^{-6} \text{ m}^2$$

Step 3 Calculate s_u.

$$s_u = \frac{(P_z)_f}{2A} = \frac{225}{2 \times 1277 \times 10^{-6}} = 88 \times 10^3 \text{ Pa} = 88 \text{ kPa}$$

Step 4: Check reasonableness of answer and determine the consistency.

From Table 8.6 with $s_u = 88$ kPa is with the ranges of undrained shear strengths for soils. The soil can be described as stiff. It is uncertain whether s_u is a peak value or a critical state value. Because the maximum load remains approximately constant after a vertical displacement of 8 mm and the total volume of the soil sample in the UC test is constant, a reasonable judgment is that s_u is a critical state value.

EXAMPLE 8.7 *Interpreting CU Triaxial Test Data*

A CU test was conducted on a stiff, saturated clay soil by isotropically consolidating the soil using a cell pressure of 200 kPa and then incrementally applying loads on the plunger while keeping the cell pressure constant. At large axial strains ($\approx$ 15%), the axial stress exerted by the plunger was approximately constant at 230 kPa and the excess porewater pressure recorded was constant at 77 kPa. Determine (a) s_u and (b) ϕ'. Illustrate your answer by plotting Mohr's circle for total and effective stresses.

Strategy You can calculate the effective strength parameters by using the Mohr–Coulomb failure criterion, or you can determine them from plotting Mohr's circle. Remember that the axial stress imposed by the plunger is not the major (axial) principal total stress, σ_1, but the deviatoric stress, $(\sigma_1 - \sigma_3) = (\sigma_1' - \sigma_3')$. Since the axial stress and the excess porewater pressure remain approximately constant at large strains, we can assume that critical state condition has been achieved. The Tresca failure criterion must be used to determine s_u.

Solution 8.7

Step 1: Calculate the stresses at critical state.

$$\frac{P_z}{A} = (\sigma_1 - \sigma_3)_{cs} = 230 \text{ kPa}$$

$$(\sigma_1)_{cs} = \frac{P_z}{A} + \sigma_3 = 230 + 200 = 430 \text{ kPa}$$

$$(\sigma_1')_{cs} = (\sigma_1)_{cs} - \Delta u_{cs} = 430 - 77 = 353 \text{ kPa}$$

$$(\sigma_3)_{cs} = 200 \text{ kPa}, \quad (\sigma_3')_{cs} = (\sigma_3)_{cs} - \Delta u_{cs} = 200 - 77 = 123 \text{ kPa}$$

Step 2: Determine the undrained shear strength.

$$s_u = (s_u)_{cs} = \frac{(\sigma_1 - \sigma_3)_{cs}}{2} = \frac{230}{2} = 115 \text{ kPa}$$

Step 3: Determine ϕ_{cs}'.

$$\sin \phi_{cs}' = \frac{(\sigma_1 - \sigma_3)_{cs}}{(\sigma_1' + \sigma_3')_{cs}} = \frac{230}{353 + 123} = 0.483$$

$$\phi_{cs}' = 28.9°$$

or

$$\sin \phi_{cs}' = \left(\frac{P_z/A}{(P_z/A) + 2(\sigma_3 - \Delta u)} \right)_{cs} = \left(\frac{230}{230 + 2(220 - 77)} \right) = 0.483$$

$$\phi_{cs}' = 28.9°$$

Step 4: Draw Mohr's circle.

See Figure E8.7.

$$\phi_{cs}' = 29°$$

$$(s_u)_{cs} = 115 \text{ kPa}$$

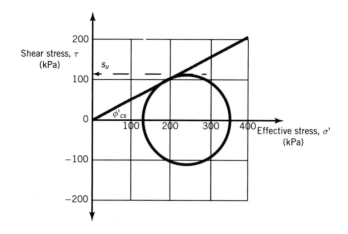

Figure E8.7

Step 5: Check reasonableness of answer and determine the consistency.

From Table 8.4, the critical state friction angle is within the range for clays. From Table 8.6, s_u is within the range for very stiff clays. The answer is reasonable.

EXAMPLE 8.8 *Interpreting CD Triaxial Test Data Using Mohr–Coulomb Failure Criterion*

The results of three CD tests on 38 mm diameter and 76 mm long samples of a dense sand are as follows:

Test number	σ'_3 (kPa)	Deviatoric stress (kPa) $(\sigma'_1 - \sigma'_3)_f$
1	72.5	179.6 (peak at axial strain of 3.8%)
2	130.5	262.5 (peak at axial strain of 4.1%)
3	217.5	409.0 (no peak observed; axial strain when test was stopped was 12%)

The detailed results for test 1 are as follows: The height of the soil after consolidation was 75 mm. The negative sign indicates expansion.

Height change ΔH_o (mm)	Volume change ΔV (cm³)	Axial load P_z (N)
0	0	0
0.152	0.02	44.3
0.228	0.03	68.4
0.38	−0.09	89.9
0.76	−0.5	146.1
1.52	−1.29	186.7
2.28	−1.98	212.4
2.66	−2.24	216.8
3.04	−2.41	216.1
3.8	−2.55	202.5
4.56	−2.59	194.6
5.32	−2.67	183.1
6.08	−2.62	172.6
6.84	−2.64	166.4
7.6	−2.66	161.9
8.36	−2.63	162.7

(a) Determine the friction angle for each test.

(b) Determine τ_p and τ_{cs} for test 1.

(c) Determine ϕ'_{cs}.

(d) Determine α_p for test 1.

Strategy From a plot of axial stress versus axial strain for test 1, you will get τ_p and τ_{cs}. The friction angles can be calculated or estimated graphically using Mohr–Coulomb failure criterion.

Solution 8.8

Step 1: Determine the friction angles.

Use a table to do the calculations.

Test no.	A σ'_3 (kPa)	B $(\sigma'_1 - \sigma'_3)_f$ (kPa)	C = A + B $(\sigma'_1)_f$ (kPa)	D = C + A $(\sigma'_1 + \sigma'_3)_f$ (kPa)	$\phi' = \sin^{-1}\left(\dfrac{\sigma'_1 - \sigma'_3}{\sigma'_1 + \sigma'_3}\right)_f$
Test 1	72.5	179.6	252.1	324.6	33.6° (peak)
Test 2	130.5	262.5	393.0	523.5	30.1° (peak)
Test 3	217.5	409.0	626.5	844.0	29°

Alternatively, plot Mohr's circles of effective stresses and determine the friction angles, as shown in Figure E8.8a.

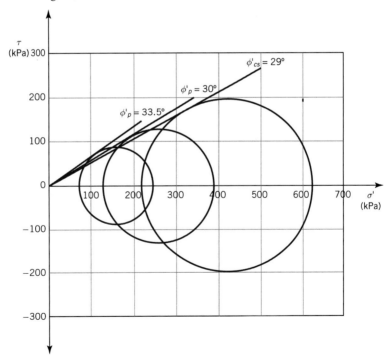

Figure E8.8a

Step 2: Determine τ_p and τ_{cs} from a plot of deviatoric stress versus axial strain response for test 1.

The initial area is $A_o = \dfrac{\pi D_o^2}{4} = \dfrac{\pi \times 0.038^2}{4} = 11.34 \times 10^{-4} \text{ m}^2$

After consolidation, $V_o = A_o H_o = 11.34 \times 7.5 = 85.1 \text{ cm}^3$

$$A = \frac{A_o[1 - (\Delta V/V_o)]}{1 - (\Delta H_o/H_o)}$$

The axial stress is the axial load (P_z) divided by the cross-sectional area of the sample.

\multicolumn{6}{c}{$A_o = 11.34 \times 10^{-4}$ m²}					
ΔH_o (mm)	$\varepsilon_1 = \dfrac{\Delta z}{H_o}$	ΔV (cm³)	$\varepsilon_p = \dfrac{\Delta V}{V_o}$	A (m²)	$\sigma_1 - \sigma_3 = \dfrac{P_z}{A}$ (kPa)
0	0	0	0.000	0.001134	0.0
0.152	0.002	0.02	0.000	0.001136	39.0
0.228	0.003	0.03	0.000	0.001137	60.1
0.38	0.005	−0.09	−0.001	0.001141	78.8
0.76	0.010	−0.5	−0.006	0.001152	126.8
1.52	0.020	−1.29	−0.015	0.001175	158.9
2.28	0.030	−1.98	−0.023	0.001197	177.5
2.66	0.035	−2.24	−0.026	0.001207	179.6
3.04	0.041	−2.41	−0.028	0.001216	177.8
3.8	0.051	−2.55	−0.030	0.001230	164.5
4.56	0.061	−2.59	−0.030	0.001244	156.4
5.32	0.071	−2.67	−0.031	0.001259	145.4
6.08	0.081	−2.62	−0.031	0.001272	135.7
6.84	0.091	−2.64	−0.031	0.001287	129.3
7.6	0.101	−2.66	−0.031	0.001301	124.4
8.36	0.111	−2.63	−0.031	0.001316	123.6

See Figure E8.8b for a plot of the results.

Extract τ_p and τ_{cs}.

The axial stress and the volumetric change appear to be constant from about $\varepsilon_1 \approx 10\%$. We can use the result at $\varepsilon_1 \approx 11\%$ to determine τ_{cs}.

$$\tau_p = \frac{(\sigma_1' - \sigma_3')_p}{2} = \frac{179.6}{2} = 89.8 \text{ kPa}, \quad \tau_{cs} = \frac{(\sigma_1' - \sigma_3')_{cs}}{2} = \frac{124}{2} = 62 \text{ kPa}$$

Step 3: Determine ϕ_{cs}'.

$$(\sigma_3')_{cs} = 72.5 \text{ kPa}, \quad (\sigma_1')_{cs} = 124 + 72.5 = 196.5 \text{ kPa}$$

$$\phi_{cs}' = \sin^{-1}\left(\frac{\sigma_1' - \sigma_3'}{\sigma_1' + \sigma_3'}\right)_{cs} = \sin^{-1}\left(\frac{124}{196.5 + 72.5}\right) = 27.5°$$

Step 5: Determine α_p

Use the peak friction angle from Test 1 (see Step 1)

$$\alpha_p = \phi_p' - \phi_{cs}' = 33.6 - 27.5 = 6.1°$$

Step 6: Check reasonableness of results.

Both the calculated and the graphical methods of estimating the friction angles gave approximately the same results. The frictions angles are within the range for dense sand (Table 8.4), but the peak dilation angle of 6.1° appears to be low compared with the range of 10°–15° for dense sand (Table 8.5). Although tables of values such as Table 8.4 and Table 8.5 are not intended to justify test results, they provide guidance to observed range of values for particular soil parameters. Your test results can fall outside of these ranges but yet plausible or explainable from observation made during testing. No information was provided on the type of failure observed. It is quite possible that this low dilation angle results from the formation of a single failure plane in the sand. If that was the case, the mass of sand

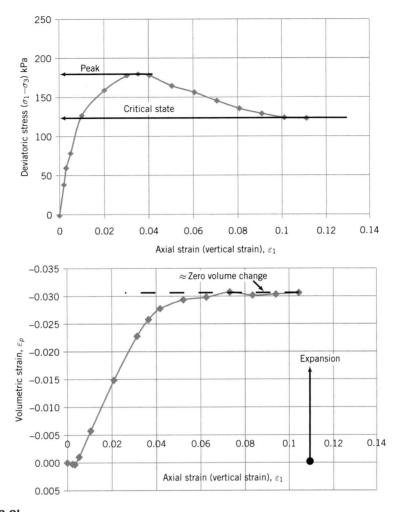

Figure E8.8b

above and below the failure plane would have behaved as rigid bodies. The sand within and just adjacent to the failure plane would have been at critical state. As more axial displacements are applied, more sand mass near this failure plane would come into critical state. However, the sand sample would become unstable well before the whole mass of sand comes into critical state. One mode of instability that is normally observed in this type of failure is the top part of the soil mass above the failure plane would slide along it pushing the membrane sideways and producing a local bulge. Another possibility is that the confining pressure (cell pressure in this test) is large enough to suppress the dilation significantly.

The strains calculated from the displacements at the exterior of the sand sample would not represent the strains that are actually occurring within the sand sample. The critical state friction angle is reasonable, but the peak dilation angle and, consequently, the peak friction angle are suspicious. This example, developed from test data from a real sand, demonstrates the uncertainty of the peak friction angle. In addition, it shows that more information on the test such as the mode of failure is necessary to interpret the test results.

EXAMPLE 8.9   *Undrained Shear Strength from a UU Triaxial Test*

A UU test was conducted on stiff, saturated clay. The cell pressure was 200 kPa, and the peak axial (deviatoric stress) was 220 kPa. Determine the undrained shear strength.

Strategy Use the Tresca failure criterion.

Solution 8.9

Step 1: Determine the undrained shear strength.

$$(s_u)_p = \frac{(\sigma_1 - \sigma_3)_p}{2} = \frac{220}{2} = 110 \text{ kPa}$$

8.7.4   *Direct simple shear*

The direct simple shear (DSS) apparatus subjects a soil sample to plane strain loading condition (the strain in one direction is zero). This test closely reproduces stress conditions for many geoengineering structures such as excavations, pile foundations, and embankment. Commercial versions of direct simple shear devices consist of a cylindrical sample with the vertical side enclosed by a wire-reinforced rubber membrane (Figure 8.17a). Rigid, rough metal plates (platens) are placed at the top and bottom of the sample. Displacing the top of the sample relative to the bottom deforms the sample (Figure 8.17b). The vertical and horizontal loads (usually on the top boundary) as well as displacements on the boundaries are measured, and thus the average normal and shear stresses and boundary strains can be deduced. The top platen can be maintained at a fixed height for constant volume tests or allowed to move vertically to permit volume change to occur (constant load test).

The direct simple shear apparatus do not subject the sample as a whole to uniform stresses and strains. However, the stresses and strains in the central region of the sample are uniform. In simple shear, the strains are $\varepsilon_x = \varepsilon_y = 0$, $\varepsilon_z = \Delta z/H_o$, and $\gamma_{zx} = \Delta x/H_o$. The stresses calculated are the normal stress, $\sigma_z = P_z/A$, and the shear stress, $\tau_{zx} = P_x/A$, where A is the cross-sectional area of the soil sample. The undrained shear strength, $s_u = \tau_{zx}$ and the friction angle is

$$\phi_i' = tan^{-1} \frac{(\tau_{zx})_i}{\sigma_z - \Delta u_i},$$

where the subscript i denotes either peak or critical state, and Δu is the excess porewater pressure.

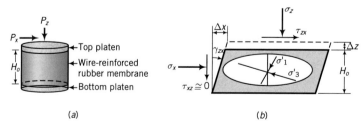

(a)   *(b)*

Figure 8.17   Direct simple shear.

EXAMPLE 8.10 *Interpreting Direct Simple Shear Test Data*

A soft clay soil sample 50 mm diameter and 20 mm high was tested in a direct simple shear device. The soil volume was maintained constant by adjusting the vertical load. At failure (critical state), the vertical load (P_z) was 500 N and the horizontal load or shear (P_x) was 152 N. The excess porewater pressure developed was 100 kPa.

(a) Determine the undrained shear strength.

(b) Determine the critical state friction angle.

Strategy You must use effective stresses to calculate the friction angle.

Solution 8.10

Step 1: Determine the total and effective stresses.

$$A = \frac{\pi}{4}D^2 = \frac{\pi}{4} \times (0.05)^2 = 19.6 \times 10^{-4}\ \text{m}^2$$

$$\sigma_z = \frac{P_z}{A} = \frac{500 \times 10^{-3}}{19.6 \times 10^{-4}} = 255\ \text{kPa}$$

$$\tau_{zx} = \frac{P_x}{A} = \frac{152 \times 10^{-3}}{19.6 \times 10^{-4}} = 77.6\ \text{kPa}$$

$$\sigma_z' = \sigma_z - \Delta u = 255 - 100 = 155\ \text{kPa}$$

Step 2: Determine ϕ_{cs}' and s_u.

$$(s_u)_{cs} = \tau_{zx} = 77.6\ \text{kPa}$$

$$\phi_{cs}' = \tan^{-1}\left(\frac{\tau_{zx}}{\sigma_z'}\right) = \tan^{-1}\left(\frac{77.6}{155}\right) = 26.6°$$

Step 3: Determine reasonableness of results.

The results are within the range of values for soft clays (Table 8.4 and Table 8.6). The results are reasonable.

What's next ... In the next section, the types of laboratory strength tests to specify for typical practical situations are presented.

8.8 SPECIFYING LABORATORY STRENGTH TESTS

It is desirable to test soil samples under the same loading and boundary conditions that would likely occur in the field. Often, this is difficult to accomplish because the loading and boundary conditions in the field are uncertain. Even if they were known to a high degree of certainty, it would be difficult and perhaps costly to devise the required laboratory apparatus. We then have to specify lab tests using conventional devices that best simulate the field conditions. A few practical cases are shown in Figure 8.18 with the recommended types of tests.

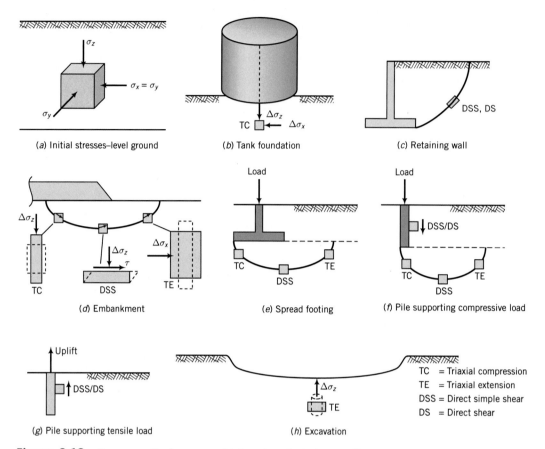

Figure 8.18 Some practical cases and laboratory tests to specify.

What's next ... A variety of field tests have been developed to identify soils and to obtain design parameters such as shear strength parameters by testing soils in situ. Field test results are often related to laboratory test results using empirical factors. In the next section, the popular field tests to estimate the shear strength parameters are briefly described.

8.9 ESTIMATING SOIL PARAMETERS FROM IN SITU (FIELD) TESTS

8.9.1 Vane shear test (VST)

The undrained shear strength from a vane shear test is calculated from

$$s_u = \frac{2T}{\pi d^3 (h/d + \frac{1}{3})} \tag{8.29}$$

where $T = T_{max}$ is the maximum torque to give the peak undrained shear strength or $T = T_{res}$ is the residual torque to give the residual undrained shear strength, h is the height and d is the diameter of the vane. Usually, $h/d = 2$, and Equation (8.29) becomes

$$(s_u)_p = 0.273 \frac{T}{d^3} \tag{8.30}$$

EXAMPLE 8.11 *Calculating s_u from VST*

The peak torque from a VST on medium-stiff blue clay was 30 N m. The test was repeated at the same depth on the remolded (residual) clay and the peak torque recorded was 16 N m. The blade diameter was 65 mm and $h/d = 2$.

(a) Determine the peak undrained shear strength.

(b) Determine the remolded undrained shear strength.

(c) Determine the sensitivity.

Strategy
The solution is an application of Equation (8.30)

Solution 8.11

Step 1: For (a), determine the peak undrained shear strength.

$$(s_u)_p = 0.273 \frac{T_{max}}{d^3} = 0.273 \frac{30}{0.065^3} = 29,822 \text{ Pa} \approx 30 \text{ kPa}$$

Step 2: For (b), determine the remolded undrained shear strength.

$$(s_u)_{residual} \approx (s_u)_{cs} = 0.273 \frac{T_{res}}{d^3} = 0.273 \frac{16}{0.065^3} = 15,905 \text{ Pa} \approx 16 \text{ kPa}$$

Step 3: For (c), determine the sensitivity.

$$\text{Equation (3.1):} \quad S_t = \frac{T_{max}}{T_{res}} = \frac{30}{16} \approx 1.9$$

8.9.2 *Standard penetration test (SPT)*

The SPT has been correlated to most soil parameters. Many of the correlations have low regression coefficients (R squared < 0.6).

Table 8.8 Correlation of Friction Angle with SPT.

N	N_{60}	Compactness	ϕ' (degrees)
0–4	0–3	Very loose	26–28
4–10	3–9	Loose	29–34
10–30	9–25	Medium	35–40*
30–50	25–45	Dense	38–45*
>50	>45	Very dense	>45*

*Values correspond to ϕ'_p.
Source: Modified from Kulhawy and Mayne (1990).

8.9.3 Cone penetrometer test (CPT)

The CPT results have been correlated to several soil parameters. Some of the important correlations are shown in Table 8.9.

Table 8.9   Some correlations of CPT with soil parameters.

Parameter	Relationship
Peak friction angle (triaxial)	$$\phi'_p = 17.6 + 11 \log \left[\frac{\frac{q_c}{p_{atm}}}{\sqrt{\left(\frac{\sigma'_{zo}}{p_{atm}} \right)}} \right] \text{ degrees}$$ σ'_{zo} is the initial or current vertical effective stress, p_{atm} is atmospheric pressure (101 kPa)
Peak undrained shear strength	$(s_u)_p = \dfrac{q_c - \sigma_z}{N_k}$ N_k is a cone factor that depends on the geometry of the cone and the rate of penetration. Average values of N_k as a function of plasticity index can be estimated from $N_k = 19 - \dfrac{PI - 10}{5}; \quad PI > 10$
Past maximum vertical effective stress	$\sigma'_{zc} = 0.33(q_c - \sigma'_{zo})^m$ Intact clays: $m = 1$; organic clays: $m = 0.9$; silts: $m = 0.85$; silty sands: $m = 0.8$; clean sands: $m = 0.72$
Bulk unit weight	$\gamma = 1.95 \gamma_w \left(\dfrac{\sigma_{zo}}{p_{atm}} \right)^{0.06} \left(\dfrac{f_s}{p_{atm}} \right)^{0.06}$ f_s is the average cone sleeve resistance over the depth of interest.

Source: Mayne et al. (2009).

Key points

1. Various field tests are used to estimate soil strength parameters.
2. You should be cautious in using correlations of field test results, especially SPT, with soil strength parameters in design.

What's next ... Several empirical relationships have been proposed to obtain soil strength parameters from laboratory tests, for example, the Atterberg limits, or from statistical analyses of field and laboratory test results. Some of these relationships together with some theoretical ones are presented in the next section.

8.10 SOME EMPIRICAL AND THEORETICAL RELATIONSHIPS FOR SHEAR STRENGTH PARAMETERS

Some suggested empirical relationships for the shear strength of soils are shown in Table 8.10. These relationships should only be used as a guide and in preliminary design calculations.

Table 8.10 Some empirical and theoretical soil strength relationships.

Soil type	Equation	Reference
	Triaxial test or axisymmetric condition	
Normally consolidated clays	$\left(\dfrac{s_u}{\sigma'_z}\right)_{cs} = 0.11 + 0.0037\, PI$; PI is plasticity index (%)	Skempton (1944)
	$\left(\dfrac{s_u}{\sigma'_{zo}}\right)_{cs} \approx 0.5 \sin\phi'_{cs}$ (see notes 1 and 2)	Wroth (1984) Budhu (2011)
	$\left[\left(\dfrac{s_u}{\sigma'_{zo}}\right)_{cs}\right]_{ic} = \dfrac{3\sin\phi'_{cs}}{3-\sin\phi'_{cs}}(0.5)^{\Lambda}$ (see note 3)	
	$\Lambda = 1 - \dfrac{C_c}{C_r}, \quad \Lambda \approx 0.8$	
Overconsolidated clays	$\left(\dfrac{s_u}{\sigma'_{zo}}\right)_{cs} \approx 0.5 \sin\phi'_{cs}(OCR)^{\Lambda}$ (see notes 1 and 2)	Wroth (1984)
		Budhu (2011)
	$\left(\dfrac{s_u}{\sigma'_{zo}}\right)_{p} \approx 0.5 \sin\phi'_{cs}\alpha_y(OCR)^{\Lambda}$ (see notes 1 and 2)	
	$\alpha_y = \dfrac{\sqrt{1.45 OCR^{0.66}-1}}{(0.725 OCR^{0.66})^{\Lambda}}, \quad OCR \leq 10$	
	$\alpha_y = \dfrac{\sqrt{1.6 OCR^{0.62}-1}}{(0.8 OCR^{0.62})^{\Lambda}}, \quad OCR > 10$	
	$\Lambda = 1 - \dfrac{C_c}{C_r}, \quad \Lambda \approx 0.8$	
	$\dfrac{s_u}{\sigma'_{zo}} = (0.23 \pm 0.04)OCR^{0.8}$ (see note 1)	Jamiolkowski et al. (1985)
Normally consolidated clays	$\phi'_{cs} = \sin^{-1}\left[0.35 - 0.1\ln\left(\dfrac{PI}{100}\right)\right]$; PI is plasticity index (%) (see note 2)	Wood (1990)
Clean quartz sand	$\phi'_p = \phi'_{cs} + 3D_r(10 - \ln p'_f) - 3$ where p'_f is the mean effective stress at failure (in kPa) and D_r is relative density. This equation should only be used if $12 > (\phi'_p - \phi'_{cs}) > 0$.	Bolton (1986)

Table 8.10 *Continued*

Soil type	Equation	Reference
Direct simple shear or plane strain condition		
Normally consolidated and overconsolidated clays	$\left[\left(\dfrac{s_u}{\sigma'_{zo}}\right)_{cs}\right]_{dss} = \dfrac{(3 - \sin\phi'_{cs})}{2\sqrt{3}}\left[\left(\dfrac{s_u}{\sigma'_{zo}}\right)_{cs}\right]_{triaxial}$ (see notes 1 and 2)	Budhu (2011)

Subscripts: *ic* = isotropic consolidation, *cs* = critical state, *p* = peak, *y* = yield, *dss* = direct simple shear.
Note 1: These are applicable to direct simple shear (DSS) tests. The estimated undrained shear strength from triaxial compression tests would be about 1.4 times greater.
Note 2: These are theoretical equations derived from critical state soil mechanics.
Note 3: These are for isotropically consolidated clays at critical state in the triaxial test.

8.11 SUMMARY

The strength of soils is interpreted using various failure criteria. Each criterion is suitable for a certain class of problems. For example, the Coulomb failure criterion is best used in situations where planar slip planes may develop. All soils, regardless of their initial state of stress, will reach a critical state characterized by continuous shearing at constant shear to normal-effective-stress ratio and constant volume. The initial void ratio of a soil and the normal effective stresses determine whether the soil will dilate. Dilating soils often exhibit (1) a peak shear stress and then strain-soften to a constant shear stress, and (2) initial contraction followed by expansion toward a critical void ratio. Nondilating soils (1) show a gradual increase of shear stress, ultimately reaching a constant shear stress, and (2) contract toward a critical void ratio. The shear strength parameters are the friction angles (ϕ'_p and ϕ'_{cs}) for drained conditions and s_u for undrained conditions. Only ϕ'_{cs} is a fundamental soil strength parameter.

A number of laboratory and field tests are available to estimate the shear strength parameters. All these tests have shortcomings. You should use careful judgment in deciding what test should be used for a particular project. Also, you should select the appropriate failure criterion to interpret the test results.

8.11.1 Practical examples

EXAMPLE 8.12 *Estimation of s_u*

You have contracted a laboratory to conduct soil tests for a site, which consists of a layer of sand, 6 m thick, with $\gamma_{sat} = 16.8$ kN/m³. Below the sand is a deep, soft, bluish clay with $\gamma_{sat} = 20.8$ kN/m³ (Figure E8.12). The site is in a remote area. Groundwater level is located at 2 m below the surface. You specified a consolidation test and a triaxial consolidated undrained test for samples of the soil taken at 10 m below ground surface. The consolidation test shows that the clay is lightly overconsolidated with an $OCR = 1.5$. The undrained shear strength at a cell pressure approximately equal to the initial vertical stress is 72 kPa. Do you think the undrained shear strength value is reasonable, assuming the OCR of the soil is accurate? Show calculations to support your thinking. Assume that the sand above the groundwater level is saturated. Use 9.8 kN/m³ as the unit weight of water.

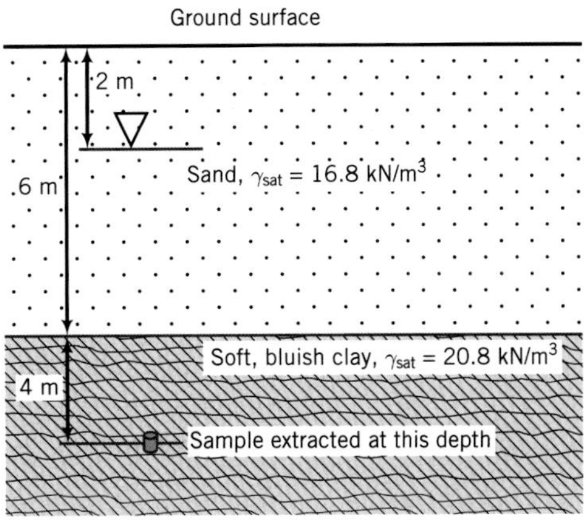

Figure E8.12

Strategy Because the site is in a remote area, it is likely that you may not find existing soil results from neighbouring constructions. In such a case, you can use empirical relationships as guides, but you are warned that soils are notorious for having variable strengths.

Solution 8.12

Step 1: Determine the initial effective stresses of the sample in the field.

$$\sigma_{zo} = (16.8 \times 6) + (20.8 \times 4) = 184 \text{ kPa}$$

$$u_o = 8 \times 9.8 = 78.4 \text{ kPa}$$

$$\sigma'_{zo} = 184 - 78.4 = 105.6 \text{ kPa}$$

$$\sigma'_{zc} = \sigma'_{zo} \times OCR = 105.6 \times 1.5 = 158.4 \text{ kPa}$$

Step 2: Determine s_u/σ'_{zo}.

$$\frac{s_u}{\sigma'_{zo}} = \frac{72}{105.6} = 0.68$$

Step 3: Use empirical equations from Table 8.10.

Jamiolkowski et al. (1985): $\dfrac{s_u}{\sigma'_{zo}} = (0.23 \pm 0.04)OCR^{0.8}$

Range of:

$$\frac{s_u}{(\sigma'_{zo})} = 0.19(OCR)^{0.8} \text{ to } 0.27(OCR)^{0.8}$$

$$= 0.19(1.5)^{0.8} \text{ to } 0.27(1.5)^{0.8}$$

$$= 0.26 \text{ to } 0.37 < 0.68$$

The Jamiolkowski et al. (1985) expression is applicable to DSS. Triaxial compression tests usually give s_u values higher than DSS. Using a factor of 1.4 (see Table 8.10, note 1), we get

$$\left(\frac{s_u}{\sigma'_{zo}}\right) = 0.36 \text{ to } 0.52 < 0.68$$

The differences between the reported results and the empirical relationships are substantial. The undrained shear strength is therefore suspicious. One possible reason for such high shear strength is that the water content at which the soil was tested is lower than the natural water content. This could happen if the soil sample extracted from the field was not properly sealed to prevent moisture loss. You should request a repeat of the test.

EXAMPLE 8.13   *Interpreting CU Triaxial Test Data*

The results of a CU test on a stiff, bluish gray clay from an alluvial deposit of soil is shown in the table below. The degree of saturation of the soil was 86%, so a back pressure was applied to saturate the soil. The final degree of saturation was 97%. The initial size of the sample was 38 mm × 76 mm high. The sample was first isotropically consolidated using a cell pressure of 230 kPa. The height of the sample and the average cross-sectional area after consolidation were 74.9 mm and 1110 mm², respectively. After isotropic consolidation, incremental axial displacement was applied on the plunger while keeping the cell pressure constant (the shear phase). The results of the shearing phase are shown in the table below. Failure mode 2 was observed just before the peak shear stress was recorded. (a) Determine the shear strength parameters for short-term and long-term loading. (b) Is any value of these parameters suspicious and, if so, explain why?

Axial load (N)	Axial displacement (mm)	u (kPa)
0	0	200.0
23.10	0.08	205.9
39.42	0.18	210.9
47.58	0.25	213.9
51 64	0.36	214.9
50.30	0.43	214.9
48.92	0.53	215.9
48.92	0.64	216.9
47.58	0.71	217.9
46.20	0.79	217.9
44.86	0.89	217.9
44.86	0.97	218.9
43.48	1.07	218.9
43.48	1.14	218.9
43.48	1.24	218.9
42.14	1.32	218.9
42.14	1.42	218.9
40.76	1.50	219.9
42.14	1.60	219.9
40.76	1.68	219.9
40.76	1.78	219.9
40.76	1.85	219.9
40.76	1.96	219.9

Strategy You can calculate the long-term strength parameters by using the Mohr–Coulomb failure criterion and the short-term strength parameter, s_u, by using the Tresca failure criterion, or you can determine them by plotting Mohr's circle.

Solution 8.13

Step 1: Calculate the stresses and strains, and plot stress–strain response.

Set up a spreadsheet to do the calculations as shown in the table below.

Radius	19	mm
Initial height	76	mm
Height after consolidation	74.9	mm
Initial area after consolidation	1110	mm²
Cell pressure	230	kPa

Axial load P_z (N)	Axial displacement (mm)	u (kPa)	$\varepsilon_z =$ $\Delta H_o/$ H_o	A (mm²)	$\sigma_1 - \sigma_3$ $= P_z/A$ (kPa)	$\Delta u =$ $u - u_o$ (kPa)	$\sigma_1 =$ $\sigma_1 - \sigma_3 + \sigma_3$ (kPa)	$\sigma_1' =$ $\sigma_1 - \Delta u$ (kPa)	$\sigma_3' =$ $\sigma_3 - \Delta u$ (kPa)
0	0	200.0	0	1110	0	0.0	230.0	30.0	30.0
23.10	0.08	205.9	0.0010	1111.1	20.8	5.9	250.8	44.9	24.1
39.42	0.18	210.9	0.0024	1112.6	35.4	10.9	265.4	54.5	19.1
47.58	0.25	213.9	0.0034	1113.8	42.7	13.9	272.7	58.8	16.1
51.64	0.36	214.9	0.0047	1115.3	46.3	14.9	276.3	61.4	15.1
50.30	0.43	214.9	0.0058	1116.4	45.1	14.9	275.1	60.1	15.1
48.92	0.53	215.9	0.0071	1118.0	43.8	15.9	273.8	57.9	14.1
48.92	0.64	216.9	0.0085	1119.5	43.7	16.9	273.7	56.8	13.1
47.58	0.71	217.9	0.0095	1120.6	42.5	17.9	272.5	54.6	12.1
46.20	0.79	217.9	0.0105	1121.8	41.2	17.9	271.2	53.3	12.1
44.86	0.89	217.9	0.0119	1123.3	39.9	17.9	269.9	52.1	12.1
44.86	0.97	218.9	0.0129	1124.5	39.9	18.9	269.9	51.0	11.1
43.48	1.07	218.9	0.0142	1126.0	38.6	18.9	268.6	49.7	11.1
43.48	1.14	218.9	0.0153	1127.2	38.6	18.9	268.6	49.7	11.1
43.48	1.24	218.9	0.0166	1128.8	38.5	18.9	268.5	49.6	11.1
42.14	1.32	218.9	0.0176	1129.9	37.3	18.9	267.3	48.4	11.1
42.14	1.42	218.9	0.0190	1131.5	37.2	18.9	267.2	48.3	11.1
40.76	1.50	219.9	0.0200	1132.7	36.0	19.9	266.0	46.1	10.1
42.14	1.60	219.9	0.0214	1134.2	37.2	19.9	267.2	47.3	10.1
40.76	1.68	219.9	0.0224	1135.4	35.9	19.9	265.9	46.0	10.1
40.76	1.78	219.9	0.0237	1137.0	35.8	19.9	265.8	46.0	10.1
40.76	1.85	219.9	0.0248	1138.2	35.8	19.9	265.8	45.9	10.1
40.76	1.96	219.9	0.0261	1139.8	35.8	19.9	265.8	45.9	10.1

See Figure E8.13 for plots of the data.

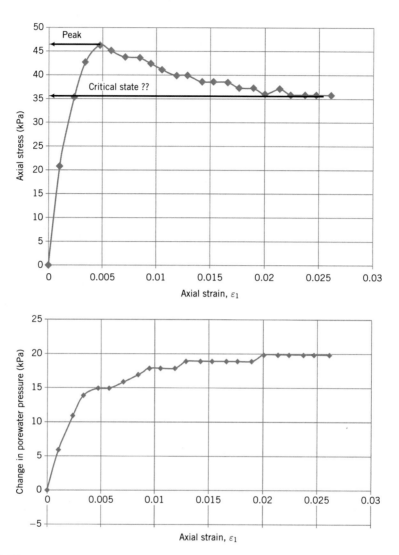

Figure E8.13

Step 2: Extract peak and critical state stresses.

Peak: $(\sigma_1 - \sigma_3)_p = 46.3$ kPa; $\Delta u_p = 14.9$ kPa; $\sigma_1' = 61.4$ kPa; $\sigma_3' = 15.1$ kPa

Critical state: $(\sigma_1 - \sigma_3)_{cs} = 35.8$ kPa; $\Delta u_{cs} = 19.9$ kPa; $\sigma_1' = 46.0$ kPa; $\sigma_3' = 10.1$ kPa

Step 3: Determine the undrained shear strength.

$$(s_u)_p = \frac{(\sigma_1 - \sigma_3)_p}{2} = \frac{46.2}{2} = 23.2 \text{ kPa}$$

$$(s_u)_{cs} = \frac{(\sigma_1 - \sigma_3)_{cs}}{2} = \frac{35.8}{2} = 17.9 \text{ kPa}$$

Step 4: Determine ϕ'_p and ϕ'_{cs}.

$$\sin\phi'_p = \frac{(\sigma_1 - \sigma_3)_p}{(\sigma'_1 + \sigma'_3)_p} = \frac{46.3}{61.4 + 15.1} = 0.605$$

$$\phi'_p = 37.2°$$

$$\sin\phi'_{cs} = \frac{(\sigma_1 - \sigma_3)_{cs}}{(\sigma'_1 + \sigma'_3)_{cs}} = \frac{35.8}{46.0 + 10.1} = 0.638$$

$$\phi'_{cs} = 39.6$$

This value of ϕ'_{cs} is exorbitant and indicates that the results after the peak shear stress are unreliable. Since failure mode 2 was observed near and after peak shear stress, the measured boundary forces and porewater pressures do not represent conditions at the failure plane. Also, the undrained shear strength at critical state may not be correct for the same reasoning.

EXERCISES

Concept understanding

8.1   What is meant by the shear strength of soils?

8.2   Some soils show a peak shear strength. Why and what type(s) of soil do so?

8.3   What is critical state for soils?

8.4   Which friction angle, peak or critical state, is more reliable? Why?

8.5   Which failure criterion is used to interpret the results from a direct shear test? Why?

8.6   Can you draw a Mohr's circle of stress using the data from a direct shear test? Explain your answer.

8.7   Which failure criterion is used to interpret triaxial test results for long-term loading condition? What is basis for this criterion?

8.8   Which failure criterion is used to interpret triaxial test results for short-term loading condition? What is basis for this criterion? Is this criterion used for clays or sand, or both? Justify your answer.

8.9   What is soil cohesion and is it the same as the undrained shear strength? Explain your answer.

8.10 Why does a sample need to be saturated to interpret the results from triaxial undrained tests?

8.11 The confining pressure on a loose saturated sand under groundwater is 100 kPa. The sand is shaken by a seismic event. What is the value of the maximum excess porewater pressure that can develop in the sand? What is the corresponding effective stress?

8.12 The confining pressure on a dense (relative density = 85%) saturated sand under groundwater is 100 kPa. Shearing stresses are applied to this sand. Would the excess porewater pressure increase or decrease as a result of the shearing? What is the minimum porewater pressure that can develop in this sand? Justify your answer.

Problem solving

8.13 The normal and shear stresses at critical state on a horizontal sliding (failure) plane for an uncemented sand is 100 kPa and 57.7 kPa, respectively. (a) Which failure criterion is best used to determine the friction angle of this sand? Justify your answer. (b) Determine the friction angle.

8.14 The following results were obtained from three direct shear (shear box) tests on a sample of uncemented sandy clay. The cross section of the shear box is 50 mm × 50 mm:

Normal force (N)	1337	892	446	223
Shearing force (N)	580	388	276	169

(a) Which failure criterion is appropriate to interpret this data set? Justify your answer. (b) Plot a graph of normal force versus shear force for the data given. (c) Determine the critical state friction angle. (d) Is the soil dilatant? (e) Determine the friction angle at a normal force of 446 N. Is this a peak or critical state friction angle? If this angle is a peak friction angle, what is the value of the dilation angle?

8.15 The results of a direct shear test on a dense uncemented sand using a vertical force of 200 N are shown in the table below. Determine ϕ'_p, ϕ'_{cs}, and α_p. The sample area was 65 mm × 65 mm and the sample height was 20 mm. In the table, Δx, Δz, and P_x are the horizontal and vertical displacements, and horizontal (shear) force, respectively.

Δx (mm)	Δz (mm)	P_x (N)
0	0	0
0.249	0.000	17.8
0.381	0.020	19.2
0.759	0.041	40.6
1.519	0.030	83.4
2.670	-0.041	127.5
3.180	-0.089	137.8
4.059	-0.150	156.5
5.080	-0.221	170.3
6.101	-0.259	177.5
6.599	-0.279	178.4
7.109	-0.279	173.5
8.131	-0.279	161.9
9.139	-0.279	160.5
10.160	-0.279	159.2

8.16 An unconfined, undrained compression test was conducted on a compacted clayey, uncemented sand classified as SC. The sample size had a diameter of 38 mm and a length of 76 mm The peak axial force recorded was 230 N, and the sample shortened by 2 mm. (a) Which failure criterion is appropriate to interpret the test data? (b) Determine the undrained shear strength.

8.17 CU triaxial tests were carried out on a sample of a saturated clay classified as CH. The sample was isotropically consolidated using a cell pressure of 400 kPa before the axial displacements were incrementally applied. The results at peak and critical states are shown in the table below.

	Peak			Critical state	
σ_3 (kPa)	$\sigma_1 - \sigma_3$ (kPa)	Δu (kPa)		$\sigma_1 - \sigma_3$ (kPa)	Δu (kPa)
400	386	200		324	221

(a) Which failure criterion is appropriate to interpret this data set? Justify your answer. (b) Draw Mohr's circles (total and effective stresses) for the test. (c) Calculate the peak and critical state friction angles and show these angles on the Mohr's circles. (d) Calculate the undrained shear strengths at peak and at critical state and show them on the Mohr's circles. (e) Determine the inclination of angle of the failure plane to the plane on which the major principal effective stress acts at peak shear stress.

8.18 A CU test on a stiff, saturated overconsolidated clay was carried out at a constant cell pressure of 90 kPa. The peak deviatoric stress recorded was 294 kPa. The corresponding excess porewater pressure was -39.6 kPa. (a) Calculate the friction angle. Is this the critical state friction angle? Explain your answer. (b) Calculate the undrained shear strength.

8.19 CU tests were conducted on three samples of a compacted clay. Each sample was saturated before shearing. The results, when no further change in excess porewater pressure or deviatoric stress occurred, are shown in the table below. Calculate (a) the friction angle at indicate if this is for peak or critical state, (b) the undrained shear strength for each test, and (c) draw Mohr's circle of total and effective stress for each test and compare your results to those calculated in (a) and (b).

σ_3 (kPa)	$(\sigma_1 - \sigma_3)$ (kPa)	Δu (kPa)
273	419	0.91
432	664	-1.8
569	865	2.73

8.20 A CU triaxial test was carried out on a silty, saturated clay that was isotropically consolidated using a cell pressure of 90 kPa. The following data were obtained:

Axial stress (kPa)	Axial strain, ε_1 (%)	Δu (kPa)
0	0	0
5.5	0.05	4.1
11.0	0.12	8.3
24.8	0.29	19.3
28.3	0.38	29.0
35.2	0.56	34.5
50.3	1.08	40.7
84.8	2.43	49.6
104.8	4.02	55.8
120.7	9.15	59.3
121.4	10.1	59.3

(a) Plot the deviatoric (axial) stress against axial strain and excess porewater pressure against axial strain. (b) Determine the undrained shear strength and the friction angle. Are these critical state or peak values? Justify your answer.

8.21 The peak deviatoric stress result of a UU test on a sample of a stiff, saturated clay classified as CL is $(\sigma_1 - \sigma_3)_p = 70$ kPa. The initial size of the sample was 38 mm × 76 mm long, and it compressed by 2 mm Determine $(s_u)_p$.

8.22 Plot the corrected N value ($N_{1,60}$), the estimated bulk unit weight, and the estimated friction angle with depth for the data given in Figure 3.16.

8.23 Plot the estimated undrained shear strength with depth for the CPT results shown in Figure 3.18.

Critical thinking and decision making

8.24 You are in charge of designing a retaining wall. What laboratory tests would you specify for the backfill soil? Give reasons.

8.25 A CU test on an extremely stiff, overconsolidated clay was carried out. The sample was back saturated to a degree of saturation of 98%. It was then isotropically consolidated incrementally under a constant cell pressure of 140 kPa, and then incrementally unloaded to a cell pressure of 35 kPa. The initial porewater pressure was zero. The deviatoric stress remains approximately constant at 55 kPa from an axial strain of 11% to 15%, at which stage the test was stopped. The corresponding excess porewater pressure also remained approximately constant at 1.4 kPa over the same axial strain range. (a) Calculate the friction angle and state if this is likely the peak or critical state. (b) Calculate the ratio of the undrained shear strength to the effective cell pressure. Is this value reasonable? Use an appropriate relationship shown in Table 8.10 to determine the reasonableness of your answer. The value of $\Lambda = 0.8$. (c) What strength parameters would you recommend for this clay?

8.26 The results of triaxial CU tests on saturated, uncemented, overconsolidated CL soil from a commercial laboratory are shown by the Mohr's peak effective stress circles in Figure P8.26. All three tests were stopped as the soon as the peak deviatoric stress started dropping. (a) Replot the Mohr's circles of stress. (b) Determine the shear strength parameter or parameters. (c) The confining effective stress on the soil in the field for which the shear strength parameters is (are) required is 300 kPa. What shear strength parameter or parameters would you recommend, and why? (d) If you cannot recommend any shear strength parameter from the test results, explain why? (e) If you cannot recommend any shear strength parameter from the test results, and samples of the soil are available, would you recommend re-testing and what instructions would you give to the laboratory?

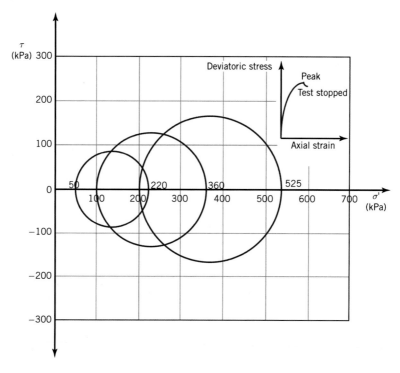

Figure P8.26

Appendix A
Derivation of the One-Dimensional Consolidation Theory

Consider a soil element shown in Figure A.1. The volume inflow of water is $Q_{in} = v\, dA$ where v is the flow velocity and dA is the cross sectional area of the soil element normal to the flow. The volume outflow over the elemental thickness dz is $Q_{out} = [v + (\partial v/\partial z)dz]dA$. The change in flow is

$$Q_{out} = Q_{out} - Q_{in} = \left[v + \left(\frac{\partial v}{\partial z}\right)dz\right]dA - vdA = \left(\frac{\partial v}{\partial z}\right)dz\, dA \tag{A.1}$$

The rate of change in volume of water expelled, $\partial V/\partial t$, which is equal to the rate of change of volume of the soil, must equal the change in flow. That is,

$$\frac{\partial V}{\partial t} = \frac{\partial v}{\partial z}dz\, dA \tag{A.2}$$

Since consolidation is assumed to take place only in the vertical direction, the vertical strain is the same as the volumetric strain. That is, $\varepsilon_z = \varepsilon_p$. The modulus of volume change is $m_v = \partial \varepsilon_z / \partial \sigma_z'$ where $\partial \sigma_z'$ is the change in vertical effective stress. Therefore, $\partial \varepsilon_z = m_v \partial \sigma_z'$, and since $\partial \varepsilon_z = \partial V/V$, where $V = dz\, dA$ is the volume of the soil element, we get that $\partial V = V \partial \varepsilon_z = V m_v \partial \sigma_z'$. The change in vertical effective stress, $\partial \sigma_z'$, is equal to the excess porewater pressure dissipated, ∂u. We can now write as $\partial V = V m_v \partial u = dz\, dA\, m_v \partial u$.

By substituting $\partial V = V m_v \partial u$ into Equation (A.2), we get

$$dz\, dA\, m_v \frac{\partial u}{\partial t} = \frac{\partial v}{\partial z}dz\, dA \tag{A.3}$$

Simplifying and rearranging Equation (A.3), we obtain

$$\frac{\partial v}{\partial z} = \frac{\partial u}{\partial t}m_v \tag{A.4}$$

Soil Mechanics Fundamentals, First Edition. Muni Budhu.
© 2015 John Wiley & Sons, Ltd. Published 2015 by John Wiley & Sons, Ltd.
Companion website: www.wiley.com\go\budhu\soilmechanicsfundamentals

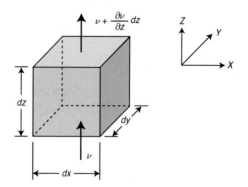

Figure A.1 Vertical flow through a soil element.

The one-dimensional flow of water from Darcy's law is

$$v = k_z i = k_z \frac{\partial h}{\partial z} \tag{A.5}$$

where k_z is the hydraulic conductivity in the vertical direction and h is the porewater pressure (u) head.

Partial differentiation of Equation (A.5) with respect to z gives

$$\frac{\partial v}{\partial z} = k_z \frac{\partial^2 h}{\partial z^2} \tag{A.6}$$

The porewater pressure at any time t is

$$u = h\gamma_w \tag{A.7}$$

Partial differentiation of Equation (A.7) with respect to z gives

$$\frac{\partial^2 h}{\partial z^2} = \frac{1}{\gamma_w} \frac{\partial^2 u}{\partial z^2} \tag{A.8}$$

By substitution of Equation (A.8) into Equation (A.6), we get

$$\frac{\partial v}{\partial z} = \frac{k_z}{\gamma_w} \frac{\partial^2 u}{\partial z^2} \tag{A.9}$$

Equating Equation (A.4) and Equation (A.9), we obtain

$$\frac{\partial u}{\partial t} = \frac{k_z}{m_v \gamma_w} \frac{\partial^2 u}{\partial z^2} \tag{A.10}$$

We can replace $k_z/m_v\gamma_w$ by a coefficient C_v called the coefficient of consolidation.

The units for C_v are length²/time, for example, cm²/min. Rewriting Equation (A.10) by substituting C_v, we get the general equation for one-dimensional consolidation as

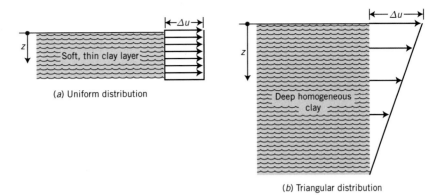

(a) Uniform distribution

(b) Triangular distribution

Figure A.2   Two types of excess porewater pressure distribution with depth: (a) uniform distribution within a thin layer and (b) triangular distribution within a thick layer.

$$\frac{\partial u}{\partial t} = C_v \frac{\partial^2 u}{\partial z^2} \qquad\qquad (A.11)$$

Equation (A.11) was developed by Terzaghi (1925). It is a common equation in many branches of engineering. For example, the heat diffusion equation commonly used in mechanical engineering is similar to Equation (A.11) except that temperature, T, replaces u and heat factor, K, replaces C_v. The values of k_z and m_v were assumed to be constants. In general, this assumption is violated in soils. As the soil consolidates, the void spaces are reduced and k_z decreases. Also, m_v is not linearly related to σ'_z. The consequence of k_z and m_v not being constants is that C_v is not a constant. In practice, C_v is assumed to be a constant, and this assumption is reasonable only if the stress changes are small enough such that k_z and m_v do not change significantly.

The solution of Equation (A.11) requires the specification of the initial distribution of excess porewater pressures at the boundaries. Various distributions of porewater pressures within a soil layer are possible. Two of these are shown in Figure A.2. One of these is a uniform distribution of initial excess porewater pressure with depth (Figure A.2a). This may occur in a thin layer of fine-grained soils. The other (Figure A.2b) is a triangular distribution. This may occur in a thick layer of fine-grained soils.

The boundary conditions for a uniform distribution of initial excess porewater pressure in which double drainage occurs are

When $t = 0$: $\Delta u = \Delta u_o = \Delta \sigma_z$.
At the top boundary: $z = 0$, $\Delta u_z = 0$.
When t tends to infinity and the excess porewater pressure has dissipated:
At the bottom boundary: $z = 2H_{dr}$, $\Delta u_z = 0$,

where Δu_o is the initial excess porewater pressure, Δu_z is the excess porewater pressure at a depth, z, within the soil and H_{dr} is the length of the drainage path.

Different analytical techniques for the general one-dimensional consolidation equation (and heat diffusion equation) including using Fourier's method and Heaviside's method have been developed for simple boundary conditions. For complex boundary conditions, numerical methods such as finite difference method and finite element method are used. The analytic solution for uniform initial excess porewater pressure, Δu_o, and double drainage using Heaviside's method is

$$\frac{\Delta u_z}{\Delta u_o} = \frac{4}{\pi} \sum_{m=1}^{\infty} \left\{ \frac{(-1)^{m-1}}{2m-1} \cos\left[(2m-1)\frac{\pi}{2}\left(\frac{H_{dr}-z}{H_{dr}}\right)\right] \exp\left[-(2m-1)^2 \frac{\pi^2}{4} T_v\right] \right\} \tag{A.12}$$

where m is a positive integer and

$$T_v = \frac{C_v t}{H_{dr}^2} \tag{A.13}$$

where T_v is known as the time factor; it is a dimensionless term.

Appendix B
Mohr's Circle for Finding Stress States

Access www.wiley.com/college/budhu, and click Chapter 8 and then Mohrcircle.zip to download an application to plot, interpret, and explore a variety of stress states interactively.

Suppose that a cuboidal sample of soil is subjected to the stresses shown in Figure B.1a. We would like to know what the stresses are at a point X within the sample due to the applied stresses. One approach to find the stresses at X, called the stress state at X, is to use Mohr's circle. The stress state at a point is the set of stress vectors corresponding to all planes passing through that point. The procedure to draw Mohr's circle to find the stress state is as follows.

1. Draw two orthogonal axes in which the x-axis (abscissa) is normal stress (usually normal effective stress for soils), σ, and the y-axis (ordinate) is the shear stress, τ.
2. Choose a sign convention. In soil mechanics, compressive stresses are positive. We will assume that counterclockwise shear is positive.
3. Plot the two coordinates of the circle as (σ_z, τ_{zx}) and (σ_x, τ_{xz}). The stresses (σ_z, τ_{zx}) act on the horizontal plane while the stresses (σ_x, τ_{xz}). act on the vertical plane. Recall from your strength of materials course that, for equilibrium, $\tau_{xz} = -\tau_{zx}$; these are called complementary shear stresses and are orthogonal to each other. For the stresses in Figure B.1a, A and B in Figure B.1b represent the two coordinates.
4. Draw a circle with AB as the diameter and O as the center.
5. The circle crosses the normal stress axis at 1 and 3. The stresses at these points are the major principal stress, σ_1, and the minor principal stress, σ_3.
6. If you want to find the stresses on any plane inclined at an angle θ from the horizontal plane, as depicted by MN in Figure B.1a, you need to identify the pole of the stress circle as follows: The stress σ_z acts on the horizontal plane and the stress σ_x acts on the vertical plane for our case. Draw these planes in Mohr's circle; where they intersect at a point P, that represents the pole of the stress circle. It is a special point because any line passing through the pole will intersect Mohr's circle at a point that represents the stresses on a plane parallel to the line. Let us see how this works. Once you locate pole P, draw a line parallel to MN through P as shown by $M'N'$ in Figure B.1b. The line $M'N'$ intersects the circle at N' and the coordinates of N', $(\sigma_\theta, \tau_\theta)$, represent the normal and shear stresses on MN.

Rather than drawing Mohr's circle, you could determine the stresses using the following equations:

Soil Mechanics Fundamentals, First Edition. Muni Budhu.
© 2015 John Wiley & Sons, Ltd. Published 2015 by John Wiley & Sons, Ltd.
Companion website: www.wiley.com\go\budhu\soilmechanicsfundamentals

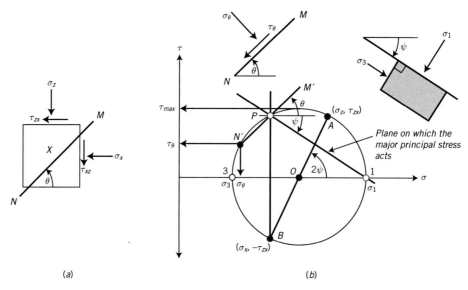

Figure B.1 Mohr's circle to determine the stress state.

The principal stresses are related to the stress components σ_z, σ_x, τ_{zx} by

$$\sigma_1 = \frac{\sigma_z + \sigma_x}{2} + \sqrt{\left(\frac{\sigma_z - \sigma_x}{2}\right)^2 + \tau_{zx}^2} \tag{B.1}$$

$$\sigma_3 = \frac{\sigma_z + \sigma_x}{2} - \sqrt{\left(\frac{\sigma_z - \sigma_x}{2}\right)^2 + \tau_{zx}^2} \tag{B.2}$$

The angle between the major principal stress plane and the horizontal plane (ψ) is

$$\tan\psi = \frac{\tau_{zx}}{\sigma_1 - \sigma_x} \tag{B.3}$$

The stresses on a plane oriented at an angle θ from the major principal stress plane are

$$\sigma_\theta = \frac{\sigma_1 + \sigma_3}{2} + \frac{\sigma_1 - \sigma_3}{2}\cos 2\theta \tag{B.4}$$

$$\tau_\theta = \frac{\sigma_1 - \sigma_3}{2}\sin 2\theta \tag{B.5}$$

The stresses on a plane oriented at an angle θ from the horizontal plane are

$$\sigma_\theta = \frac{\sigma_z + \sigma_x}{2} + \frac{\sigma_z - \sigma_x}{2}\cos 2\theta + \tau_{zx}\sin 2\theta \tag{B.6}$$

$$\tau_\theta = \tau_{zx}\cos\theta - \frac{\sigma_z - \sigma_x}{2}\sin 2\theta \tag{B.7}$$

In these equations, θ is positive for counterclockwise orientation.

The maximum (principal) shear stress is at the top of the circle with magnitude

$$\tau_{max} = \frac{\sigma_1 - \sigma_3}{2} \tag{B.8}$$

Appendix C
Frequently Used Tables of Soil Parameters and Correlations

Table 1.1 Soil types, descriptions, and average grain sizes.

Category	Soil type	Symbol	Description	Grain size, D
Coarse-grained	Gravel	G	Rounded and/or angular bulky hard rock, coarsely divided	Coarse: >75 mm Fine: 4.75 mm–19 mm
	Sand	S	Rounded and/or angular hard rock, finely divided	Coarse: 2.0 mm–4.75 mm Medium: 0.425 mm–2.0 mm Fine: 0.075 mm–0.425 mm
Fine-grained (also called fines)	Silt	M	Particle size between clay and sand; nonplastic or very slightly plastic; exhibits little or no strength when dried; easily brushed off when dried	0.002 mm–0.075 mm
	Clay	C	Particles are smooth and mostly clay minerals; greasy and sticky when wet; exhibits plasticity and significant strength when dried; water reduces strength	<0.002 mm

Table 2.1 Typical values of unit weight for soils.

Soil type	γ_{sat} (kN/m³)	R_d	γ_d (kN/m³)	R_d
Gravel	20–22	2.04–2.24	15–17	1.52–1.73
Sand	18–20	1.84–2.04	13–16	1.33–1.63
Silt	18–20	1.84–2.04	14–18	1.43–1.84
Clay	16–22	1.63–2.24	14–21	1.43–2.15

Soil Mechanics Fundamentals, First Edition. Muni Budhu.
© 2015 John Wiley & Sons, Ltd. Published 2015 by John Wiley & Sons, Ltd.
Companion website: www.wiley.com\go\budhu\soilmechanicsfundamentals

Table 2.2　Description of coarse-grained soils based on relative density and porosity.

D_r (%)	Porosity, n (%)	Description
0–20	100–80	Very loose
20–40	80–60	Loose
40–70	60–30	Medium dense or firm
70–85	30–15	Dense
85–100	<15	Very dense

Table 2.3　Ranges of free swell for some clay minerals.

Clay minerals	Free swell (%)
Calcium montmorillonite (Ca-smectite)	45–145
Sodium montmorillonite (Na-smectite)	1400–1600
Illite	15–120
Kaolinite	5–60

Table 2.4　Description of the strength of fine-grained soils based on liquidity index.

Values of LI	Description of soil strength
$LI < 0$	Semisolid state: high strength, brittle, (sudden) fracture is expected
$0 < LI < 1$	Plastic state: intermediate strength, soil deforms like a plastic material
$LI > 1$	Liquid state: low strength, soil deforms like a viscous fluid

Table 2.5　Typical Atterberg limits for soils.

Soil type	LL (%)	PL (%)	PI (%)
Sand		Nonplastic	
Silt	30–40	20–25	10–15
Clay	40–150	25–50	15–100
Minerals			
Kaolinite	50–60	30–40	10–25
Illite	95–120	50–60	50–70
Montmorillonite	290–710	50–100	200–660

Table 2.6 Description of fine-grained soils based on consistency index.

Description	CI
Very soft (ooze out of finger when squeezed)	<0.25
Soft (easily molded by finger)	0.25–0.50
Firm or medium (can be molded using strong finger pressure)	0.50–0.75
Stiff (finger pressure dents soil)	0.75–1.00
Very stiff (finger pressure barely dents soil, but soil cracks under significant pressure)	>1

Table 2.7 Activity of clay-rich soils.

Description	Activity, A
Inactive	<0.75
Normal	0.75–1.25
Active	1.25–2
Very (highly) active (e.g., montmorillonite or bentonite)	>6
Minerals	
Kaolinite	0.3–0.5
Illite	0.5–1.3
Na-montmorillonite	4–7
Ca-montmorillonite	0.5–2.0

Table 3.2 Guidelines for the minimum number of boreholes for buildings and subdivisions based on area.

Buildings		Subdivisions	
Area (m²)	No. of boreholes (min.)	Area (m²)	No. of boreholes (min.)
100	2	4000	2
250	3	8000	3
500	4	20,000	4
1000	5	40,000	5
2000	6	80,000	7
5000	7	400,000	15
6000	8		
8000	9		
10,000	10		

Table 3.3 Guidelines for the minimum number or frequency and depths of boreholes for common geostructures.

Geostructure	Minimum number of boreholes	Minimum depth
Shallow foundation for buildings	1, but generally boreholes are placed at node points along grids of sizes varying from 15 m × 15 m to 40 m × 40 m	5 m or 1B to 3B, where B is the foundation width
Deep (pile) foundation for buildings	Same as shallow foundations	25–30 m; if bedrock is encountered, drill 3 m into it
Bridge	Abutments: 2 Piers: 2	25–30 m; if bedrock is encountered, drill 3 m into it
Retaining walls	length < 30 m: 1 length > 30 m: 1 every 30 m, or 1 to 2 times the height of the wall	1 to 2 times the wall height Walls located on bedrock: 3 m into bedrock
Cut slopes	Along length of slope: 1 every 60 m; if the soil does not vary significantly, 1 every 120 m On slope: 3	6 m below the bottom of the cut slope
Embankments including roadway (highway, motorway)	1 every 60 m; if the soil does not vary significantly, 1 every 120 m	The greater of 2× height or 6 m

Table 3.4 Correction factors for rod length, sampler type, and borehole size.

Correction factor	Item	Correction factor
C_R	Rod length (below anvil)	$C_R = 0.8$; $L \leq 4$ m $C_R = 0.05L + 0.61$; 4 m $< L \leq 6$ m $C_R = -0.0004L^2 + 0.017L + 0.83$; 6 m $< L < 20$ m, $C_R = 1$; $L \geq 20$ m L = rod length
C_S	Standard sampler US sampler without liners	$C_S = 1.0$ $C_S = 1.2$
C_B	Borehole diameter: 65 mm to 115 mm 152 mm 200 mm	 $C_B = 1.0$ $C_B = 1.05$ $C_B = 1.15$
C_E	Equipment: Safety hammer (rope, without Japanese "throw" release) Donut hammer (rope, without Japanese "throw" release) Donut hammer (rope, with Japanese "throw" release) Automatic trip hammer (donut or safety type)	 $C_E = 0.7$ to 1.2 $C_E = 0.5$ to 1.0 $C_E = 1.1$ to 1.4 $C_E = 0.8$ to 1.4

Sources: Youd et al. (2001)

Table 3.5 Compactness of coarse-grained soils based on N values.

N	γ (kN/m³)	D_r (%)	Compactness
0–4	11–13	0–20	Very loose
4–10	13–16	20–40	Loose
10–30	16–19	40–70	Medium
30–50	19–21	70–85	Dense
>50	>21	>85	Very dense

Source: Terzaghi and Peck (1948).

Table 3.6 Consistency of saturated fine-grained soils based on SPT.

Description	N
Very soft	0–2
Soft	2–4
Medium stiff	4–8
Stiff	8–15
Very stiff	15–30
Hard	>30

Source: Terzaghi and Peck (1948).

Table 4.1 Hydraulic conductivity for common saturated soil types.

Soil type	k_z (cm/s)	Description	Drainage
Clean gravel (GW, GP)	>1.0	High	Very good
Clean sands, clean sand and gravel mixtures (SW, SP)	1.0 to 10^{-3}	Medium	Good
Fine sands, silts, mixtures comprising sands, silts, and clays (SM-SC) Weathered and fissured clays	10^{-3} to 10^{-5}	Low	Poor
Silt, silty clay (MH, ML)	10^{-5} to 10^{-7}	Very low	Poor
Homogeneous clays (CL, CH)	<10^{-7}	Practically impervious	Very poor

Table 5.1 Comparison of field compactors for various soil types.

		Compaction type				
		Static		Dynamic		
		Pressure with kneading	Kneading with pressure	Vibration	Impact	
Material	Lift thickness (mm)	Static sheeps-foot grid roller; scraper	Scraper; rubber-tired roller; loader; grid roller	Vibrating plate compactor; vibrating roller; vibrating sheepsfoot roller	Vibrating sheepsfoot rammer	Compactability
Gravel	300±	Not applicable	Very good	Good	Poor	Very easy
Sand	250±	Not applicable	Good	Excellent	Poor	Easy
Silt	150±	Good	Excellent	Poor	Good	Difficult
Clay	150±	Very good	Good	No	Excellent	Very difficult

Table 7.1 Typical values of E'_{sec} and G_{sec}.

Soil type	Description	E'_{sec} (MPa)	G_{sec} (MPa)
Clay	Soft	1–15	0.5–5
	Medium	15–50	5–15
	Stiff	50–100	15–40
Sand	Loose	10–20	2–10
	Medium	20–50	10–15
	Dense	50–100	15–40
Gravel	Loose	20–75	2–20
	Medium	75–100	20–40
	Dense	100–200	40–75

Estimated from standard penetration test (SPT)*	E'_{sec} (MPa)	G_{sec} (MPa)
Gravels and gravels with sands	$1.2N_{60}$*	$0.45N_{60}$
Coarse sands, sands with gravel (<10%)	$0.95N_{60}$	$0.35N_{60}$
Fine and medium sands, clean and with fines (<10%)	$0.7N_{60}$	$0.25N_{60}$
Silts and sandy silt	$0.4N_{60}$	$0.15N_{60}$

Estimated from cone penetrometer test (CPT)*	E'_{sec} (kPa)	G_{sec} (kPa)
Fine and medium sands, clean and with fines (<10%)	$3q_c^*$	$1.2q_c$
Clayey silt and silty sand	$5q_c$	$2q_c$
Clays	$7q_c$	$2.5q_c$

Note: *N_{60} is the SPT N values corrected for 60% energy (see Chapter 3); q_c is the cone tip resistance in kPa.

Table 7.2   Typical values of Poisson's ratio.

Soil type	Description	ν'
Clay	Soft	0.35–0.4
	Medium	0.3–0.35
	Stiff	0.2–0.3
Sand	Loose	0.15–0.25
	Medium	0.25–0.3
	Dense	0.25–0.35

Note: For all soils at constant volume, $\nu = 0.5$ (total stress condition).

Table 7.3   Typical range of values of C_c and C_r.

$C_c = 0.1$ to 0.8
$C_r = 0.015$ to 0.35; also, $C_r \approx C_c/5$ to $C_c/10$
$C_\alpha/C_c = 0.03$ to 0.08

Table 7.4   Some empirical relationships for C_c and C_r.

Empirical relationships	Reference
$C_c = 0.009(LL - 10)$	Terzaghi and Peck, 1948
$C_c = 1.35$ PI (remolded clays)	Schofield and Wroth, 1968
$C_c = 0.40(e_o - 0.25)$	Azzouz et al., 1976
$C_c = 0.01(w - 5)$	Azzouz et al., 1976
$C_c = 0.37(e_o + 0.003LL - 0.34)$	Azzouz et al., 1976
$C_r = 0.15(e_o + 0.007)$	Azzouz et al., 1976
$C_r = 0.003(w + 7)$	Azzouz et al., 1976
$C_r = 0.126(e_o + 0.003LL - 006)$	Azzouz et al., 1976
$C_r = 0.000463LL\ G_s$	Nagaraj and Murthy, 1985

Note: w is the natural water content (%), LL is the liquid limit (%), e_o is the initial void ratio, and PI is the plasticity index.

Table 7.5   Typical values of C_v.

Soil	c_v (cm²/s × 10⁻⁴)	c_v (m²/yr)
Boston blue clay (CL)	40 ± 20	12 ± 6
Organic silt (OH)	2–10	0.6–3
Glacial lake clays (CL)	6.5–8.7	2.0–2.7
Chicago silty clay (CL)	8.5	2.7
Swedish medium sensitive clays (CL-CH)		
1. laboratory	0.4–0.7	0.1–0.2
2. field	0.7–3.0	0.2–1.0
San Francisco Bay mud (CL)	2–4	0.6–1.2
Mexico City clay (MH)	0.9–1.5	0.3–0.5

Source: Modified from Carter and Bentley (1991).

Table 8.3 Summary of equations for the three failure criteria.

Name	Peak	Critical state
Coulomb	$\tau_p = (\sigma'_n)_p \tan(\phi'_{cs} + \alpha_p) = (\sigma'_n)_p \tan\phi'_p$ unsaturated, cemented soils: $\tau_p = C + (\sigma'_n)_p \tan(\xi_o)$ $C = c_o + c_t + c_{cm}$	$\tau_{cs} = (\sigma'_n)_{cs} \tan\phi'_{cs}$
Mohr–Coulomb	$\sin\phi'_p = \left(\dfrac{\sigma'_1 - \sigma'_3}{\sigma'_1 + \sigma'_3} \right)_p$ $\dfrac{(\sigma'_3)_p}{(\sigma'_1)_p} = \dfrac{1 - \sin\phi'_p}{1 + \sin\phi'_p} = \tan^2\left(45° - \dfrac{\phi'_p}{2}\right)$ Cemented soils: $\sin\xi_o = \dfrac{(\sigma'_1 - \sigma'_3)}{2C \cot\xi_o + (\sigma'_1 + \sigma'_3)}$ $C = c_o + c_t + c_{cm}$	$\sin\phi'_{cs} = \left(\dfrac{\sigma'_1 - \sigma'_3}{\sigma'_1 + \sigma'_3} \right)_{cs}$ $\dfrac{(\sigma'_3)_{cs}}{(\sigma'_1)_{cs}} = \dfrac{1 - \sin\phi'_{cs}}{1 + \sin\phi'_{cs}} = \tan^2\left(45° - \dfrac{\phi'_{cs}}{2}\right)$
	Inclination of the failure plane to the plane on which the major principal effective stress acts. $\theta_p = 45° + \dfrac{\phi'_p}{2}$	Inclination of the failure plane to the plane on which the major principal effective stress acts. $\theta_{cs} = 45° + \dfrac{\phi'_{cs}}{2}$
Tresca	$(s_u)_p = \dfrac{(\sigma_1 - \sigma_3)_p}{2}$	$(s_u)_{cs} = \dfrac{(\sigma_1 - \sigma_3)_{cs}}{2}$

Table 8.4 Ranges of friction angles for soils (degrees).

Soil type	ϕ'_{cs}	ϕ'_p	ϕ'_r
Gravel	30–35	35–50	
Mixtures of gravel and sand with fine-grained soils	28–33	30–40	
Sand	27–37[a]	32–50	
Silt or silty sand	24–32	27–35	
Clays	15–30	20–30	5–15

[a]Higher values (32°–37°) in the range are for sands with significant amount of feldspar (Bolton, 1986). Lower values (27°–32°) in the range are for quartz sands.

Table 8.5 Typical ranges of dilation angles for soils.

Soil type	α_p (degrees)
Dense sand	10–15°
Loose sand	<10°
Normally consolidated clay	0°

Table 8.6 Typical values of s_u for saturated fine-grained soils.

Description	s_u (kPa)
Very soft (extremely low)	<10
Soft (low)	10–25
Medium stiff (medium)	25–50
Stiff (high)	50–100
Very stiff (very high)	100–200
Extremely stiff (extremely high)	>200

Table 8.8 Correlation of friction angle with SPT.

N	N_{60}	Compactness	ϕ' (degrees)
0–4	0–3	Very loose	26–28
4–10	3–9	Loose	29–34
10–30	9–25	Medium	35–40*
30–50	25–45	Dense	38–45*
>50	>45	Very dense	>45*

Source: Modified from Kulhawy and Mayne (1990).
Note: Values marked by an * correspond to ϕ'_p.

Table 8.9 Some correlations of CPT with soil parameters.

Parameter	Relationship
Peak friction angle (triaxial)	$$\phi'_p = 17.6 + 11\log\left[\frac{\dfrac{q_c}{p_{atm}}}{\sqrt{\left(\dfrac{\sigma'_{zo}}{p_{atm}}\right)}}\right] \text{ degrees}$$ σ'_{zo} is the initial or current vertical effective stress, p_{atm} is atmospheric pressure (101 kPa)
Peak undrained shear strength	$(s_u)_p = \dfrac{q_c - \sigma_z}{N_k}$ N_k is a cone factor that depends on the geometry of the cone and the rate of penetration. Average values of N_k as a function of plasticity index can be estimated from $N_k = 19 - \dfrac{PI - 10}{5}; \quad PI > 10$
Past maximum vertical effective stress	$\sigma'_{zc} = 0.33(q_c - \sigma'_{zo})^m$ Intact clays: $m = 1$; organic clays: $m = 0.9$; silts: $m = 0.85$; silty sands: $m = 0.8$; clean sands: $m = 0.72$
Bulk unit weight	$\gamma = 1.95\gamma_w\left(\dfrac{\sigma_{zo}}{p_{atm}}\right)^{0.06}\left(\dfrac{f_s}{p_{atm}}\right)^{0.06}$ f_s is the average cone sleeve resistance over the depth of interest.

Source: Mayne et al. (2009).

Table 8.10   Some empirical and theoretical soil strength relationships.

Soil type	Equation	Reference
Normally consolidated clays	Triaxial test or axisymmetric condition $\left(\dfrac{s_u}{\sigma'_z}\right)_{cs} = 0.11 + 0.0037\, PI$; PI is plasticity index	Skempton (1944)
	$\left(\dfrac{s_u}{\sigma'_{zo}}\right)_{cs} \approx 0.5 \sin\phi'_{cs}$ (see notes 1 and 2)	Wroth (1984) Budhu (2011)
	$\left[\left(\dfrac{s_u}{\sigma'_{zo}}\right)_{cs}\right]_{ic} = \dfrac{3\sin\phi'_{cs}}{3 - \sin\phi'_{cs}}(0.5)^{\Lambda}$ (see note 3)	
	$\Lambda = 1 - \dfrac{C_c}{C_r}$, $\Lambda \approx 0.8$	
Overconsolidated clays	$\left(\dfrac{s_u}{\sigma'_{zo}}\right)_{cs} \approx 0.5 \sin\phi'_{cs}(OCR)^{\Lambda}$ (see notes 1 and 2)	Wroth (1984)
	$\left(\dfrac{s_u}{\sigma'_{zo}}\right)_{p} \approx 0.5 \sin\phi'_{cs}\alpha_y(OCR)^{\Lambda}$ (see notes 1 and 2)	Budhu (2011)
	$\alpha_y = \dfrac{\sqrt{1.45 OCR^{0.66} - 1}}{(0.725 OCR^{0.66})^{\Lambda}}$, $OCR \leq 10$	
	$\alpha_y = \dfrac{\sqrt{1.6 OCR^{0.62} - 1}}{(0.8 OCR^{0.62})^{\Lambda}}$, $OCR > 10$	
	$\Lambda = 1 - \dfrac{C_c}{C_r}$, $\Lambda \approx 0.8$	
	$\dfrac{s_u}{\sigma'_{zo}} = (0.23 \pm 0.04) OCR^{0.8}$ (see note 1)	Jamiolkowski et al. (1985)
Normally consolidated clays	$\phi'_{cs} = \sin^{-1}\left[0.35 - 0.1\ln\left(\dfrac{PI}{100}\right)\right]$; PI is plasticity index (%) (see note 2)	Wood (1990)
Clean quartz sand	$\phi'_p = \phi'_{cs} + 3D_r(10 - \ln p'_f) - 3$ where p'_f is the mean effective stress at failure (in kPa) and D_r is relative density. This equation should only be used if $12 > (\phi'_p - \phi'_{cs}) > 0$.	Bolton (1986)

Direct simple shear or plane strain condition

Soil type	Equation	Reference
Normally consolidated and overconsolidated clays	$\left[\left(\dfrac{s_u}{\sigma'_{zo}}\right)_{cs}\right]_{dss} = \dfrac{(3 - \sin\phi'_{cs})}{2\sqrt{3}}\left[\left(\dfrac{s_u}{\sigma'_{zo}}\right)_{cs}\right]_{triaxial}$ (see notes 1 and 2)	Budhu (2011)

Subscripts: ic = isotropic consolidation, cs = critical state, p = peak, y = yield, dss = direct simple shear.
Note 1: These are applicable to direct simple shear (DSS) tests. The estimated undrained shear strength from triaxial compression tests would be about 1.4 times greater.
Note 2: These are theoretical equations derived from critical state soil mechanics.
Note 3: These are for isotropically consolidated clays at critical state in the triaxial test.

Appendix D
Collection of Equations

CHAPTER 1: COMPOSITION AND PARTICLE SIZES OF SOILS

Determination of particle size of soils

Particle size of coarse-grained soils

Percentage retained

$$\text{\% retained on } i\text{th sieve} = \frac{W_i}{W} \times 100$$

Percentage finer

$$\text{\% finer than } i\text{th sieve} = 100 - \sum_{i=1}^{n} (\text{\% retained on } i\text{th sieve})$$

Particle size of fine-grained soils

Diameter of the particle at time t_D

$$D = \sqrt{\frac{30\mu z}{980(G_s - 1)t_D}} = K\sqrt{\frac{z}{t_D}} = K\sqrt{v_{set}}$$

where μ is the viscosity of water [0.01 Poise at 20°C; 10 Poise = 1 Pascal second (Pa.s) = 1000 centiPoise], z is the effective depth (cm) of the hydrometer, G_s is the specific gravity of the soil particles. At a temperature of 68°F and $G_s = 2.7$, $K = 0.01341$.

Soil Mechanics Fundamentals, First Edition. Muni Budhu.
© 2015 John Wiley & Sons, Ltd. Published 2015 by John Wiley & Sons, Ltd.
Companion website: www.wiley.com\go\budhu\soilmechanicsfundamentals

Characterization of soils based on particle size

Uniformity coefficient

$$Cu = \frac{D_{60}}{D_{10}}$$

Coefficient of curvature

$$CC = \frac{(D_{30})^2}{D_{10}D_{60}}$$

CHAPTER 2: PHYSICAL SOIL STATES AND SOIL CLASSIFICATION

Phase relationships

Total volume of the soil: $V = V_s + V_w + V_a = V_s + V_v$
Weight of the soil: $W = W_d + W_w$

Water content: $w = \dfrac{W_w}{W_d} \times 100\%$

Void ratio: $e = \dfrac{V_v}{V_s}$

Porosity: $n = \dfrac{V_v}{V}$

Relationship between porosity and void ratio: $n = \dfrac{e}{1+e}$

Specific gravity: $G_s = \dfrac{W_s}{V_s \gamma_w}$

Degree of saturation: $S = \dfrac{V_w}{V_v} = \dfrac{wG_s}{e}$ or $Se = wG_s$

Unit weight (bulk unit weight): $\gamma = \dfrac{W}{V} = \left(\dfrac{G_s + Se}{1+e}\right)\gamma_w$

Saturated unit weight ($S = 1$): $\gamma_{sat} = \left(\dfrac{G_s + e}{1+e}\right)\gamma_w$

Dry unit weight: $\gamma_d = \dfrac{W_s}{V} = \left(\dfrac{G_s}{1+e}\right)\gamma_w = \dfrac{\gamma}{1+w}$

Effective or buoyant unit weight: $\gamma' = \gamma_{sat} - \gamma_w = \left(\dfrac{G_s - 1}{1+e}\right)\gamma_w$

Relative density: $D_r = \dfrac{e_{max} - e}{e_{max} - e_{min}}$

$$D_r = \frac{\gamma_d - (\gamma_d)_{min}}{(\gamma_d)_{max} - (\gamma_d)_{min}} \left\{ \frac{(\gamma_d)_{max}}{\gamma_d} \right\}$$

Density index: $I_d = \dfrac{\gamma_d - (\gamma_d)_{min}}{(\gamma_d)_{max} - (\gamma_d)_{min}}$

Plasticity index: $PI = LL - PL$

Liquidity index: $LI = \dfrac{w - PL}{PI}$

Shrinkage index: $SI = PL - SL$

Estimate of SL: $SL = 46.4 \left(\dfrac{LL + 45.5}{PI + 46.4} \right) - 43.5$

where LL and PI are in percent.

Activity that describes the importance of the clay fractions on the plasticity index

$$A = \frac{PI}{\text{Clay fraction (\%)}}$$

Soil classification schemes

A-line in plasticity chart: $PI = 0.73(LL - 20)(\%); \quad PI > 4\%$

CHAPTER 3: SOILS INVESTIGATION

VST

Sensitivity: $S_t = \dfrac{T_{max}}{T_{res}}$

SPT

Correction for N_{60}: $N_{60} = N \left(\dfrac{ER_r}{60} \right) = NC_E$

SPT composite correction factor:

$$C_{RSBEN} = C_R C_S C_B C_E C_N$$

$$C_N = \sqrt{\frac{98.5}{\sigma'_{zo}}}, \quad C_N \leq 2; \quad \text{the unit for } \sigma'_{zo} \text{ is kPa}$$

Corrected N value, $N_{1,60}$: $N_{1,60} = C_{RSBEN} N$

Correction factor	Item	Correction factor
C_R	Rod length (below anvil)	$C_R = 0.8$; $L \leq 4\,m$ $C_R = 0.05L + 0.61$; $4\,m < L \leq 6\,m$ $C_R = -0.0004L^2 + 0.017L + 0.83$; $6\,m < L < 20\,m$, $C_R = 1$; $L \geq 20\,m$ L = rod length
C_S	Standard sampler US sampler without liners	$C_S = 1.0$ $C_S = 1.2$
C_B	Borehole diameter: 65 mm to 115 mm 152 mm 200 mm	 $C_B = 1.0$ $C_B = 1.05$ $C_B = 1.15$
C_E	Equipment: Safety hammer (rope, without Japanese "throw" release) Donut hammer (rope, without Japanese "throw" release) Donut hammer (rope, with Japanese "throw" release) Automatic trip hammer (donut or safety type)	 $C_E = 0.7$ to 1.2 $C_E = 0.5$ to 1.0 $C_E = 1.1$ to 1.4 $C_E = 0.8$ to 1.4

N	γ (kN/m³)	D_r (%)	Compactness
0–4	11–13	0–20	Very loose
4–10	13–16	20–40	Loose
10–30	16–19	40–70	Medium
30–50	19–21	70–85	Dense
>50	>21	>85	Very dense

Description	N
Very soft	0–2
Soft	2–4
Medium stiff	4–8
Stiff	8–15
Very stiff	15–30
Hard	>30

CPT

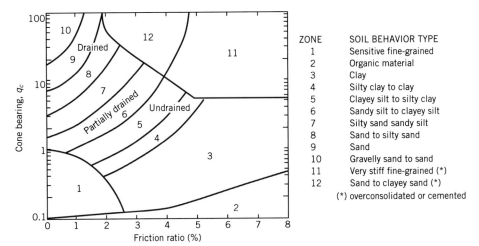

1 MPa = 20.88 ksf = 20,880 psf

CHAPTER 4: ONE- AND TWO-DIMENSIONAL FLOW OF WATER THROUGH SOILS

Darcy's law: $v_z = k_z \dfrac{\Delta H}{l} = k_z i$

Seepage velocity: $v_s = \dfrac{k_z}{n} i$

Volume rate of flow: $q_z = v_z A = A k_z i$

Empirical relationships for k_z proposed by Hazen: $k_z = C D_{10}^2$ cm/s

Flow parallel to soil layers: $k_{x(eq)} = \dfrac{1}{H_o}(z_1 k_{x1} + z_2 k_{x2} + \cdots + z_n k_{xn})$

Flow normal to soil layers: $k_{z(eq)} = \dfrac{H_o}{(z_1/k_{z1}) + (z_2/k_{z2}) + \cdots + (z_n/k_{zn})}$

Equivalent hydraulic conductivity: $k_{eq} = \sqrt{k_{x(eq)} k_{z(eq)}}$

Constant head test

Hydraulic conductivity in vertical direction: $k_z = \dfrac{q}{Ai} = \dfrac{QL}{tAh}$

Hydraulic conductivity corrected to a baseline temperature of 20°C: $k_{20°C} = k_{T°C} \dfrac{\mu_{T°C}}{\mu_{20°C}} = k_{T°C} R_T$

where $R_T = 2.42 - 0.475 \ln(T)$; T is degrees Celsius.

Falling head test

Hydraulic conductivity in the vertical direction: $k_z = \dfrac{aL}{A(t_2 - t_1)} \ln\left(\dfrac{h_1}{h_2}\right)$

Flow rate

Head loss (Δh) between each consecutive pair of equipotential lines: $\Delta h = \Delta H / N_d$

Flow rate: $q = k \sum_{i=1}^{N_f} \left(\dfrac{\Delta H}{N_d} \right)_i = k \Delta H \dfrac{N_f}{N_d}$

Critical hydraulic gradient: $i_{cr} = \dfrac{G_s - 1}{1 + e}$

Porewater pressure distribution

Porewater pressure head at any point j within the flow domain (flownet): $(h_p)_j = \Delta H - (N_d)_j \Delta h - h_z$

Porewater pressure: $u_j = (h_p)_j \gamma_w$

Simpson's rule: $P_w = \dfrac{\Delta x}{3} \left(u_1 + u_n + 2 \sum_{\substack{i=3 \\ \text{odd}}}^{n} u_i + 4 \sum_{\substack{i=2 \\ \text{even}}}^{n} u_i \right)$

CHAPTER 5: SOIL COMPACTION

Dry unit weight: $\gamma_d = \left(\dfrac{G_s}{1 + e} \right) \gamma_w = \dfrac{\gamma}{1 + w}$

Degree of saturation at any value of dry unit weight: $S = \dfrac{w G_s}{(G_s \gamma_w / \gamma_d) - 1}$

Theoretical maximum dry unit weight: $(\gamma_d)_{\max} = \left(\dfrac{G_s}{1 + e_{\min}} \right) \gamma_w = \left(\dfrac{G_s}{1 + w G_s} \right) \gamma_w$

Degree of compaction (DC): $DC = \dfrac{\text{Measured dry unit weight}}{\text{Desired dry unit weight}}$

CHAPTER 6: STRESSES FROM SURFACE LOADS AND THE PRINCIPLE OF EFFECTIVE STRESS

Stress increases in soil from surface loads

Point load: $\Delta \sigma_z = \dfrac{Q}{z^2} I \, ; \, I = \dfrac{3}{2\pi} \left(\dfrac{1}{1 + (r/z)^2} \right)^{5/2}$

Line load: $\Delta \sigma_z = \dfrac{2Q}{\pi} \dfrac{z^3}{(x^2 + z^2)^2}$

Strip loaded area transmitting a uniform stress: $\Delta \sigma_z = \dfrac{q_s}{\pi} \{ \alpha + \sin \alpha \cos(\alpha + 2\beta) \}$

Uniformly loaded circular area: $\Delta\sigma_z = q_s\left[1 - \left\{\dfrac{1}{1+(r_o/z)^2}\right\}^{3/2}\right]$

Uniformly loaded rectangular area

Below the corner of a rectangular area of width B and length L

$$\Delta\sigma_z = \frac{q_s}{2\pi}\left[\tan^{-1}\frac{LB}{zR_3} + \frac{LBz}{R_3}\left(\frac{1}{R_1^2} + \frac{1}{R_2^2}\right)\right]$$

where $R_1 = (L^2 + z^2)^{1/2}$, $R_2 = (B^2 + z^2)^{1/2}$, and $R_3 = (L^2 + B^2 + z^2)^{1/2}$
or $\Delta\sigma_z = q_sI_z$ where

$$I_z = \frac{1}{4\pi}\left[\frac{2mn\sqrt{m^2 + n^2 + 1}}{m^2 + n^2 + m^2n^2 + 1}\left(\frac{m^2 + n^2 + 2}{m^2 + n^2 + 1}\right) + \tan^{-1}\left(\frac{2mn\sqrt{m^2 + n^2 + 1}}{m^2 + n^2 - m^2n^2 + 1}\right)\right];$$
$$(mn)^2 \geq (m^2 + n^2 + 1)$$

and $m = B/z$ and $n = L/z$

Approximate method for rectangular loads

Center of the load: $\Delta\sigma_z = \dfrac{q_sBL}{(B+z)(L+z)}$

Total and effective stresses

Terzaghi's principle of effective stress: $\sigma' = \sigma - u$
Effective stress for unsaturated soils: $\sigma' = \sigma - u_a + \chi(u_a - u_w)$
Capillary depth: $z_c = \dfrac{4T\cos\theta}{d\gamma_w}$

Effective stresses due to Geostatic Stress Fields

Total vertical stress: $\sigma = \gamma_{sat}z$
Porewater pressure: $u = \gamma_w z$
Effective vertical stress: $\sigma' = \sigma - u = \gamma_{sat}z - \gamma_w z = (\gamma_{sat} - \gamma_w)z = \gamma'z$

Effects of seepage

Seepage force per unit volume: $j_s = \dfrac{\Delta H\gamma_w}{L} = i\gamma_w$

Resultant vertical effective stress for seepage downward:

$$\sigma'_z = \gamma'z + iz\gamma_w = \gamma'z + j_sz$$

Resultant vertical effective stress for seepage upward:

$$\sigma'_z = \gamma'z - iz\gamma_w = \gamma'z - j_sz$$

Lateral earth pressure at rest

Lateral earth pressure coefficient: $K_o = \sigma_3'/\sigma_1'$

$$K_o^{nc} \approx 1 - \sin\phi_{cs}'$$

$$K_o^{oc} = K_o^{nc}(OCR)^{1/2} = (1 - \sin\phi_{cs}')(OCR)^{1/2}$$

CHAPTER 7: SOIL SETTLEMENT

Settlement of coarse-grained soils

Rectangular flexible loaded area: $\rho_e = \dfrac{q_s B\left(1 - (v')^2\right)}{E'}I_s$

Center of the rectangle: $I_s \approx 0.62\ln\left(\dfrac{L}{B}\right) + 1.12$

Center of the rectangle: $I_s \approx 0.31\ln\left(\dfrac{L}{B}\right) + 0.56$

Circular flexible loaded area: $\rho_e = \dfrac{q_s D\left(1 - (v')^2\right)}{E'}I_{ci}$

Center of the circular area: $I_{ci} = 1$

Edge of circular area: $I_{ci} = \dfrac{2}{\pi}$

Settlement of fine-grained soils

Length of drainage path: $H_{dr} = \dfrac{H_{av}}{2} = \dfrac{H_o + H_f}{4}$

Vertical strain: $\varepsilon_z = \dfrac{\Delta z}{H} = \dfrac{\Delta e}{1 + e_o}$

Soil settlement: $\Delta z = H\dfrac{\Delta e}{1 + e_o}$

Void ratio at the end of consolidation under load: $e = e_o - \Delta e = e_o - \dfrac{\Delta H}{H}(1 + e_o)$

Coefficient of compression or compression index: $C_c = -\dfrac{e_2 - e_1}{\log\left[(\sigma_z')_2/(\sigma_z')_1\right]}$

Recompression index: $C_r = -\dfrac{e_2 - e_1}{\log\left[(\sigma_z')_2/(\sigma_z')_1\right]}$

Modulus of volume re-compressibility: $m_{vr} = -\dfrac{(\varepsilon_z)_2 - (\varepsilon_z)_1}{(\sigma_z')_2 - (\sigma_z')_1}$

Young's modulus of elasticity: $E_c' = \dfrac{\Delta\sigma_z'}{\Delta\varepsilon_z} = \dfrac{1}{m_{vr}}$

Overconsolidation ratio: $OCR = \dfrac{\sigma_{zc}'}{\sigma_{zo}'}$

Primary consolidation settlement of fine-grained soils, OCR = 1

$$\rho_{pc} = H\frac{\Delta e}{1+e_o} = \frac{H}{1+e_o}C_c\log\frac{\sigma'_{fin}}{\sigma'_{zo}}, \quad OCR = 1$$

where $\Delta e = C_c\log\sigma'_{fin}/\sigma'_{zo}$

Primary consolidation settlement of overconsolidated fine-grained soils

Case 1: $\sigma'_{fin} = \sigma'_{zo} + \Delta\sigma_z < \sigma'_{zc}$.

$$\rho_{pc} = \frac{H}{1+e_o}C_r\log\frac{\sigma'_{fin}}{\sigma'_{zo}}, \quad \sigma'_{fin} < \sigma'_{zc}$$

Case 2: $\sigma'_{fin} = \sigma'_{zo} + \Delta\sigma_z > \sigma'_{zc}$.

$$\rho_{pc} = \frac{H}{1+e_o}\left\{C_r\log(OCR) + C_c\log\frac{\sigma'_{fin}}{\sigma'_{zc}}\right\}$$

Primary consolidation settlement using m_v: $\rho_{pc} = Hm_v\Delta\sigma_z$

Thick soil layers: $\Delta\sigma_z = \dfrac{n(\Delta\sigma_z)_1 + (n-1)(\Delta\sigma_z)_2 + (n-2)(\Delta\sigma_z)_3 + \cdots + (\Delta\sigma_z)_n}{n + (n-1) + (n-2) + \cdots + 1}$

One-dimensional consolidation theory

$$T_v = \frac{\pi}{4}\left(\frac{U}{100}\right)^2 \quad \text{for } U < 60\%$$

$$T_v = 1.781 - 0.933\log(100 - U) \quad \text{for } U \geq 60\%$$

$$C_v = \frac{T_v H_{dr}^2}{t}$$

Relationship between laboratory and field consolidation

$$\frac{t_{field}}{t_{lab}} = \frac{(H_{dr}^2)_{field}}{(H_{dr}^2)_{lab}}$$

Secondary compression settlement

Secondary compression index:

$$C_\alpha = -\frac{(e_t - e_p)}{\log(t/t_p)}$$

Secondary consolidation settlement:

$$\rho_{sc} = \frac{H}{(1+e_p)}C_\alpha\log\left(\frac{t}{t_p}\right)$$

CHAPTER 8: SOIL STRENGTH

Coulomb's frictional law

Uncemented soils

$$\tau_{cs} = \left(\sigma'_n\right)_{cs} \tan \phi'_{cs}$$

$$\tau_p = \left(\sigma'_n\right)_p \tan \phi'_p = \left(\sigma'_n\right)_p \tan \left(\phi'_{cs} + \alpha_p\right)$$

Dilation angle if a soil mass is constrained in the lateral directions:

$$\alpha_p = \tan^{-1}\left(\frac{-\Delta H_o}{\Delta x}\right)_p$$

Cemented soils

$$\tau_p = C + \left(\sigma'_n\right)_p \tan \left(\xi_o\right)$$

$$C = c_o + c_t + c_{cm}$$

Mohr–Coulomb failure criterion

Uncemented soils

$$\sin \phi'_{cs} = \left(\frac{\sigma'_1 - \sigma'_3}{\sigma'_1 + \sigma'_3}\right)_{cs}$$

$$\tau_{cs} = \frac{\sigma'_1 - \sigma'_3}{2} \cos \phi'_{cs}$$

$$\theta_{cs} = 45° + \frac{\phi'_{cs}}{2}$$

$$\sin \phi'_p = \left(\frac{\sigma'_1 - \sigma'_3}{\sigma'_1 + \sigma'_3}\right)_p$$

$$\tau_p = \left(\frac{\sigma'_1 - \sigma'_3}{2}\right)_p \cos \phi'_p = \left(\frac{\sigma'_1 - \sigma'_3}{2}\right)_p \cos \left(\phi'_{cs} + \alpha_p\right)$$

$$\theta_p = 45° + \frac{\phi'_p}{2}$$

Cemented soils

$$\sin \xi_o = \frac{\left(\sigma'_1 - \sigma'_3\right)}{2C \cot \xi_o + \left(\sigma'_1 + \sigma'_3\right)}$$

$$\tau_p = C + \frac{1}{2} \tan \xi_o \left[\sigma'_1 (1 - \sin \xi_o) + \sigma'_3 (1 + \sin \xi_o)\right]$$

Tresca failure criterion

Undrained shear strength

$$\left(s_u\right)_{cs} = \frac{\left(\sigma_1\right)_{cs} - \left(\sigma_3\right)_{cs}}{2}$$

$$\left(s_u\right)_p = \frac{(\sigma_1)_p - (\sigma_3)_p}{2}$$

Laboratory tests to determine shear strength parameters

Shear box or direct shear test

Peak: $\tau_p = \dfrac{(P_x)_p}{A}$; $\phi_p' = \tan^{-1}\dfrac{(P_x)_p}{P_z}$

Critical state: $\tau_{cs} = \dfrac{(P_x)_{cs}}{A}$; $\phi_{cs}' = \tan^{-1}\dfrac{(P_x)_{cs}}{P_z}$

$$\alpha_p = \tan^{-1}\left(\frac{-\Delta H_o}{\Delta x}\right)_p$$

$$\alpha_p = \phi_p' - \phi_{cs}'$$

Conventional triaxial apparatus

Major principal total stress: $\sigma_1 = \dfrac{P_z}{A} + \sigma_3$

Deviatoric stress: $\sigma_1 - \sigma_3 = \dfrac{P_z}{A}$

Axial strain: $\varepsilon_1 = \varepsilon_a = \dfrac{\Delta H_o}{H_o}$

Radial strain: $\varepsilon_3 = \dfrac{\Delta r}{r_o}$

Volumetric strain: $\varepsilon_p = \dfrac{\Delta V}{V_o} = \varepsilon_1 + 2\varepsilon_3$

Shear strain: $\gamma = (\varepsilon_1 - \varepsilon_3)$

Area of sample at any given instance as it changes during loading

$$A = \frac{V}{H} = \frac{V_o - \Delta V}{H_o - \Delta H_o} = \frac{V_o\left(1 - \dfrac{\Delta V}{V_o}\right)}{H_o\left(1 - \dfrac{\Delta H_o}{H_o}\right)} = \frac{A_o(1 - \varepsilon_p)}{1 - \varepsilon_1} = \frac{A_o(1 - \varepsilon_1 - 2\varepsilon_3)}{1 - \varepsilon_1}$$

Unconfined compression (UC) test

Undrained shear strength:

$$s_u = \frac{P_z}{2A}$$

Field test

Shear vane: $\left(s_u\right)_p = 0.273\dfrac{T}{d^3}$

References

Atterberg, A. (1911). Über die physikalishe Bodenuntersuchung und über die Plastizität der Tone. *Int. Mitt. Boden*, **1**, 10–43.

Azzouz, A. S., Krizek, R. J., and Corotis, R. B. (1976). Regression analysis of soil compressibility. *Soils and Fnds.*, **16** (2), 19–29.

Bardet, Jean-Pierre. (1997). *Experimental soil mechanics*. Prentice Hall, Upper Saddle River, NJ.

Becker, D. E., Crooks, J. H., Been, K., and Jefferies, H. G. (1987). Work as a criterion for determining in situ and yield stresses in clays. *Can. Geotech. J.*, **24** (4), 549–564.

Bishop, A. W., Alpan, I., Blight, G. E., and Donald, I. B. (1960). Factors controlling the shear strength of partly saturated cohesive soils. *ASCE Res. Conf. on Shear Strength of Cohesive Soils*, University of Colorado, Boulder, 503–532.

Blanchet, R., Tavenas, F., and R. Garneau (1980). Behaviour of friction piles in soft sensitive clays. *Can. Geotech. J.*, **17** (2), 203–224.

Bolton, M. D. (1986). The strength and dilatancy of sands. *Geotechnique*, **36** (1), 65–78.

Boussinesq, J. (1885). *Application des potentiels a l'étude de l'équilibre et du mouvement des solides élastiques*. Gauthier-Villars, Paris.

Budhu, M. (2011). *Soil Mechanics and Foundations*, 3rd ed., Wiley, Hoboken, NJ.

Burland, J. B. (1973). Shaft friction piles in clay—a simple fundamental approach. *Ground Eng.*, **6** (3), 30–42.

Carter, M., and Bentley, S. P. (1991). *Correlations of Soil Properties*. Pentech, London.

Casagrande, A. (1932). Research on the Atterberg limits of soils. *Public Roads*, **13** (8), 121–136.

Casagrande, A. (ed.) (1936). The determination of the precon-solidation load and its practical significance. *1st Int. Conf. on Soil Mechanics and Foundation Engineering*, Cambridge, MA, **3**, 60–64.

Casagrande, A., and Fadum, R. E. (1940). Notes on soil testing for engineering purposes. *Soil Mechanics Series*. Graduate School of Engineering, Harvard University, Cambridge, MA, **8** (268), 37.

Coulomb, C. A. (1776). Essai sur une application des regles de maximia et minimis a quelques problèmes de statique relatifs a l'architecture. *Mémoires de la Mathématique et de Physique, présentés a l'Académie Royale des Sciences, par divers savants, et lus dans ces Assemblées*. L'Imprimérie Royale, Paris, 3–8.

Darcy, H. (1856). *Les Fontaines publiques de la ville de Dijon*. Dalmont, Paris.

Fancher, G. H., Lewis, J. A., and Barnes, K. B. (1933). Mineral Industries Experiment Station. *Bull.*, **12**. Penn State College, University Park.

Soil Mechanics Fundamentals, First Edition. Muni Budhu.
© 2015 John Wiley & Sons, Ltd. Published 2015 by John Wiley & Sons, Ltd.
Companion website: www.wiley.com\go\budhu\soilmechanicsfundamentals

Feng, T. W. (2000). Fall-cone penetration and water content relationship of clays. *Geotechnique*, 50(2), 181–187.

Foster, C., and Ahlvin, R. (1954). Stresses and deflections induced by a uniform circular load. In *Highway Research Board Proceedings* 33, National Academy of Sciences, Washington, DC.

Fredlund, D. G., and Xing, A. (1994). Equations for the soil-water characteristic curve. *Can. Geotech. J.*, **31**, 521–532.

Giroud, J. P. (1968). Settlement of a linearly loaded rectangular area. *J. Soil Mech. Found. Div.*, ASCE, **94** (SM4), 813–831.

Hazen, A. (1892). Some physical properties of sand and gravels with special reference to the use in filtration. *Massachusetts State Board of Health, 24th Ann. Rep.*, Boston.

Hazen, A. (1930). Water supply. *American Civil Engineers Handbook*. Wiley, New York, 1444–1518.

Jaky, J. (1944). The coefficient of earth pressure at rest. *J. Soc. Hungarian Architects Eng.*, 7, 355–358.

Jamiolkowski, M., Ladd, C. C., Germaine, J. T., and Lancellotta, R. (eds.) (1985). New developments in field and laboratory testing of soils. *11th Int. Conf. on Soil Mechanics and Foundation Engineering*, San Francisco, **1**, 57–154.

Jurgenson, L. (1934). The application of theories of elasticity and plasticity to foundation problems. *Journal of the Boston Society of Civil Engineers*, 21/1934, Boston Society of Civil Engineers, Boston.

Koerner, R. M. (1970). Effect of particle characteristics on soil strength. *J. Soil Mech. Found. Div.*, ASCE, **96** (SM4), 1221–1234.

Kulhawy, F. H., and Mayne, P. W. (1990). Manual on estimating soil properties for foundation design. EL-6800, Research project 1493-6, Cornell University, Ithaca, NY.

Liao, S. S. C., and Whitman, R. V. (1985). Overburden correction factors for SPT in sand. *J. Geotech. Eng. Div.*, ASCE, **112** (3), 373–377.

Lunne, T. Robertson, P. K., and Powell, J. J. M. (1997). *Cone Penetration Testing in Geotechnical Practice*. Chapman Hall, London.

McCammon, N. R., and Golder, H. Q. (1970). Some loading tests on long pipe piles. *Geotechnique*, **20** (2), 171–184.

Mayne, P. W., Coop, M. R., Springman, S. M., Huang, An-Bin, and Zornberg, J. G. (2009). Geomaterial behavior and testing. *Proc. 17th Int. Conf. on Soil Mechanics and Geotechnical Engineering*, Alex-andria, Egypt, OIS Press. Theme Lecture, **4**, 2777–2872.

Meyerhof, G. G. (1976). Bearing capacity and settlement of pile foundations. *J. Geotech. Eng. Div.* ASCE, **102** (GT3), 195–228.

Nagaraj, T. S., and Srinivasa Murthy, B. R. (1985). Prediction of the preconsolidation pressure and recompression index of soils. *Geo-tech. Testing J. ASTM*, 8 (4), 199–202.

Newmark, N. M. (1942). *Influence Charts for Computation of Stresses in Elastic Foundations*, 338. University of Illinois Engineering Experiment Station, Urbana.

Pacheco Silva, F. (1970). A new graphical construction for determination of the pre-consolidation stress of a soil sample. *Proc. of 4th Brazilian Conf. on Soil Mechanics and Foundation Engineering*, Rio de Janeiro, **2** (1), 225–232.

Poulos, H. G., and Davis, E. H. (1974). *Elastic Solutions for Soil and Rock Mechanics*. Wiley, New York.

Proctor, R. R. (1933). Fundamental principles of soil compaction. *Eng. News Record*, **111**, 9–13.

Robertson, P. K. (1990). Soil classification using the cone penetration test. *Can. Geotech. J.*, **27** (1), 151–158.

Schmertmann, J. H. (1953). The undisturbed consolidation behavior of clay. *Trans. ASCE*, 120, 1201.

Schofield, A., and Wroth, C. P. (1968). *Critical State Soil Mechanics*. McGraw-Hill, London.

Seed, R. B., Cetin, K. O., Moss, R. E. S., Kammerer, A. M., Wu, J., et al. (2003). Recent advances in soil liquefaction engineering: A unified and consistent framework. *26th Annual ASCE Los Angeles Geotechnical Spring Seminar*, keynote presentation, H.M.S. Queen Mary, Long Beach, CA, April 30, pp. 1–71.

Skempton, A. W. (1944). Notes on the compressibility of clays. *Q. J. Geol. Soc. London*, **100** (C: pts. 1–2), 119–135.

Skempton, A. W. (ed.) (1953). The colloidal activity of clays. *Proc. 3rd Int. Conf. on Soil Mechanics and Foundation Engineering*, Zurich, **1**, 57–61.

Taylor, D. W. (1942). *Research on consolidation of clays*. Serial No. 82. MIT, Cambridge.

Taylor, D. W. (1948). *Fundamentals of Soil Mechanics*. Wiley, New York.

Terzaghi, K. (1925). *Erdbaumechanik*. Franz Deuticke, Vienna.

Terzaghi, K. (1943). *Theoretical Soil Mechanics*. Wiley, New York.

Terzaghi, K., and Peck, R. B. (1948). *Soil Mechanics in Engineering Practice*. Wiley, New York.

Van Genuchten, M. Th. (1980). A closed-form equation for predicting the hydraulic conductivity of unsaturated soils. *Soil Soc. Am. J.*, **44**, 892–898.

Wagner, A. A. (1957). The use of the unified soil classification system by the Bureau of Reclamation. *Proc. 4th Int. Conf. on Soil Mechanics and Foundation Engineering*, London, **1**, 125.

Wood, D. M. (1990). *Soil Behavior and Critical State Soil Mechanics*. Cambridge University Press, Cambridge, UK.

Wroth, C. P. (1984). The interpretation of in situ soil tests. *Geotechnique*, **34** (4), 449–489.

Youd, T. L., Idriss, I. M., Andrus, R. D., Arango, I., Castro, G., Christian, J. T., Dobry, R., Liam Finn, W. D., Harder, L. F., Jr., Hynes, M. E., Ishihara, K., Koester, J. P., Liao, S. S. C., Marcuson, W. F. III, Martin, G. R., Mitchell, J. K., Moriwaki, Y., Power, M. S., Robertson, P. K., Seed, R. B., and Stokoe, K. H. II. (2001). Liquefaction resistance of soils: Summary report from the 1996 NCEER and 1998 NCEER/NSF workshops on evaluation of liquefaction resistance of soils. *J. Geotechnical and Geoenvironmental Eng.*, **124** (10), 817–833.

Index

Note: Page entries in *italics* refer to figures; tables are noted with a "t."

Soil Mechanics Fundamentals, First Edition. Muni Budhu.
© 2015 John Wiley & Sons, Ltd. Published 2015 by John Wiley & Sons, Ltd.
Companion website: www.wiley.com\go\budhu\soilmechanicsfundamentals